PVP-Vol. 415

RECENT ADVANCES IN SOLIDS AND STRUCTURES — 2000 —

presented at
THE 2000 ASME INTERNATIONAL MECHANICAL ENGINEERING CONGRESS AND EXPOSITION
NOVEMBER 5–10, 2000
ORLANDO, FLORIDA

sponsored by
THE PRESSURE VESSELS AND PIPING DIVISION, ASME

edited by

HOWARD H. CHUNG
CHICAGO MANUFACTURING CENTER

YOUNG W. KWON
NAVAL POSTGRADUATE SCHOOL

THE AMERICAN SOCIETY OF MECHANICAL ENGINEERS
Three Park Avenue / New York, N.Y. 10016

ISBN No. 0-7918-1905-1

FOREWORD

This special publication is the collection of papers presented at the symposium entitled "Recent Advances in Solids and Structures" as a part of the program at the International Mechanical Engineering Congress & Exposition held in Orlando, Florida. This symposium is the tenth year continuation of the symposia on the same subject.

The papers included in this volume cover various subjects in solids and structures. The topics include fracture, fatigue, damage, and failure in homogeneous or multiphase composite materials; stability; thermal and residual stress; vibration and dynamics; computational techniques; stress analysis; inelastic material behavior, etc. The papers highlight state-of-the-art technology and research in these subjects. The editors hope this symposium volume will serve as a useful reference for researchers and practitioners in the area of solid and structural mechanics.

The editors wish to express sincere thanks to the authors for their contributions to the symposium. The editors also would like to thank the reviewers for their critical comments and assistance in the publication of this symposium volume.

Howard H. Chung
Chicago Manufacturing Center
Chicago, Illinois

Young W. Kwon
Naval Postgraduate School
Monterey, California

CONTENTS

PVP-Vol. 415, Recent Advances in Solids and Structures — 2000
ASME 2000

STRESS INTENSITY FACTORS FOR CRACKS WITHIN AND NEAR TO BONDLINES IN SOFT INCOMPRESSIBLE MATERIALS

C. W. Smith*, K. T. Gloss*,
D. M. Constantinescu † and C. T. Liu‡

* Department of Engineering Science and Mechanics
Virginia Polytechnic Institute and State University
Blacksburg, VA 24061

† On Sabbatical from Bucharest Polytechnic University

‡ Air Force Research Laboratory, PRSM
10 E. Saturn Blvd.
Edwards, AFB, CA 93524-7680

ABSTRACT

Using a polyurethane photoelastic material, thick test specimens of several configurations with bonded end tabs are examined for measuring stress intensity factors (SIFs) for cracks within and near to bondlines in bonded photoelastic models. Effects of specimen height, glued end tabs, bondline and crack size and location are studied and analyzed using a two parameter model for extracting the SIFs and results are compared with cracked, homogeneous model results.

INTRODUCTION

Large volumes of particulate composites are finding use in many applications in the commercial and military transportation industry. One of these is solid propellant, which consists of hard polyhedral particles embedded in a soft rubber matrix which stiffens at low temperatures. The presence of defects, mainly cracks, in structures composed of these materials has become important as a result of efforts to extend the life of these structures for economic reasons.

When a crack occurs in such a structure during storage or transport, it may extend under a given or changing environment until it becomes sufficiently large to affect the structural and operational integrity of the unit. When a crack grows in such a material, the matrix ahead of the crack tip stretches, separating into strands and producing severe blunting and moving the hard particles above and below the crack plane until the crack breaks the strands to advance through the stretch zone. This blunt-growth-process is repeated, producing a highly non-linear phenomena. At very low temperatures, the stretching of the matrix is suppressed, producing an embrittling effect.

In order to bypass the complications described above, designers often treat the composite as a homogeneous material using average properties. In establishing average properties for cracks in soft bonded structures, glued end tabs must be used with relatively simple test geometries. The present study uses this approach in order to measure effects of specimen geometry, end tabs, crack location and bondlines on the stress intensity factors for cracks within and near to the bondlines and compares results with unbonded specimen results.

THE EXPERIMENTAL METHOD

The photoelastic method was used to analyze all specimens. The method uses stress fringe patterns obtained from tension tests of photoelastic models. The fringes are loci of constant maximum shear stress magnitude. In order to convert the test data into fracture parameters, in this case stress intensity factors,(SIF^s), it is necessary to have an algorithm for this conversion. The authors used a two parameter algorithm developed some time ago (Smith and Kobayashi, 1993) for homogeneous materials. This method is valid in this work since the interface fracture equations (Hutchinson and Suo, 1992) reduce to homogeneous form for bondline cracks in thick, incompressible bodies. The algorithm for Mode I loading is briefly described in the Appendix.

THE EXPERIMENTS

The experiments can be divided into two basic categories:

1. Bonded specimens containing double edge bondline cracks and
2. Bonded specimens with single edge cracks near to and parallel to the bondline.

1. Bondline Cracks

A series of polyurethane test specimens (Fig. 1) was prepared with double edge cracks in the bondline for both square specimens and some with the specimen height reduced by half. Specimen ends were inserted into grooves in aluminum plates and bonded to simulate methods used for bonding solid propellant material specimens. The extent to which this method achieves a uniform axial stress field in the direction of the load will be shown by global stress fringe photographs of the test specimens under load. The analytical models also assume zero transverse displacement of the test specimen at the boundaries. Companion tests were run for different crack lengths. These specimens were made by fitting the adherends to teflon coated razor blades on each side of the specimen and then pouring an adhesive with a modulus higher than the polyurethane adherends into the opening in between and then curing at 160°F for 12 hrs., after which the razor blades were removed. It has been previously found (Smith, Finlayson and Liu, 1998) that modulus mismatch between adherends exerted negligible effect upon the SIF values for the cracks within the bondline for modulus ratios not exceeding about 3.5 so no modulus mismatch was introduced into the models.

Figs. 2a and 2b are global fringe photos from a parallel circular polariscope using monochromatic light of the square and short specimens respectively under load. They reveal some irregularities along the fixed edges possibly due to adhesive slippage. This results in slight dysymmetry of the near tip fringe patterns but the effect on K_1 is well within the expected experimental scatter of ±5%. This effect was accommodated by averaging the K_1 values for the two cracks in each specimen. Figs 3a and 3b show that the near tip fringe patterns show no rotation of the fringe loops which suggest the absence of any shear mode K_2. The data zones for K_1 determination in Figure 3 are taken along a line normal to the crack plane and passing through the tip of the crack. The data zone locations are indicated. The dark bands surrounding the crack opening in Fig. 3 are part of the crack surface which turns outward due to increased crack opening at the near surface of the specimen.

2. Cracks Parallel to the Bondline

In this experimental study, a series of photoelastic experiments were conducted on bonded plates consisting of small cracks of various lengths parallel to the bondline between adherends of the same modulus bonded with an adhesive of somewhat higher modulus to effect a fairly strong bond. Both square and rectangular specimens were analyzed, the latter consisting of heights of half of the square specimens as above.

The studies were conducted under dead weight loading with uniform displacements along the edge of the grips. The specimens were glued to slots in aluminum grips without additional restraints in order to study such a gripping arrangement.

A typical specimen is pictured in Fig. 4. After machining and bonding the adherends, the crack was introduced by using a small saw with a razor blade as the cutting edge. After inserting into aluminum grips and bonding to them, the specimens were dead loaded, with the load on the lower grip being adjusted to keep the lower grip perfectly horizontal. Tests comparing a clamped upper grip with the pin hole arrangement yielded identical results so the pin hole arrangement at the upper grip was used. Room temperature values of modulus were measured to be 4.7 MPa and the value of Poisson's Ratio was 0.50. The nominal value of the adhesive modulus was given by the manufacturer to be 6.9 MPa.

The specimens were loaded and photographed in a crossed circular polariscope using monochromatic light. The near tip fringe patterns were analyzed using the two parameter method described in the Appendix. Typical global and local photographs were shown in Fig. 5, for a square specimen (SP11). The global fringe pattern (Fig. 5a) suggests some local slippage near the center of the specimen in the upper grips and an unsymmetric pattern near the lower grips due to the crack. The local fringe pattern (Fig. 5b) clearly reveals a pure Mode 1 pattern near the crack tip (i.e. no rotation of the near tip fringe loops) but a clear presence (i.e. loop folding into a butterfly shape) of the non-singular stress parallel to the crack.

Zones from which test data for computing the SIF values were taken are indicated in (Fig. 5b). SIF values were determined both above and below the crack tip and averaged to minimize error due to the crack tip location, and then normalized with respect to $K_0 = \bar{\sigma}(\pi a)^{1/2}$ where $\bar{\sigma}$ was the remote average stress.

In these thick plate experiments, for both bondline cracks and those parallel to the bondline, some transverse constraints, tip blunting, and edge curling along the upper and lower crack surfaces occurred, so a single edge crack photoelastic test on a homogeneous thick plate was run and the results compared with an accurate two dimensional solution (Gross, Srawley and Brown, 1964) obtained by boundary collocation. This comparison suggested that the test results were 7.8% higher than predicted and this factor was used to correct all of the test data for the above noted three dimensional effects. The effect of the bondline was to constrict the size of the data zone between the bondline and the crack tip (Fig. 5b) which also altered the value of the nonsingular stress σ^0_{ij} but not the value of K_1.

RESULTS

1) Bondline cracks

Using the near tip fringe pattern photos, the algorithm described in the Appendix was used to extract K_1 values from the local fringe pattern data. An example of the procedure is given in Fig. A-2. K_1 values were computed both above and below the bondline and agreed to well within experimental scatter of 5%. A summary of the test data and results are found in Table 1.

A plot of the results for the square and short specimen test data are presented in Fig. 6. Also shown is the theoretical solution for the double edge crack with no bondline due to Bowie (1964). The experimental data are corrected for thickness and Poisson Ratio ($\nu = 0.5$) effects for comparison with Bowie's solution.

2) Cracks Parallel to the Bondline

These results (Table II) show that the values of K/K_0 were greater for cracks which were further from the bondline than those with smaller separate distance from there, suggesting the presence of a shielding effect due to the bondline. The variation of K/K_0 with a/w for both square and half height specimens are shown in Fig. 7. Also shown is a solution (Bowie and Neal, 1965) for homogeneous square specimens for both uniform displacement and uniform stress.

The larger gradients in the experimental curves above the theoretical results are conjectured to be partly due to slippage in the grips which tended to achieve more of a uniform load effect and partly due to the effect of the bondline. The experimental data suggests that both the shorter cracks and those closer to the bondline may have benefited from a shielding effect due to the higher bondline modulus.

It was also found that, in analyzing the data for the cracks in the half height specimen, a modification of the form of the non-singular stress between the crack and the bondline as described in Smith, Finlayson, and Liu, (1998) was necessary. This is believed due to a residual stress field very close to the bondline. A second half height specimen did not show bondline residual stress and was analyzed as described in the Appendix. Results were virtually identical with the results for the specimen with residual stress.

In assessing the overall effect of these residual stresses, it was found that while the SIF values above and below the crack tip in the square specimens differed by about 2%, those in the half height specimens differed by 6%. These differences were minimized by averaging the results in Table II.

In summary, results show that:

1) Normalized SIFs increase with relative crack length for both square and half height bonded specimens more rapidly than indicated by solutions for unbonded specimens for both bondline cracks and cracks parallel to the bondline.

2) Reducing specimen height reduces normalized SIF's for all crack lengths in both bondline cracked specimens and those containing cracks parallel to the bondline. Since Torvik, (1979) indicated such an effect in unbonded edge cracked specimens, this reduction appears to be due to specimen shape rather than a bondline effect and may be conjectured to apply to the bondline cracked specimen results as well.

3) The experiments revealed

   a) No shear mode effect

   b) A shielding effect due to the bondline for short cracks and those with less separation from the bondline for the cracks parallel to the bondlines.

4) Imperfections in the glued tab arrangement were clearly shown.

ACKNOWLEDGMENTS

The authors wish to acknowledge the use of the facilities of the Department of Engineering Science and Mechanics and the support of the Air Force Research Laboratory through Sub-contract 98-522 with Sparta Inc.

REFERENCES

Bowie, O. L., "Symmetric Edge Cracks in Tensile Sheet with Constrained Ends," *Journal of Applied Mechanics,* pp. 726-728, Dec. (1964).

Gross, B., Srawley, J. E., and Brown, W. F., "Stress Intensity Factors for a Single-Edge Notch Tensile Specimen by Boundary Collocation of a Stress Function." *Technical Note D-2395 NASA,* 11 pages, Aug. (1964).

Bowie, O. L., and Neal, D. U., U. S. Army Materials Research Agency Technical Report AMRATR -65-20, (1965).

Torvik, P. J., "On the Determination of Stresses, Displacements, and Stress Intensity Factors in Edge Cracked Sheets with Mixed Boundary Conditions," *Trans. ASME Series E. J. Ap. Mech.*, Vol. 46, pp. 611-617, (1979).

Hutchinson, J. W., and Suo, H., "Mixed Mode Cracking in Layered Materials," *Advances in Applied Mechanics,* Vol. 29, New York, Academic Press, pp. 63-91, (1992).

Smith, C. W., and Kobayaski, A. S., "Experimental Fracture Mechanics," Chapter 20: *Handbook on Experimental Mechanics*, 2nd Ed., Kobayashi, A. S., Ed. VCH New York, pp. 905-968, (1993).

Smith, C. W., Finlayson, E. F., and Liu, C. T., "Influence of Material Properties on SIF's Determined by Frozen Stress," *Journal of Engineering Fracture Mechanics,* Vol. 61, No. 5/6, pp. 555-568, (1998).

APPENDIX

Beginning with the Griffith-Irwin Equations, we may write, for Mode I, for the homogeneous case

$$\sigma_{ij} = \frac{K_1}{(2\pi r)^{1/2}} f_{ij}(\theta) + \sigma^{\circ}_{ij}(i.j. = n, z) \qquad (1)$$

where: σ_{ij} are components of stress, K_1 is SIF, r, θ are measured from crack tip (Fig. A-1), σ°_{ij} are non-singular stress components.

Then, along $\theta = \pi/2$, after truncating σ_{ij}

$$(\tau_{nz})_{max} = \frac{K_1}{(8\pi r)^{1/2}} + \tau^{\circ} = \frac{K_{AP}}{(8\pi r)^{1/2}} \qquad (2)$$

where $\tau^{\circ} = f(\sigma^{\circ}_{ij})$ and is constant over the data range, K_{AP} = apparent SIF, $(\tau_{nz})_{max}$ = maximum shear stress in nz plane

$$\therefore \frac{K_{AP}}{\bar{\sigma}(\pi a)^{1/2}} = \frac{K_1}{\bar{\sigma}(\pi a)^{1/2}} + \frac{\sqrt{8}\tau^{\circ}}{\bar{\sigma}}\left(\frac{r}{a}\right)^{1/2} \qquad (3)$$

where (Fig. A-1) a = crack length, and $\bar{\sigma}$ = remote normal stress i.e.

$$\frac{K_{AP}}{\bar{\sigma}(\pi a)^{1/2}} \text{ vs. } \sqrt{\frac{r}{a}} \text{ is linear}$$

Since from the Stress-Optic Law $(\tau_{nz})_{max} = nf/2t$ where, n = stress fringe order, f = material fringe value, t = specimen thickness, then from Eq. 2, $K_{AP} = (8\pi r)^{1/2}(\tau_{nz})_{max} = (8\pi r)^{1/2}\, nf/2t$. A typical plot of normalized K_{AP} vs. $\sqrt{r/a}$ for a homogeneous specimen is shown in Fig. A-2.

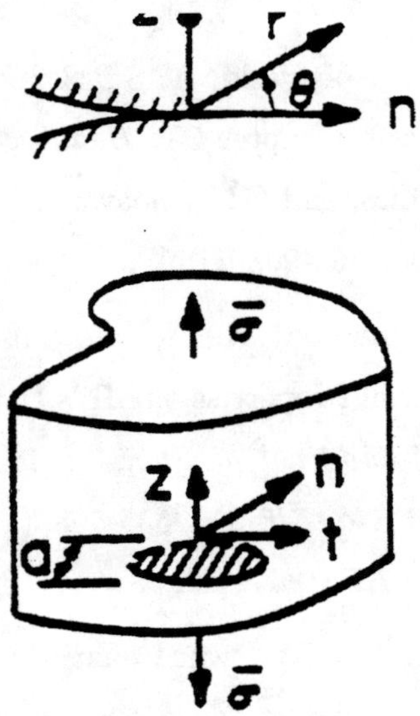

Figure A-1: Mode I Near-Tip Notation

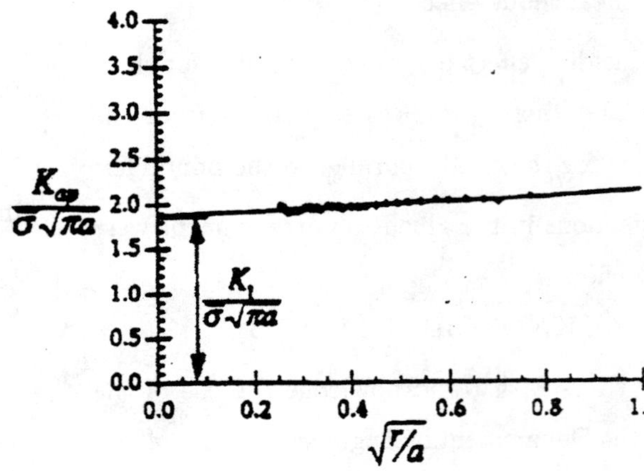

Figure A-2 - Estimating Normalized SIF from Test Data

Table 1 - Data and Results for Cracks Within Bondline

test#	a(mm)	h(mm)	a/b	P (N)	exp K/Ko	exp. corr. ($\nu = 0.5$) K/Ko	Bowie* ($\nu = 0.3$) K/Ko
DS2	7.94	50.8	0.16	74.95	1.14	1.05	1.06
DS3	12.7	50.8	0.25	74.95	1.09	1.00	1.04
DS4	17.4	50.8	0.34	74.95	1.15	1.06	1.04
DS5	20.6	50.8	0.41	74.95	1.23	1.13	1.06
DS6	25.4	50.8	0.50	74.95	1.38	1.27	1.10
DS7	27.9	50.8	0.55	74.95	1.37	1.26	1.12
DS8	7.94	25.4	0.16	74.95	0.93	0.86	—
DS9	12.7	25.4	0.25	74.95	0.94	0.87	—
DS10	17.4	25.4	0.34	52.68	0.98	0.90	—
DS11	20.6	25.4	0.41	50.72	1.00	0.93	—
DS12	25.4	25.4	0.50	50.72	1.18	1.09	—
DS13	27.9	25.4	0.55	51.01	1.22	1.12	—

* plane stress, no bondline $K_0 = \bar{\sigma}\sqrt{\pi a}$

Table 2 - Data and Results for Cracks Parallel to Bondline

w = 101.6

Test #	a	d	h	a/w	K^*/K_0
SP6	2.78	2.58	50.80	0.027	0.976
SP7	2.98	1.19	50.80	0.029	0.885
SP8	8.33	2.58	50.80	0.082	1.139
SP9	7.95	1.19	50.80	0.078	1.017
SP10	13.09	1.19	50.80	0.129	1.173
SP11	12.70	2.58	50.80	0.125	1.365
SP12	2.58	2.78	25.4	0.025	0.806
SP13	7.54	2.78	25.4	0.074	0.876
SP14	12.70	2.78	25.4	0.125	1.028

Dimensions in mm, * Corrected for 3-D effects to a 2D solution.

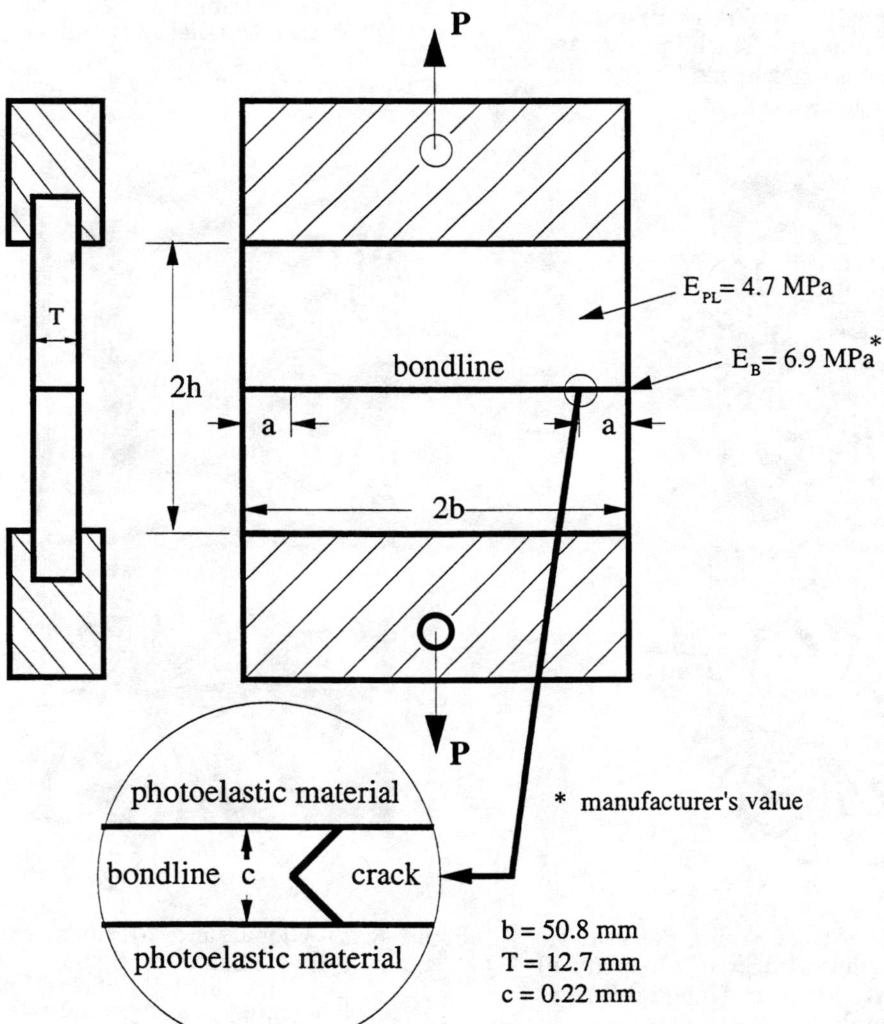

Figure 1 Bondline Specimens with Double Edge Bondline Cracks

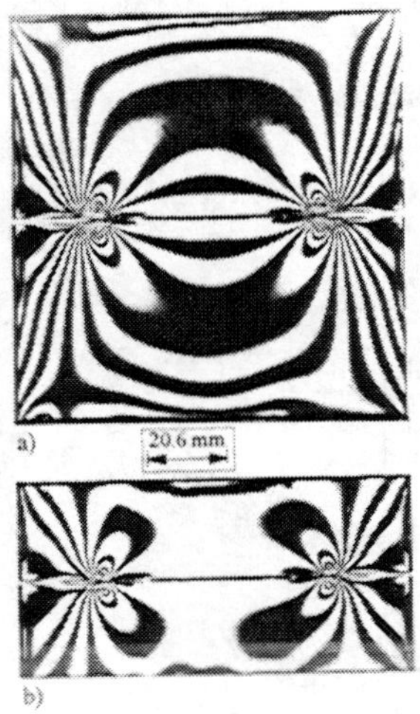

Figure 2: Global Stress Fringe Patterns for Bondline Crack in a) square b) short specimens (Integral Fringes white).

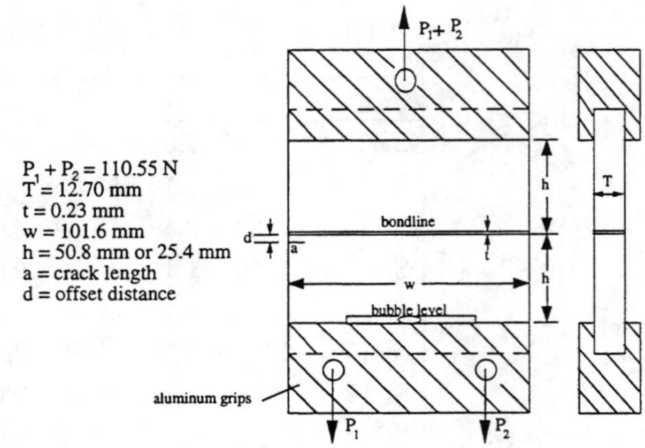

Figure 4: Test specimen and End Tabs; Crack Parallel to Bondline.

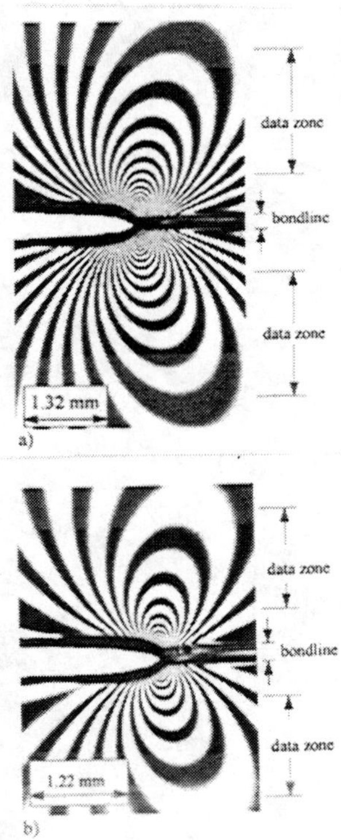

Figure 3: Local Stress Fringe Patterns for Bondline Cracks in a) square b) short specimens (Integral Fringes White).

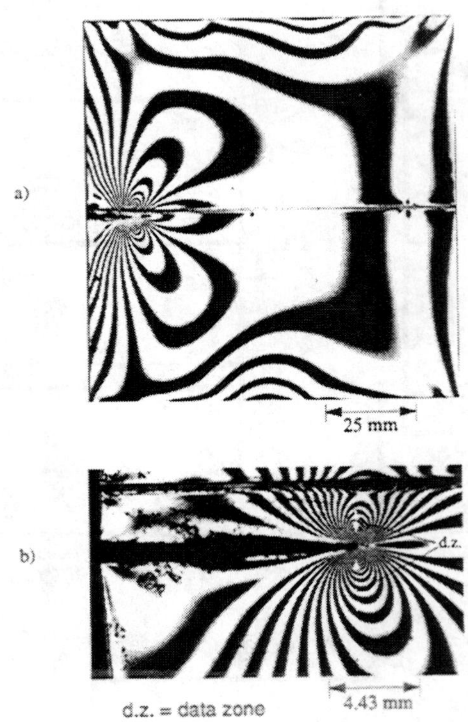

Figure 5: a) Global and b) Local Stress Fringe Photos for Square Specimen with Crack Parallel to Bondline (Integral Fringes are Dark)

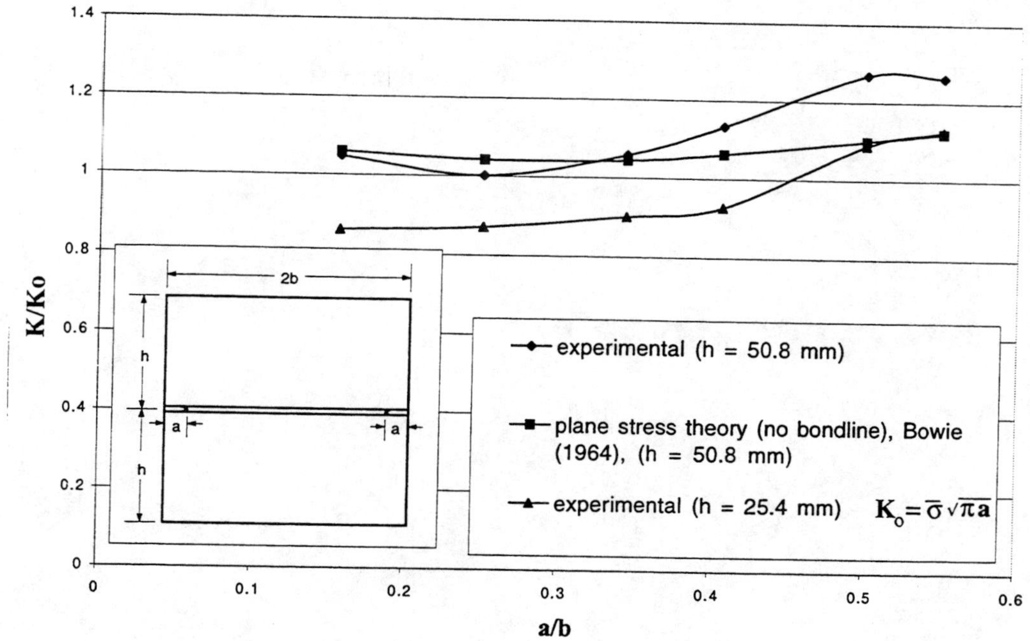

Figure 6 Normalized SIF vs. Normalized Crack Length for Bondline Cracks (b = 50.8mm).

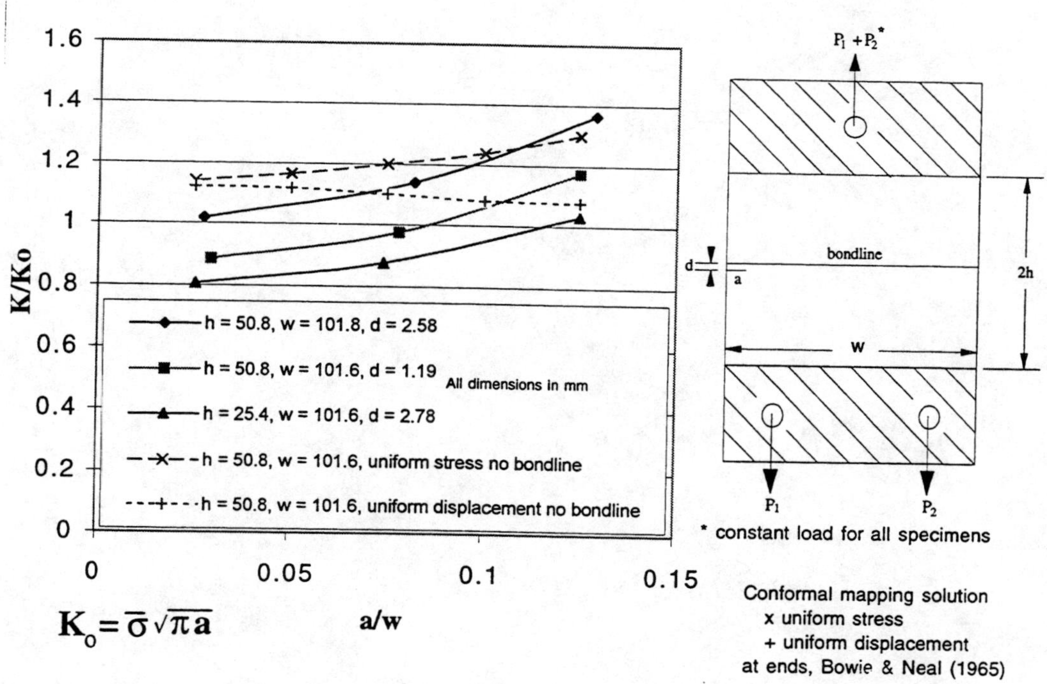

Figure 7 Normalized SIF vs. Normalized Crack Length for Cracks Parallel to Bondline.

PVP-Vol. 415, Recent Advances in Solids and Structures — 2000
ASME 2000

MODELING OF HYDROSTATIC PRESSURE EFFECT ON PROGRESSIVE DAMAGE IN PARTICULATE COMPOSITES

Y. W. Kwon
Mechanical Engineering Department
Naval Postgraduate School
Monterey, California

C. T. Liu
Air Force Research Laboratory
Edward AFB, California

ABSTRACT

Hydrostatic pressure affects the damage growth in a particulate composite. As a result, an analytical model was presented to represent the damage growth in a particulate composite under various hydrostatic pressures. The model was based on a multi-level approach with damage description at the micro-level. A damage theory was presented to describe the material behavior under hydrostatic pressure. The predicted stress-strain curves compared well with the experimental measurements. Furthermore, the initial crack size at a notch tip was studied with and without hydrostatic pressure.

INTRODUCTION

Particulate composites have beneficial properties for engineering applications. Especially, particulate composites are easy to produce. Thus, many researchers have devoted their studies to particulate composites in order to understand and predict their mechanical characteristics [Farris, 1971; Gent and Park, 1984; Schapery, 1973; Weng, 1984]. One of the important mechanical characteristics is strength and toughness. In particular, crack initiation and growth in such composite materials is an important issue for safety. As a result, progressive damage at the notch tip and its vicinity of a rectangular specimen with a center hole was studied in the past [Kwon and Baron, 1998; Kwon and Liu, 1997, 1998, 1999].

An experimental study [Farris 1968] showed that the hydrostatic pressure suppressed the damage initiation in a particulate composite. A phenomenological constitutive model was presented to describe the effect of the hydrostatic pressure on the stress-strain curve [Ravichandran and Liu, 1995].

The present study investigated the effect of the hydrostatic pressure on damage initiation and growth leading to the macro-scale crack called the initial crack. A multi-level approach was adopted for the study. Because the damage was implemented at the micro-level, a damage theory was developed for the matrix material. The particulate composite under study has very hard/stiff particles embedded in a soft/weak matrix. For example, particles are metallic materials while the binding matrix is a rubber-like polymer material. Those composites have damage in the matrix material and/or particle-matrix interface but not particle cracking.

The next section describes the multi-level approach briefly. Then, a damage theory was developed to describe the effect of the hydrostatic pressure on the progressive damage. Numerical results and discussion are given next followed by conclusions.

MULTI-LEVEL APPROACH

As an analysis tool, a multi-level approach was adopted to model progressive damage in particulate composite specimens. The approach consists of two levels of analyses: micro-level and macro-level analyses. The macro-level analysis is the structural (specimen) level analysis while the micro-level analysis is the constituent material level analysis like particle and matrix materials. The two levels are closely coupled together. The interaction between the two levels is illustrated in Fig. 1.

The damage is implemented at the micro-level and its effect on the macro-level is determined. As a result, damage can be described naturally and clearly in terms of the constituent materials. For a particulate composite, damages at the micro-level can be classified as the micro-cavity (called vacuole) in the matrix material, interface debonding (called dewetting), or particle cracking. Depending on the strength and stiffness of the constituent materials and their interface properties, one form of damage may occur first and cause other forms of damage subsequently. For the present composite material under study, the matrix material is much softer and weaker than the particle material so that particle cracking does not occur.

The detailed description for the multi-level analysis was given in the references [Kwon and Berner, 1995, Kwon and Baron, 1998; Kwon and Craugh; Kwon and Kim, 1998; Kwon and Liu, 1999]. As a result, it was omitted here. However, the damage theory will be described next. The damage theory is used in the micro-level analysis to represent the damage of the respective constituent material.

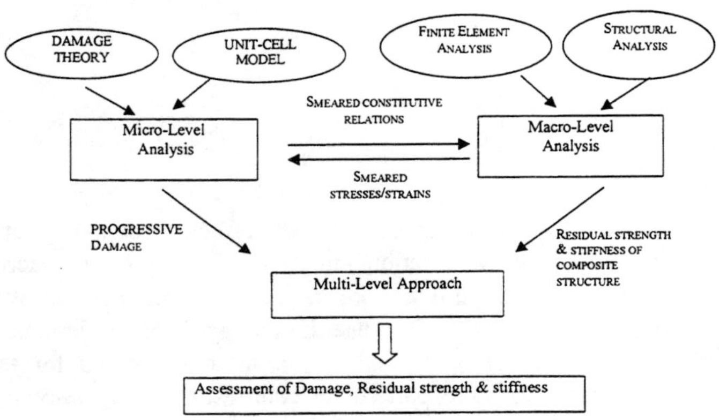

Figure 1. Multi-Level Approach

DAMAGE THEORY

The damage behavior of the particulate composite under study showed large dependency on the hydrostatic loading condition. The uniaxial stress-strain plots under various constant hydrostatic pressures indicated that the damage initiation in the composite material was highly dependent on the magnitude of the hydrostatic pressure [Farris, 1968]. Hydrostatic pressure delayed the damage initiation so that the linear portion of the stress-strain curve was longer for a greater hydrostatic pressure loading. The nonlinear portion of the stress-strain curve was considered as being caused by damage initiation and growth. The experimental data [Farris, 1968] showed that the damage initiation was significantly affected by the hydrostatic pressure but the damage growth rate (damage increment per unit strain) was much less affected by the hydrostatic pressure.

In order to represent this behavior of the composite material, the damage function of the matrix material is assumed to be

$$F = f(u_v, u_d) - \kappa \qquad (1)$$

where u_v and u_d are the dilatational and distortional strain energy density, and κ is a constant for the damage threshold value which increases with the damage. If F is less than zero, damage does neither occur nor grow. However, when $F = 0$, damage initiates or grows. For the present damage model, the function f is assumed as

$$f = a_1 u_d + a_2 \operatorname{sgn}(e) u_v \qquad (2)$$

where a_i is the material constants, and $e = e_{ii}$ is the dilatation. In addition, the sign function $\operatorname{sgn}(x)$ is 1 if x is positive and -1 if x is negative. Thus, when the dilatation is negative under a hydrostatic pressure, it delays damage.

The damage evolution function is expressed as

$$\dot{d} = \dot{\kappa} g \qquad (3)$$

in which the overdot denotes the temporal derivative. The Kuhn-Tucker conditions for damage loading and unloading can be stated as

$$\dot{\kappa} \geq 0, \; F \leq 0, \; \dot{\kappa} F = 0 \qquad (4)$$

For damage accumulation, $\dot{\kappa} > 0$, then $F = 0$ and $\dot{d} > 0$. Otherwise, $F < 0$ and $\dot{\kappa} = 0$. Then $\dot{d} = 0$ which means no more damage. The function g in Eq. (3) was also assumed to be a function of the strain energy density.

RESULTS AND DISCUSSION

Using the damage theory described in the previous section, the stress-strain curves were obtained under various constant hydrostatic pressure loads as shown in Fig. 2. The stress-strain curves agree with the experimental observation qualitatively [Farris, 1968; Ravichandran and Liu, 1995]. In other words, a higher hydrostatic pressure holds the damage initiation until a higher stress/strain. When the hydrostatic pressure is very high like 6895 kPa, the stress-strain curve was almost linear throughout the range. Figure 3 also shows that there is a delay of damage initiation in association with the hydrostatic pressure but the damage growth rate per unit strain does not vary so much for different hydrostatic pressures.

The next study considered damage accumulation at a notch tip caused by strain concentration. The saturated and localized damage would result in a macro-scale crack initiating from a notch tip. The specimen considered was 38.1mm (1.5 in.) x 38.1mm (1.5 in.) x 5.08mm (0.2 in) with a circular hole at the center. The hole diameter was 6.35mm (0.25 in.) or 12.7mm (0.5 in.). The stress-strain curve of the material is shown in Fig. 4. In the figure, the analytical stress-strain curve agrees well with the experimental data. These curves were obtained without hydrostatic pressure.

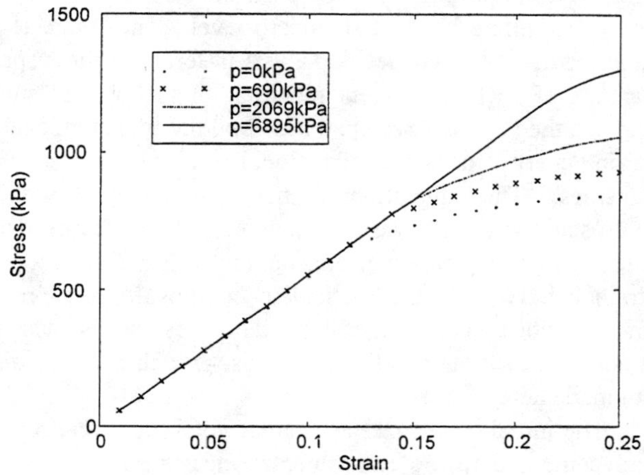

Figure 2. Stress-Strain Curves under Different Hydrostatic Pressures

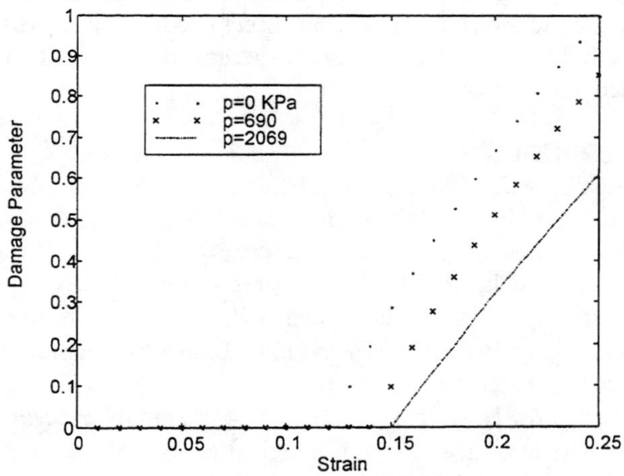

Figure 3. Damage Initiation and Growth under Different Hydrostatic Pressures

The initial, macro-scale crack size was determined based on the criterion developed in previous papers [Kwon and Liu, 2000a, 2000b]. The criterion proposed that the size of a crack initiating from a notch tip is the same as the length of an unstable material zone at the immediate front of the notch tip when there is onset of crack initiation at the notch tip. The unstable material zone is caused by accumulated damage in the zone. The onset of crack initiation is considered to be damage saturation at the notch tip so that it cannot sustain the load any more. The criterion is based on the concept that the crack initiating from a notch tip will propagate through the unstable material zone and will be arrested at the stable material zone. The unstable zone is shown in Fig. 5. Based on the criterion, the predicted initial crack size was 0.615mm for the 6.35mm diameter hole and 1.24mm for the 12.7mm diameter hole, respectively. The averages of the measurements were 0.780mm for the smaller hole and 1.33mm for the larger hole. Thus, the predicted results agreed well with the experimental data. Figures 6 through 8 shows the stress, strain, and damage distributions near the notch tip when the notch tip has the damage saturation.

The same specimen was considered under a hydrostatic pressure. Because of the pressure, the damage initiation and the formation of the initial crack occurred at a higher applied load. However, the initial crack size was almost the same as that without the hydrostatic pressure. In this study, the penetration of the pressure load into the crack face was not considered. The pressure loading on the crack face might further propagate the crack resulting in a longer initial crack or an unstable initial crack. This effect will be investigated in the subsequent study.

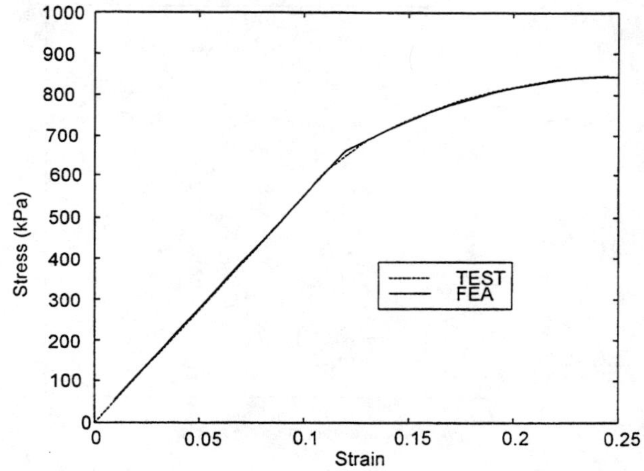

Figure 4. Comparison of Stress-Strain Curves without Hydrostatic Pressure

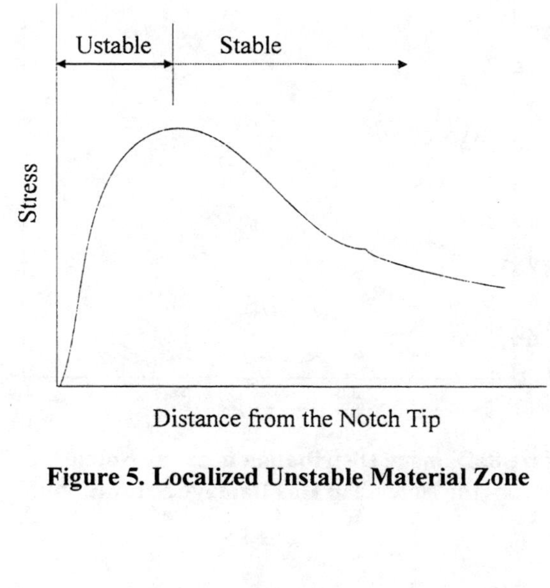

Figure 5. Localized Unstable Material Zone

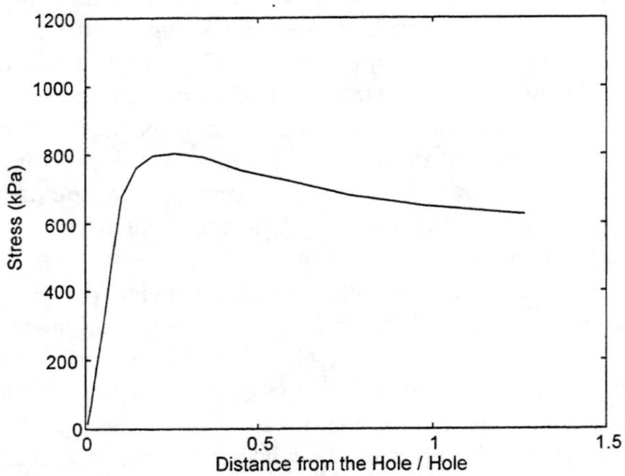

Figure 6. Stress Distribution near the Notch Tip When the Notch Tip Has Damage Saturation

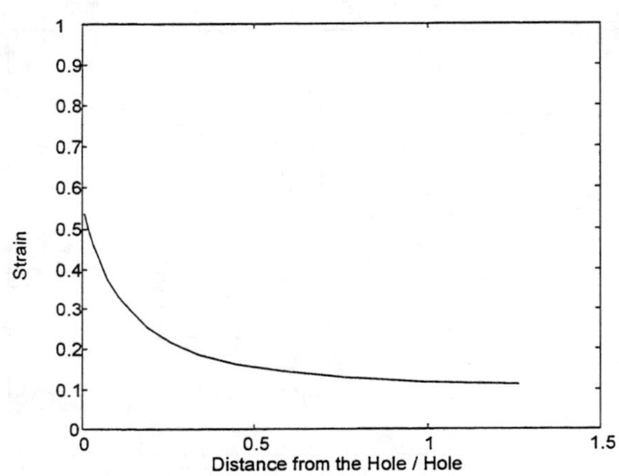

Figure 7. Strain Distribution near the Notch Tip When the Notch Tip Has Damage Saturation

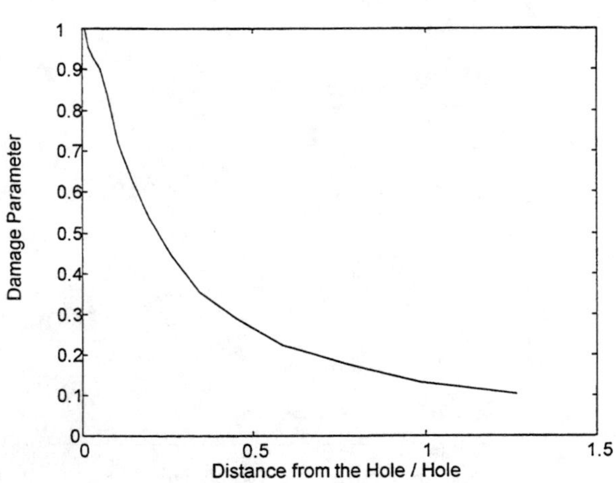

Figure 8. Damage Distribution near the Notch Tip When the Notch Tip Has Damage Saturation

CONCLUSIONS

The stress-strain behavior of a particulate composite specimen under hydrostatic pressure was modeled using the multi-level approach. The approach consists of two levels of analyses: micro-level and macro-level. The damage was described at the micro-level analysis in terms of the respective damage of each constituent material. In the present study, a damage theory was developed based on the two components of strain energy density: dilatational and deviatoric energy densities. The dilatational energy associated with the hydrostatic pressure was assumed to hold back the damage initiation. Using this concept, qualitatively acceptable stress-strain behaviors of the particulate composite were predicted under various levels of the hydrostatic pressure. Because there was no data for quantitative comparison, such comparison was not made here.

The initial crack sizes at circular notch tips were predicted and compared to the experimental measurement when there was no hydrostatic pressure. The predicted values compared very well to the test data. When the study included a hydrostatic pressure, the initial crack size did not vary even though the crack formed at a higher applied strain level. This study did not include the pressure penetration into the crack to be applied to the crack faces. This effect needs to be studied in the subsequent study, and the experimental tests are also needed with hydrostatic pressures.

REFERENCES

Farris, R.J., 1968, "The Character of the Stress-Strain Function for Highly Filled Elastomers", *Transactions of the Society of Rheology*, Vol. 12, No. 2, pp. 303-314.

Farris, R.J., 1971, "The Stress-Strain Behavior of Mechanically Degradable Polymers", *Polymer Networks: Structural and Mechanical Properties*, (Edited by Chompff and Newman, S.) Plenum, New York.

Gent, A. N. and Park, B., 1984, "Failure Process in Elastomers at or near a Rigid Spherical Inclusion", *Journal of Materials Science*, Vol. 19, pp. 1947-1956.

Gurson, A.L., 1977, "Continuum Theory of Ductile Rupture by Void Nucleation and Growth: Part I-Yield Criteria and Flow Rules for Porous Ductile Materials", *Journal of Engineering Material Technology*, Vol. 99 pp. 2-5.

Kwon, Y.W. and Baron, D.T., 1998, "Numerical Predictions of Progressive Damage Evolution in Particulate Composites", *Journal of Reinforced Plastics and Composites*, Vol. 17, No. 8, pp. 691-711.

Kwon, Y.W. and Berner, J.M., 1995, "Micromechanics Model for Damage and failure Analyses of Laminated Fibrous Composites", *Engineering Fracture Mechanics*, Vol. 52, No. 2, pp. 231-242.

Kwon, Y.W. and Craugh, L.E., "Progressive Failure Modeling in Notched Cross-ply Fibrous Composites", *Applied Composite Materials*, Accepted for Publication.

Kwon, Y.W. and Kim, C., "Micromechanical Model for Thermal Analysis of Particulate and Fibrous Composites", *Journal of Thermal Stresses*, Vol. 21, 1998, pp. 21-39.

Kwon, Y.W. and Liu, C.T., 1997, "Study of Damage Evolution in Composites Using Damage Mechanics and Micromechanics", *Composite Structures*, Vol. 38, No. 1-4, pp. 133-139.

Kwon, Y.W and Liu, C.T., 1998, "Effects of Non-Uniform Particle Distributions on Damage Evolution in Pre-Cracked, Particulate Composite Specimens", *Polymers and Polymer Composites*, Vol. 6, No. 6, pp. 387-397

Kwon, Y.W. and Liu, C.T., 1999, "Numerical Study of Damage Growth in Particulate Composites", *Journal of Engineering Materials and Technology*, Vol. 121, Oct. 1999, pp. 476-482.

Kwon, Y.W. and Liu, C.T., 2000a, "Prediction of Initial Crack Size in Particulate Composites with a Circular Hole", *Mechanics Research Communications*, Vol. 27, No. 4.

Kwon, Y.W. and Liu, C.T., 2000b, "Effect of Particle Distribution on Initial Cracks Forming from Notch Tips of Composites with Hard Particles Embedded in a Soft Matrix", *Composites, Part B: Engineering*, Accepted for Publication.

Ravichandran, G. and Liu, C.T., 1995, "Modeling Constitutive Behavior of Particulate Composites Undergoing Damage", *International Journal of Solids and Structures*, Vol. 32, No. 6/7, pp. 979-990.

Schapery, R.A., 1973, "A Theory of Crack Growth in Viscoelastic Media", Texas A&M University Report No. MM2764-73-1, March.

Weng, G.J, 1984, "Some Elastic Properties of Reinforced Solids, with Special Reference to Isotropic Ones Containing Spherical Inclusions", *Int. J. Engng Sci*, Vol. 22, No. 7, pp. 845-856.

PVP-Vol. 415, Recent Advances in Solids and Structures — 2000
ASME 2000

AN OPTIMIZED SPECIMEN FOR CRACK GROWTH STUDIES IN A CONSTANT K ENVIRONMENT

R. Cammino, M. Gosz, and S. Mostovoy
Department of Mechanical, Materials, and Aerospace Engineering
Illinois Institute of Technology
Chicago, IL 60616-3793

ABSTRACT

Techniques in computational fracture mechanics were employed to optimize the performance of a fracture specimen for use in crack growth studies in a constant K environment. The finite element method was used to model the specimen. In the numerical calculations, the mode I stress intensity factors were obtained using a domain integral approach. The specimen was optimized by systematically changing its geometry and performing finite element calculations in an iterative fashion. The procedure was carried out until a variation in the mode I stress intensity factor of less than three percent within the desired range of crack propagation was achieved.

INTRODUCTION

In the development of fatigue crack growth experiments, it was recognized early on that the stress intensity factor range is an important parameter governing the fatigue crack growth rate. This observation lead to the development of empirical formulas, such as Paris law [1], that provide relationships between crack growth rate and stress intensity factor range. One of the main difficulties in performing fatigue crack growth tests is keeping the stress intensity factor range constant during the test (the stress intensity factor range typically changes with increasing crack length when the fracture specimen is subjected to a constant cyclic loading range). To remedy this problem closed-loop computer controlled fatigue tests [2-4] have been developed. In these tests, a computer monitors the load and displacement, and the load is continuously adjusted to keep the stress intensity factor range constant during the test. Because of the relatively random nature of crack front geometry, this procedure introduces errors in calculated crack position that contribute to data scatter. An alternative approach for material fatigue crack growth analysis is to use a fracture specimen designed in such a way so that the stress intensity factor range remains constant as the crack propagates under constant cyclic loading. Such specimens are commonly called “zero K-gradient” specimens. In the literature, they are also referred to as contoured double-cantilevered beam specimens (CDCB).

Mostovoy [5] developed the original design of the CDCB specimen using a mechanics of materials approach. Since then, the CDCB specimen has been used extensively in fatigue and environmentally assisted crack growth experiments [6]. The CDCB specimen has proven useful in a recent study by Nash *et al.* [7] to determine the threshold stress intensity factor range (the stress intensity factor range at which the crack propagation rate is considered to be negligible) for high strength low alloy steels. These authors performed the fatigue crack growth experiments in a hydrogen-enriched environment to reduce the time required to obtain the threshold range (as opposed to performing the test in a non-corrosive environment).

In a purely experimental evaluation of a given CDCB specimen subjected to a constant load, the stress intensity factor corresponding to a given crack length is determined by performing a linear least squares fit through compliance data. The slope of the resulting straight line is used to calculate the stress intensity factor. Based on experiments of the original CDCB specimen, the correlation coefficient for the linear regression was of the order of 0.999. The excellent correlation coefficient leads one to conclude that the stress intensity factor is constant for the range of crack lengths monitored in the experiment.

In the present paper the finite element method is employed to perform a numerical evaluation of the original CDCB specimen. The domain integral method is employed to calculate the mode I stress intensity factors as a function of crack length under constant load. It was found that the mode I stress intensity factor distribution for this specimen deviated from constant K by up to fifteen percent within the range of crack lengths for which the specimen was designed. Such variations in the stress intensity factor cannot be detected using the experimental approach discussed above. Therefore, it can be advantageous to evaluate fracture specimens based on numerical determination of stress intensity factors over experimental procedures.

Because numerical evaluations of fracture specimens can be carried out very quickly, further optimization of the specimens can be accomplished. By modifying the geometry of the specimen in a systematic fashion, we have achieved a specimen having a stress intensity factor variation of less than three percent for a crack propagation range of 1.5 inches. The optimization procedure described in the paper also allows the development of constant K specimens having a range of constant compliance vs. crack length curves.

To outline the remainder of the paper, the mechanics of materials approach employed by Mostovoy et al. [5] to develop the original CDCB specimen is presented in the next section. The finite element approach employed to model the CDCB specimen along with some details of the domain integral method used to obtain the stress intensity factors is described in Section 3. The numerical and experimental results for the original CDCB specimen and the procedure for optimizing the specimen are provided in Section 4. Finally, a summary and some concluding remarks are given in Section 5.

MECHANICS OF MATERIALS APPROACH TO CONSTANT K DESIGN

In this section, we present the mechanics of materials approach to constant K specimen design employed by Mostovoy et al. [5]. To begin, consider a tapered double cantilever beam as shown in Figure 1. As shown in the figure, a is the crack length, P is the remote applied load, δ is the crack opening displacement, and $h(x)$ is the height of the specimen at a distance x from the left end. Treating the upper and lower arms of the specimen as cantilever beams of length a, the complementary energy U^* in the arms can be expressed as

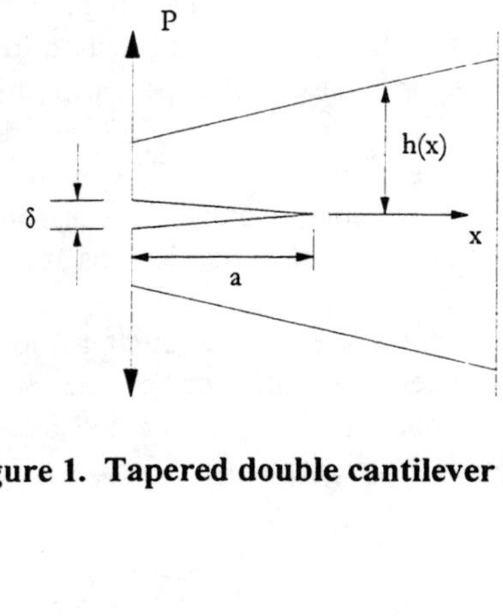

Figure 1. Tapered double cantilever beam.

$$U^* = 2\int_0^a \left(\frac{M^2}{2EI} + \frac{kV^2(1+\nu)}{EA} \right) dx \tag{1}$$

where M and V are the resultant bending moment and transverse shear force acting on the cross section, I is the moment of inertia about the neutral axis, A is the cross sectional area, k is a shear correction factor, E is Young's modulus, and ν is Poisson's ratio. Noting that $M = Px$, $V = P$, $I = bh(x)^3/12$, and $A = bh(x)$, the complementary energy expression (1) can be written as

$$U^* = \frac{12P^2}{Eb}\int_0^a \frac{x^2}{h^3}dx + \frac{2k(1+\nu)P^2}{Eb}\int_0^a \frac{1}{h}dx \tag{2}$$

where b is the width of the beam.

The energy release rate is defined as

$$G = \frac{1}{b}\frac{\partial U^*}{\partial a} \tag{3}$$

where b is the crack width. Upon differentiating equation (2) with respect to a and taking $\nu = 1/3$ and $k = 1.5$, we obtain the following expression for the energy release rate which appears in Mostovoy *et al.* [5]

$$G = \frac{P^2}{2}\frac{8}{Eb^2}\left\{ \frac{3a^2}{h^3} + \frac{1}{h} \right\} \tag{4}$$

Since the mode I stress intensity factor is proportional to the square root of G, a constant K specimen can be designed by contouring the specimen in such a way as to keep the quantity

$$m = \left\{ \frac{3a^2}{h^3} + \frac{1}{h} \right\} \tag{5}$$

equal to a constant.

Experimental validation of constant K specimens

The compliance of the CDCB specimen is defined as

$$C = \frac{\delta}{P} \tag{6}$$

where δ is the crack opening displacement and P is the applied load as shown in Figure 1. The energy release rate is related to the compliance through the relation

$$G = \frac{P^2}{2b}\frac{\partial C}{\partial a} \tag{7}$$

Hence, the compliance of a constant K specimen should vary linearly with crack length.

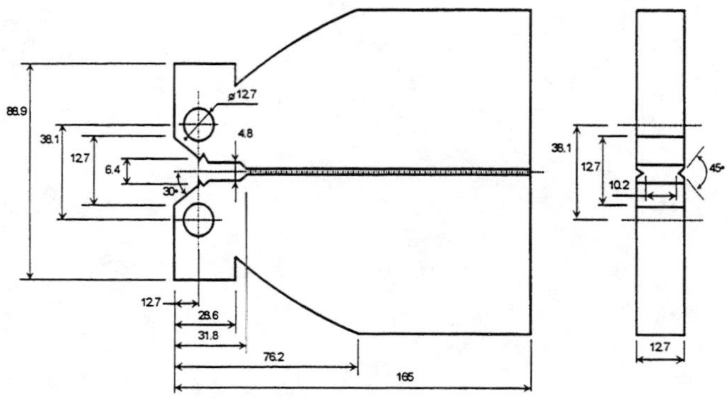

Figure 2. The geometry of the original specimen (all dimensions given in mm).

In the experimental validation of a CDBC fracture specimen, compliance versus crack length data is obtained while the specimen is subjected to cyclic loading. The compliance is obtained by measuring the crack opening displacement associated with a known externally applied load. After all of the data has been gathered, a linear least squares fit through the data is performed. If the data exhibits a correlation coefficient close to one over a sufficient range of crack lengths, the specimen is deemed to be constant K.

The correlation coefficient for the original specimen designed by Mostovoy et al. [5] is of the order of 0.999. The geometry and dimensions of this specimen are shown in Figure 2. The main problem associated with the experimental validation technique described above stems from the fact that the linear least squares curve fit through the data masks local variations in the slope of the compliance versus crack length curves. These local variations lead to significant deviations in the stress intensity factor as will be demonstrated in the next section.

FINITE ELEMENT MODELING

In this section, we describe the finite element approach used to evaluate the performance of the original CDCB fracture specimen. The geometry of this specimen was determined from the mechanics of materials approach described in the previous section. Due to symmetry, it is only necessary to model one half of the specimen. A schematic of the model employed for the upper arm is shown in Figure 3.

To describe the boundary conditions applied to the model, a point load of magnitude *P* is applied at the center of the hole located at the left end of the specimen. The line AB represents the crack front. The displacement component u_y was prescribed to be zero on the surface ABCD. In order to prevent rigid body rotation or translation of the model without losing generality, we fix the displacement components u_x, u_y, and u_z at point E located a sufficient distance away from the crack front. We also fix the displacement component u_z at point F

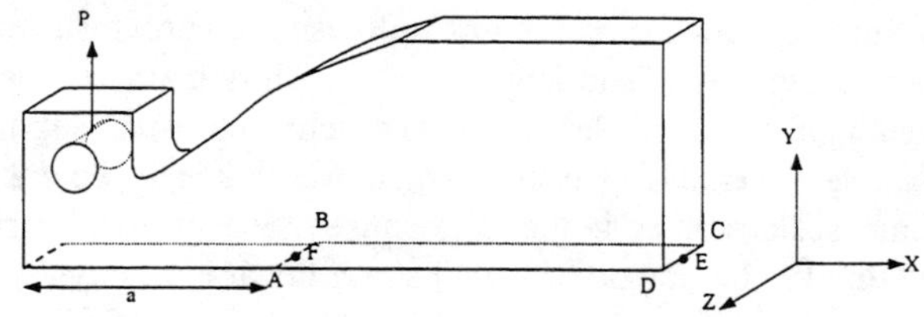

Figure 3. The model employed for the upper arm.

located at the middle of the crack front.

The domain integral method was employed to numerically extract the energy release rates and stress intensity factors along the crack front. In the domain integral method, a crack-tip contour integral is expressed as a volume integral over a finite domain surrounding the crack tip. The process of recasting the contour integral into a volume integral is advantageous for numerical purposes, because accurate fracture parameters can be obtained without having to precisely capture the details of the singular fields in the vicinity of the crack tip. See Moran and Shih [8-9] for a general discussion on crack-tip contour integrals and their associated domain integral representations.

The finite element mesh employed for the analysis of the original fracture specimen is shown in Figure 4a. The mesh is composed of approximately 10,000 eight-node brick elements. A very fine mesh of eight-node brick elements surrounds the crack tip. The details of the mesh in the vicinity of the crack-tip are shown in Figure 4b. The crack-tip mesh was constructed by sweeping a two-dimensional rectangular mesh along the crack front. The domains of integration are rectangular blocks two elements thick in the direction of the

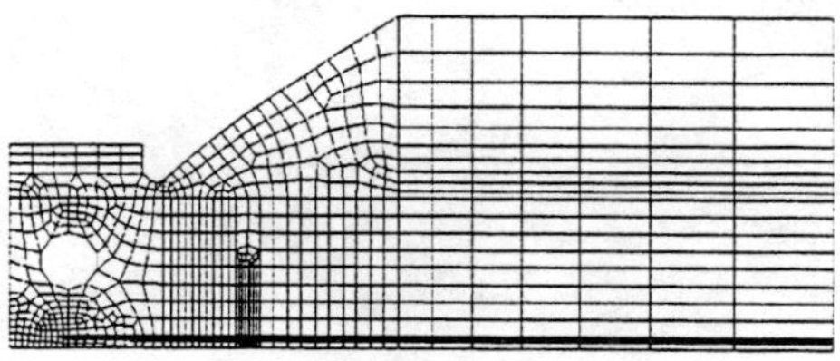

(a)

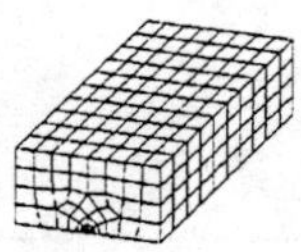

(b)

Figure 4. The finite element mesh used to model the original fracture specimen (a), and the finite element mesh in the vicinity of the crack tip (b). The mesh is composed of approximately 10,000 eight-node brick elements.

crack front. These domains can be seen in Fig. 4b. Each domain contains approximately 50 elements. The dimensions of the two-dimensional crack tip mesh are chosen such that the longest characteristic length, l, of each domain is at least one tenth of the crack length a. The characteristic length of the smallest element in each domain was taken to be $l/25$. The analyses described in the following sections were all carried out within the framework of small strain linear elasticity.

NUMERICAL RESULTS AND SHAPE OPTIMIZATION

Numerical evaluation of original specimen

The finite element procedure described in the previous section was employed to evaluate the original CDCB specimen. As a starting point, the compliance of the specimen was obtained numerically for several different crack lengths. The compliance versus crack length data obtained numerically and experimentally is plotted in Figure 5. As shown in the figure, excellent agreement is obtained between the numerical and experimental compliance data. A linear least squares fit through the data points yields a correlation coefficient of 0.999 leading one to assume that the specimen is constant K.

The mode I stress intensity factor obtained numerically by the domain integral method described in the previous section is plotted versus crack length as shown in Figure 6. In the figure, the mode I stress intensity factor is normalized with respect to the stress intensity factor value obtained for the shortest crack considered. From the figure, it is observed that the mode I stress intensity factor increases by 15% within the range of crack lengths between 38 and 76 mm. Therefore, it appears that in order to properly evaluate a constant K specimen, numerical simulation is required because of the difficulty in obtaining precise information from curve fitting through experimental data.

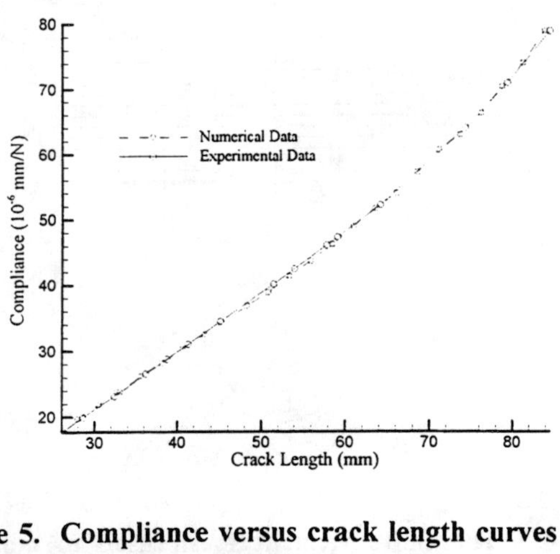

Figure 5. Compliance versus crack length curves obtained for the original fracture specimen. The circular markers represent the experimental data and the square markers represent the data obtained numerically.

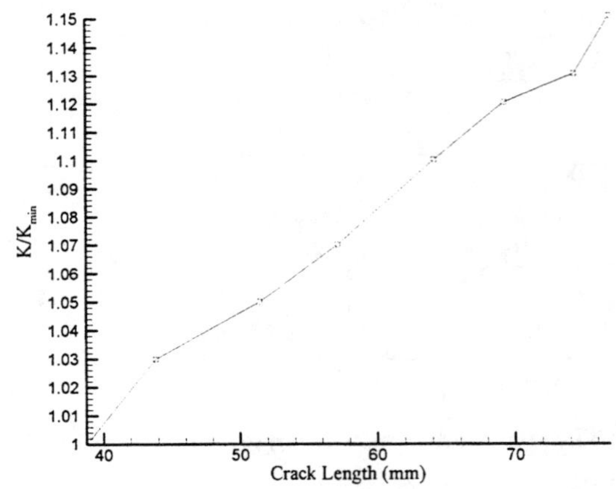

Figure 6. The mode I stress intensity factor plotted versus crack length. The stress intensity factor is normalized with respect to the K_{min}, the stress intensity factor obtained when the crack reaches the beginning of the contoured portion of the specimen.

Shape optimization

In this section, we demonstrate that it is possible to design an improved fracture specimen by systematically altering the geometry of the original specimen and evaluating the modified geometry numerically. The optimization procedure involves two phases. A pre-optimization is performed in the first phase, and the final optimization is achieved in the second phase.

Phase I

In the first phase of the optimization procedure, the contoured portion of the original specimen is approximated as a straight line. This simplified geometry is shown in Figure 7, where the leading and trailing edges of the contour are labeled as points A and B respectively. Let a_1 denote the minimum crack length to be analyzed and a_2 be the maximum crack length to be analyzed as shown in Figure 7. The next step in the first phase of the optimization procedure is to compute the mode I stress intensity factors K_{min} and K_{max} associated with these two crack lengths. The last step in this procedure depends on the desired compliance of the specimen. To produce a less compliant specimen than the original, the height, h, of the back end of the specimen is increased until the difference $K_{max} - K_{min}$ is minimized. We note that as h increases, K_{max} approaches K_{min}. Alternatively, to produce a more compliant specimen, the height of the front end of the specimen is decreased, keeping h constant, until $K_{max} - K_{min}$ is minimized. The first phase of the optimization process produces a reduction in $K_{max} - K_{min}$ and also establishes the magnitude of the compliance of the specimen.

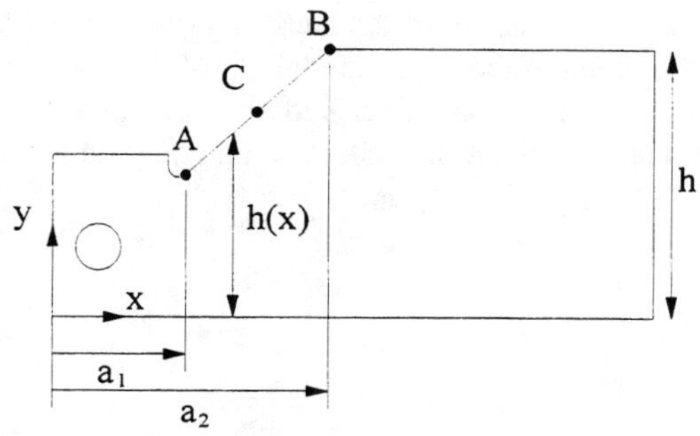

Figure 7. Original specimen after performing phase I shape optimization.

To illustrate the benefits of performing the first phase of the shape optimization process, the contoured portion of the original specimen was approximated as a straight line. The height h of the original specimen is 73.7 mm. For this specimen height, the difference between K_{max} and K_{min} was computed to be approximately 15 percent. Increasing h to 83.8 mm yields an 8 percent difference between K_{max} and K_{min}. The normalized mode I stress intensity factor is plotted versus crack length as shown in Figure 8. We note that increasing the height of the back of the specimen produces an increase in the stiffness of the specimen.

Keeping the height of the back end of the specimen fixed at the value obtained in phase I, the following iterative procedure was performed in order to further improve the design of the specimen.

Phase II

The first step in the second phase of the optimization process is to locate the midpoint C of line AB as shown in Figure 7. The next step is to choose additional control points on the line AB (equally spaced and symmetric about point C).

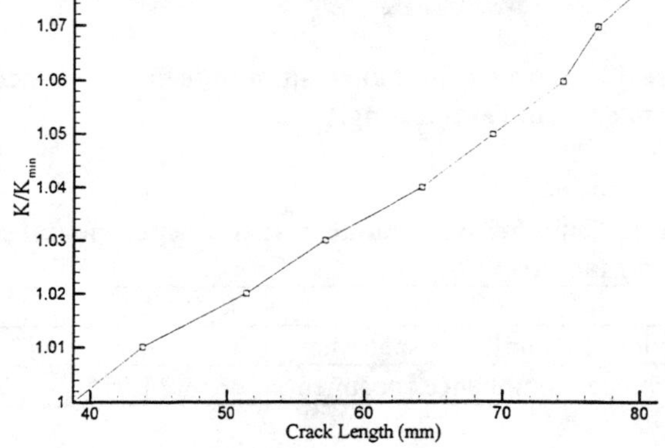

Figure 8. The mode I stress intensity factor normalized with respect to K_{min} plotted versus crack length (lower compliance specimen after phase I shape optimization).

Keeping point C fixed, the locations of the control points are then systematically varied in the y direction one pair at a time (points equidistant from C are considered pairs). After each movement of the control points, the mode I stress intensity factor is calculated as a function of crack length. This process of moving control points in the y direction and calculating the mode I stress intensity factor versus crack length is repeated until an acceptable variation is achieved. A least squares curve fit is then performed in order to determine the polynomial function $h(x)$ that provides the best fit through the final locations of the control points. We note that as the optimization process is performed step by step, increasing the height of the control points causes a localized increase in the stress intensity factor while decreasing the height causes a localized decrease.

Numerical results for the new specimens

Following the optimization procedure described above, we have designed a specimen with a variation in the mode I stress intensity factor K_I of less than three percent for a crack length range of 38 mm. The finite element mesh for the optimized specimen is shown in Figure 9. The dimensions of the front end of this specimen are identical to those given in Figure 2. The height h of the back end of the specimen is 83.8 mm. The height $h(x)$ of the contoured portion of the specimen can be described by a forth order polynomial,

$$h(x) = \alpha x^4 + \beta x^3 + \delta x^2 + \gamma x + \eta \tag{10}$$

Here in equation (10), x is measured from the left end of the specimen. The coefficients of equation (10) for this specimen are given in Table 1. This equation is valid for the range $33.02 \le x \le 83.82$ mm.

The normalized stress intensity factors are plotted versus crack length in Figure 10. We note that this modified specimen has a lower compliance and hence a higher stiffness than the original specimen.

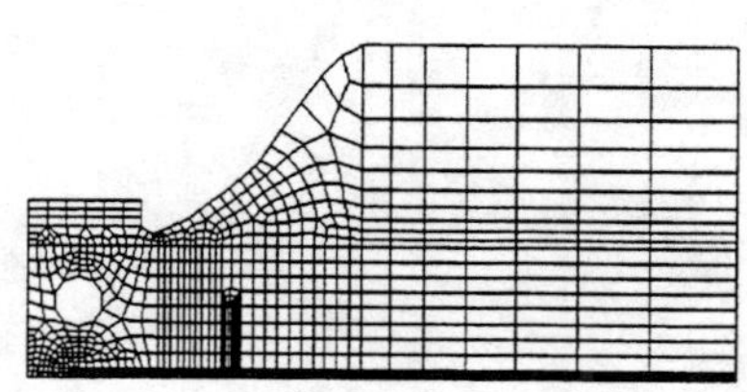

Figure 9. The finite element mesh used to model the lower compliance specimen after phase II shape optimization. The mesh is composed of approximately 10,000 eight-node brick elements.

Table 1. Coefficients describing the contoured portions of the new specimens.

	Lower compliance specimen	Higher compliance specimen
α	2.175×10^{-5}	-8.960×10^{-6}
β	4.622×10^{-2}	1.673×10^{-3}
δ	-3.487×10^{-1}	-1.000×10^{-1}
γ	11.34	3.039
η	104.2	-7.980

In order to check the validity of the numerical results, an aluminum test specimen was manufactured having the same dimensions as the model. Experimental compliance versus crack length data was then compared to the crack length versus compliance data obtained experimentally. The numerical and experimental compliance data is plotted as a function of crack length as shown in Figure 11. As shown in the figure, excellent agreement is obtained between the numerical and experimental results. A photo of the new specimen before the fracture test was conducted is shown in Figure 12.

To further establish the validity of the domain integral calculations, the ratio of the mode I stress intensity factor computed numerically versus that obtained experimentally was calculated and given in Table 2. The ratio is reported for a crack length of 57 mm. The experimentally determined stress intensity factor is obtained from the slope of the compliance curve shown in Figure 11. Excellent agreement is obtained between the numerical and experimental stress intensity factors.

Using the optimization procedure described in the previous section, a constant K specimen with a higher compliance than the original specimen was also designed. The finite element mesh for the optimized higher compliance specimen is shown in Figure 12. Again, the dimensions of the front end of this specimen are identical to those given in

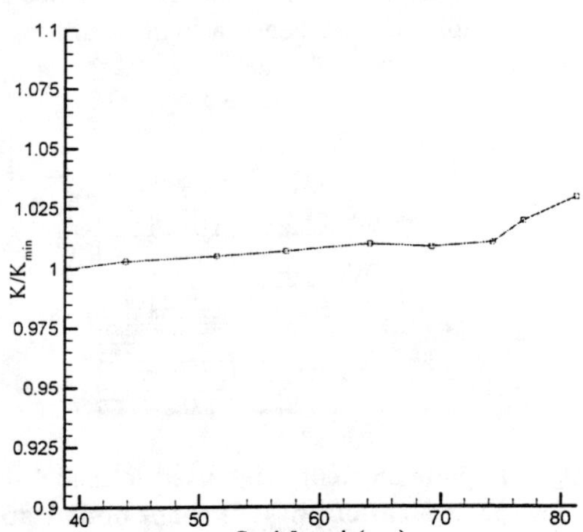

Figure 10. The mode I stress intensity factor plotted versus crack length for the optimized lower compliance specimen.

Figure 2. The height *h* of the back end of the specimen is 63.5 mm. The equation for *h(x)*, the heigth of the contoured portion of the specimen, is described by equation (10). The coefficients are given in Table 1. The equation is valid for the range $33.02 \le x \le 83.82$ mm.

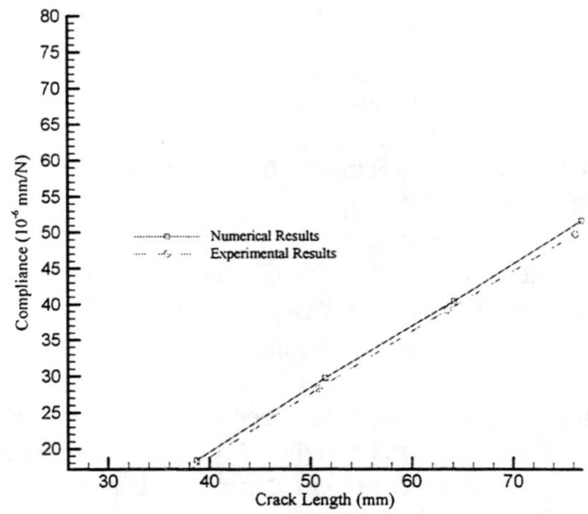

Figure 11. Compliance versus crack length curves for the optimized lower compliance specimen. The circular markers represent the experimental data and the square markers represent the data obtained numerically.

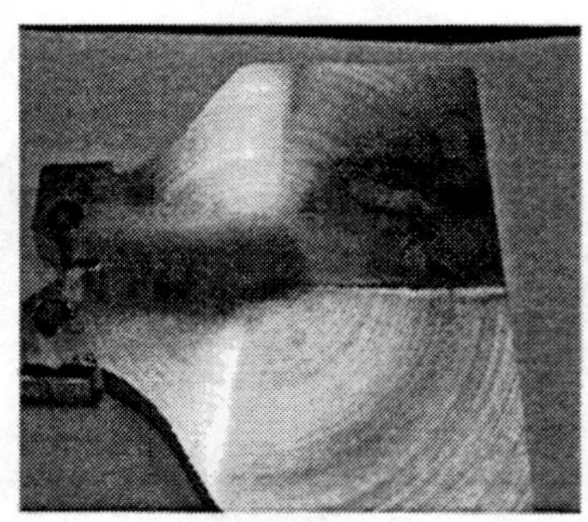

Figure 12. Lower compliance aluminum fracture specimen prior to experimental testing.

Table 2. Comparison of numerical and experimental stress intensity factors.

	$K_I^{FEM} / K_I^{experimental}$
lower compliance specimen	0.988
higher compliance specimen	1.015

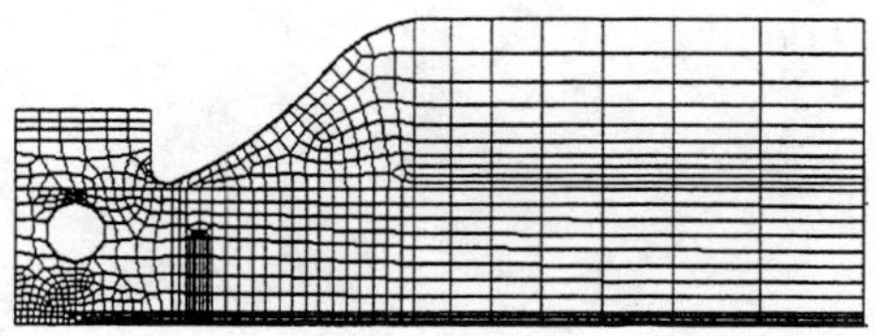

Figure 12. The finite element mesh used to model the optimized specimen with a higher compliance. The mesh is composed of approximately 10,000 eight-node brick elements.

The normalized stress intensity factors are plotted versus crack length in Figure 13. Again, in order to check the validity of the numerical results, an aluminum test specimen was manufactured having the same dimensions as the model. Experimental compliance versus crack length data was then compared to the crack length versus compliance data obtained experimentally. The numerical and experimental compliance data is plotted as a function of crack length as shown in Figure 14. As shown in the figure, excellent agreement is obtained between the numerical and experimental results. A photo of the new, higher compliance specimen before the fracture test was conducted is shown in Figure 15. The ratio of numerical versus experimental stress intensity factor for a crack length of 57 mm is reported in Table 2. Again, excellent agreement was obtained between the stress intensity factors computed experimentally and numerically.

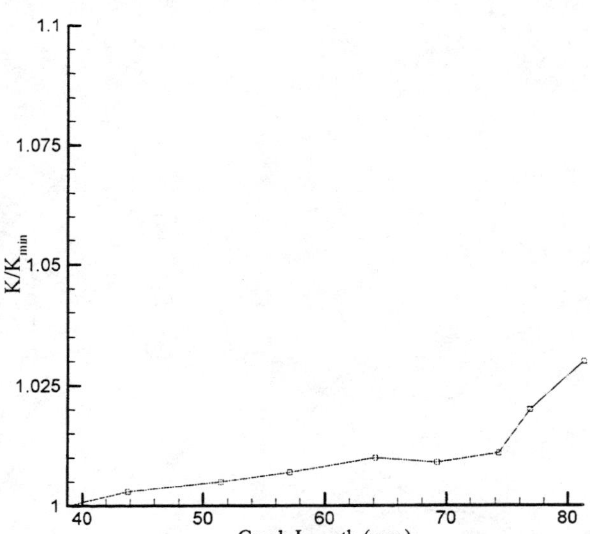

Figure 13. The mode I stress intensity factor normalized with respect to K_{min} plotted versus crack length (optimized specimen with higher compliance).

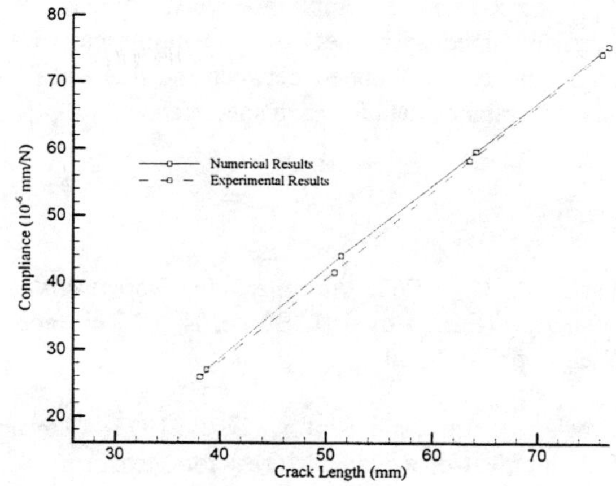

Figure 14. Compliance versus crack length curves obtained for the optimized fracture specimen with a higher compliance. The circular markers represent the experimental data and the square markers represent the data obtained numerically.

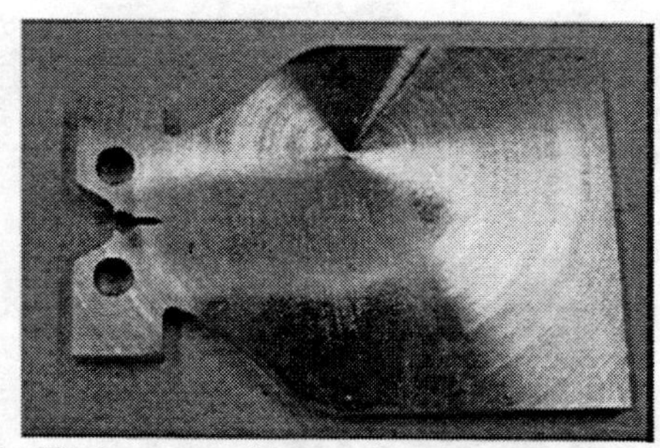

Figure 15. Higher compliance aluminum fracture specimen prior to experimental testing.

SUMMARY AND CONCLUDING REMARKS

The original CDCB fracture specimen developed by Mostovoy et al. [5] using a mechanics of materials approach was evaluated using the finite element method. A domain integral technique was employed to numerically evaluate the mode I stress intensity factor as a function of crack length. In the numerical investigations, it was found that the original CDCB specimen deviates from constant K by as much as 15 percent. The geometry of the original specimen was then optimized using an iterative finite element approach until an acceptable variation in the stress intensity factor curve was achieved. Using this optimization process, two new fracture specimens were developed. The first specimen was less compliant than the original specimen, and the second was more compliant. The stress intensity factor versus crack length curves deviated by less than 3 percent for both new specimens. In order to validate the numerical results, the new specimens were manufactured according to the dimensions determined numerically. Experimental compliance versus crack length data was then obtained as a check on the numerical results. Excellent agreement was obtained between the numerical and experimental compliance data for each specimen.

REFERENCES

[1] Paris, P. C., 1965, *Fatigue: An Interdisciplinary Approach* (Edited by J. J. Burke, N. L. Reid and V. Weiss), pp. 107-130.

[2] Crosley, P. B., and Ripling, E. J., 1977, "Towards Development of a Standard Test for Measuring Kia," *Fast Fracture and Crack Arrest, ASTM STP 627*, Eds., G.T. Hahn and M.F. Kannienen, American Society for Testing and Materials, pp. 372-391.

[3] Hertzberg, R. W., 1983, *Deformation and Fracture Mechanics of Engineering Materials*, 2nd edition, Wiley, New York, pp. 523-524.

[4] Todd, J. A., Li, Pi, Liu, G., and Raman, V., 1988, "A New Mechanism of Crack Closure in Cathodically Protected ASTM A710 Steel," *Scripta Metallurgica*, Vol. 22, pp. 745-750.

[5] Mostovoy, S., Crosley, P. B., and Ripling, E. J., 1967, "Use of Crack-Line-Loaded Specimens for Measuring Plane-Strain Fracture Toughness," *Journal of Materials*, Vol. 2, No. 3, pp. 661-681.

[6] Fransen, J. D., and Marcus, H. L., 1975, "The Correlation Between Grain Size and Plastic Zone Size for Environmental Hydrogen Assisted Fatigue Crack Propagation," *Scripta Metallurgica*, Vol. 9, pp.1089-1094.

[7] Liu-Nash, G., Todd, J. A., and Mostovoy, S., 1997, "Accelerated Test Method for Determining Near-Threshold Stress Intensity Value in HSLA Steels," *Fatigue Fracture Engineering Material Structure*, Vol. 20, No. 12, pp. 1657-1664.

[8] Moran, B., and Shih, C. F., 1987, "A general Treatment of Crack-Tip Contour Integrals," *International Journal of Fracture,* Vol. 35, pp. 295-310.

[9] Moran, B.. and Shih, C. F., 1987, "Crack-Tip and Associated Domain Integrals from Momentum and Energy Balance," *Engineering Fracture Mechanics,* Vol. 27, No. 6, pp. 615-642.

PVP-Vol. 415, Recent Advances in Solids and Structures — 2000
ASME 2000

BUCKLING VS. CRACK GROWTH: NUMERICAL SIMULATION

T. Siegmund
School of Mechanical Engineering,
Purdue University,
West Lafayette, IN 47907-1288

ABSTRACT

Thin walled structures are characterized by configurations that possess large aspect ratios, i.e. large in-plane dimensions and small thickness dimensions. The present work aims on investigations of the mechanical integrity of such structures thereby focusing on the competition and interaction between global failure due to buckling and local failure due to crack growth.

INTRODUCTION

Two specific configurations are found in prominent examples that will be investigated here:

(1) Shell structures made of sheet metal of several millimeter thickness and in-plane dimensions of several meters, and

Thin walled shell structures constructed of high strength sheet metal are frequently used if lightweight and high stiffness construction is required. These thin walled shell structures are strong but are also known to be very instability sensitive in the presence of imperfections. Failure of thin-walled shell structures due to buckling and collapse has attracted considerable attention in the past. In addition to this deformation instability behavior, the structural behavior of a thin-walled shell structure is affected by the presence of cracks. Long cracks can arise from accidental damage during use or from fatigue crack growth whereby e.g. several smaller cracks along a row of rivets might have linked up. In a design of a thin-wall shell structure aiming for structural integrity the presence of cracks has thus to be taken explicitly into account. It is thus necessary to understand the interaction between crack growth and the deformation instabilities of buckling and collapse, (Harris et al. 1998 and Starnes et al. 1998). Within the concept of the "damage tolerant"-design approach, it has to be assumed that critical cracks can be present within the structure at any time. Consequently, it is required that the structure under consideration is "tolerant" to this damage and that it can contain a crack without failing in an unstable manner. Damage tolerance and damage containment is thus most important.

(2) Thin-film deposits of (sub)-micrometer thickness deposited on substrates as used in the microelectronics industry.

A technology of central importance to the microelectronic industry is deposition of thin films onto a substrate of another material. The production of such structures leads to stresses in the thin film due to lattice mismatch, differences in coefficients of thermal expansion in addition to external loading, Evans et al. 1995. These stresses in the film may cause cracking, buckling and delamination, limiting applications of the film (Nilson et al. 1995 and Choi et al, 1999). Stresses in thin films have to be controlled to a high level of accuracy to avoid failure, yet in several applications some stresses may be necessary to optimize performance, e.g. in lasers and detectors with precisely controlled optical properties. To calibrate the amount of stress in thin films and to avoid failure it is of paramount importance to develop design tools for the analysis of their thermo-mechanical behavior and their failure. Furthermore, a set of new thin film tests has been proposed recently that takes advantage of the buckling instability in the determination of adhesion properties, (De Boer et al. 1996).

In both systems describe above the formation of cracks limits the integrity of the structure. The systems considered are different from other cracked structures in that the large aspect ratio also introduced the possibility of buckling. It is the non-interaction between the processes associated with crack initiation, growth and arrest and the buckling and post-buckling processes that will be investigated.

Regardless of the structure investigated, simulation and modeling is required since experimental testing is sometimes difficult to perform both on large as well as micro-sized structures. The expense of a full-scale testing of e.g. a complete fuselage of an airplane, the main cabin of a space station or a hazardous goods container limits the number of such tests that can be performed. Simulation tools are necessary to interpret experimental results, reduce the number of experiments that need to be performed and put focus on the testing of the most critical parameters. Current predictions of failure of thin walled structures are based primarily on fracture mechanics. Two main causes severely limit the applicability of this approach:

(1) <u>Non-linear effects</u>: Several causes contribute to non-linearity that must be fundamentally understood and accurately predicted so that safe lightweight structures and reliable thin films can be designed.

- Larger amounts of deformation normal to the shell/film are present in the case that a shell/film becomes unstable and buckling occurs, (Nemeth et al. 1998), also
- Locally, at the crack tip stresses and strains reach larger values than in areas remote from the crack. Significant out-of-plane deformations in shell/film structures are observed in the area surrounding the crack. This phenomenon is known as crack bulging. In both cases, if the out-of-plane deformation becomes large compared to the shell/film thickness a non-linear amplification of the level of local stresses at the crack tip occurs, (Riks et al. 1992, Markstroem et al. 1980).
- With increasing crack length load re-distributions occurs resulting in a continuous change of the shell/film stiffness.

Currently the nonlinear behavior is accounted for mainly by the use of *correction* factors, (Rose et al. 1999). Considering the advances in non-linear structural mechanics analysis tools, e.g. the finite element method, this approach becomes antiquated.

(2) <u>Size</u>: Thin film structures are generally of size so that any size requirements necessary for the application of fracture mechanics is violated.

Within the present proposal the goal is to overcome these restrictions. Capabilities will be introduced to analytically predict not only the onset of cracking, but also the subsequent crack propagation and the possible crack arrest coupled to the strong non-linearity of buckling and post-buckling. This task can be accomplished by the development and application of theories that allow fracture to emerge as a natural outcome of the deformation history. Such an approach accounts for the progressive nature of fracture. Hereby, the dominant microscopic mechanisms of damage leading to crack growth are captured by specific constitutive equations that incorporate the process of material separation itself. It contains two material parameters in addition to the conventional elastic-plastic parameters: a cohesive strength and a cohesive energy. Material toughness and the resulting damage tolerance of a structure are the result of interplay between the local material separation mechanism and the elastic-plastic deformation of the surrounding continuum. Previous investigations on similar problems were performed using the concept of a critical crack tip opening angle as crack growth criterion (Seshadri et al. 1998, and Gullerud et al. 1999). These investigations demonstrate that the choice of the critical angle effects the numerical predictions of buckling considerably. The present approach avoids this issue since the crack opening angles are an outcome of the analysis.

A length-scale is introduced via the use of the material separation constitutive equation. It contains the cohesive length, i.e. the ratio of cohesive energy to cohesive strength, as a material parameter.

The computing power to deal with a non-linear length scale dependent mechanics of real three-dimensional large-scale structures is becoming available. The proposed work will contribute to substitute experience based phenomenological fracture theories by a much more fundamental and universal approach. Designers will benefit from this work as their understanding of how damage initiates and develops will significantly increase and the limits of current design rules become clearly understood.

COMPUTATIONAL METHOD

In the present investigation a cohesive zone model is used to account for material separation, (Needleman 1992). Cohesive zone models have been predominately used for investigations in plane strain or axisymmetrix configurations. Recently, interest in full 3D analyses with this approach has emerged, (Pandolfi et al. 2000). The present development contributes to this development by adopting the CZ approach to be used with shell elements. The CZ elements thereby retain their 1D line element structure but incorporate arbitrary 3D kinematics. The specific for of the traction separations law used is:

$$T_n = -\sigma_{\max} e \left\{ \frac{\Delta_n}{\delta_n} \exp\left(-\frac{\Delta_t^2}{\delta_t^2} \right) \exp\left(-\frac{\Delta_n}{\delta_n} \right) \right\} ,$$

$$T_t = -\sigma_{\max} e \left\{ \frac{2\delta_n}{\delta_t} \frac{\Delta_t}{\delta_t} \left[1 + \frac{\Delta_n}{\delta_n} \right] \exp\left(-\frac{\Delta_t^2}{\delta_t^2} \right) \exp\left(-\frac{\Delta_n}{\delta_n} \right) \right\}$$

The cohesive zone material parameters are the cohesive zone strength, σ_{max}, and the cohesive lengths for normal and shear separation, δ_n and δ_t. In the following, the conditions $\delta_n = \delta_t$ was used. Cracks are constrained to grow a linear extension of the initial crack line. The cohesive zone model was implemented into the FE-code ABAQUS, (H.K.S. 1999) via the UEL interface capability.

SHEET METAL STRUCTURES

Material Parameter Determination

In the present studies on sheet metal structures Al 2024-T3 Alclad of thickness 1.0 mm is considered. The flow properties of this material were considered in the form of:

$$\varepsilon_t = \begin{cases} \sigma_t / E & \text{for} \quad \sigma_t < \sigma_0 \\ (\sigma_0 / E)(\sigma_t / \sigma_0)^n & \text{for} \quad \sigma_t > \sigma_0 \end{cases}$$

The material properties used in this equations are E=71400 MPa, ν=0.3, σ_0=285 MPa and n=8, (Deng et al. 1996).

Motivated by micromechanical considerations and continuum damage mechanics investigations the cohesive zone material parameter for this material were determined to be, (Siegmund et al. 2000):

$$\sigma_{\max} = 2\sigma_0 \quad , \quad \delta = 0.05\text{mm}$$

Crack growth experiments on center-cracked panels were reported in (Broek et al. 1994). Results comparing the experimental data to FE predictions using the above cited parameters are depicted in Fig. 1. Good agreement between experiment and numerical simulation was obtained; see also (Siegmund et al. 2000).

Crack Growth in a Cylindrical Shell

Crack growth simulations in a cylindrical shell containing a longitudinal through crack under a combined loading of internal pressure and axial compression were performed. The geometry of the finite element model is depicted in Figure 2.

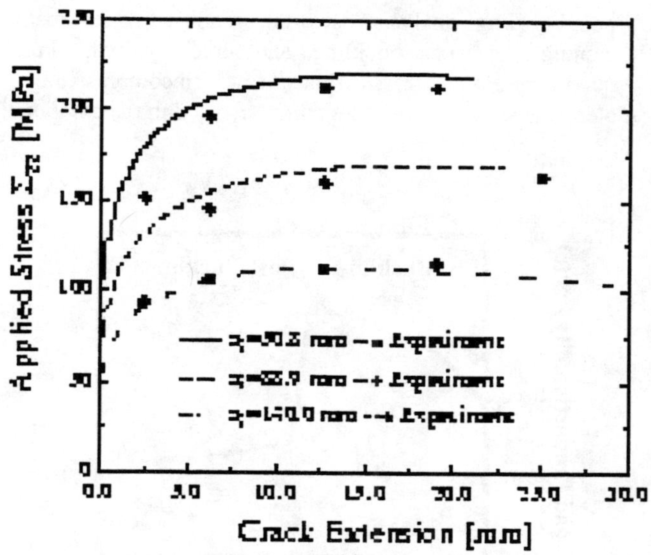

Figure 1: - Applied stress vs. crack extension for the center crack panels of Al 2024-T3 Alclad. Three different crack lengths are considered. Comparison of numerically determined values to the experimental data.

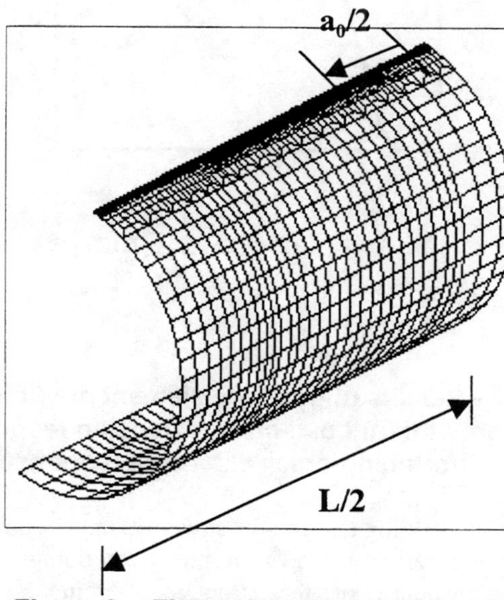

Figure 2: - Finite element model for a cracked cylindrical shell.

As in all other following investigations the computations are performed by first running a linear perturbations analysis without the cohesive elements. Small perturbations based on the FE-mesh containing the cohesive elements. The subsequent post-buckling analysis uses the modified Riks method to capture the post buckling and crack growth. Four-node quadrilaterals, stress/displacement incompressible shell elements with reduced integration are used in the computations.

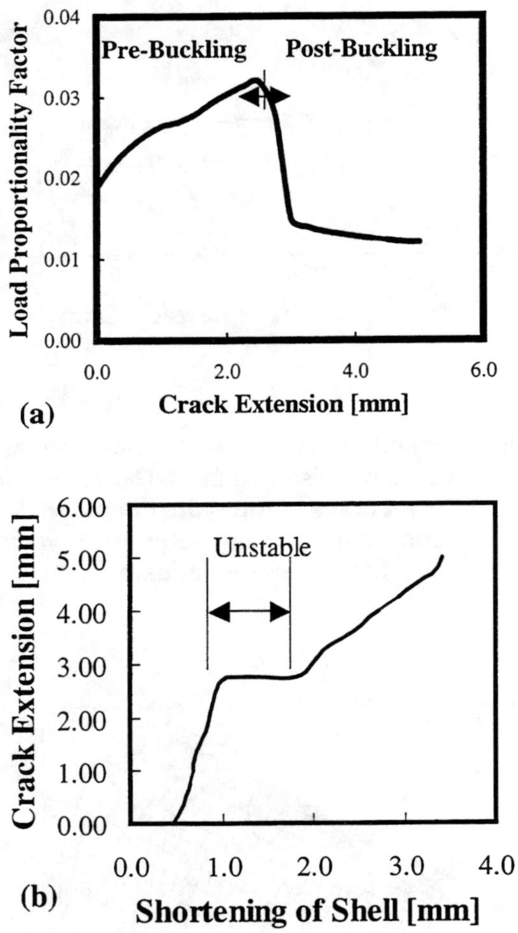

Figure 3: - Results of the finite element predictions of crack growth: (a) Load-crack extension response; (b) Shortening-crack extension response.

Results from buckling analyses of structures under compressive loads are generally presented in plots describing the load-shortening response of the structure. Within the present approach it is possible to extent this evaluation into load-crack growth plots. A typical representative load-crack growth plot for a structure in which buckling and crack growth occur simultaneously is depicted in Figure 3a. Initially the crack propagates under increasing load, at the buckling event the load drops and subsequent crack growth occurs under a significantly reduced applied stress level. The crack growth rate is predicted to be large initially, during the instability event crack growth stops and the crack temporarily arrests. Subsequently after the buckling event, the crack reinitiates but propagates at a much lower rate; see Figure 3b.

Details of the hoop stress distribution as computed for the cylindrical shells for 5mm crack extension are depicted in Figure 4.

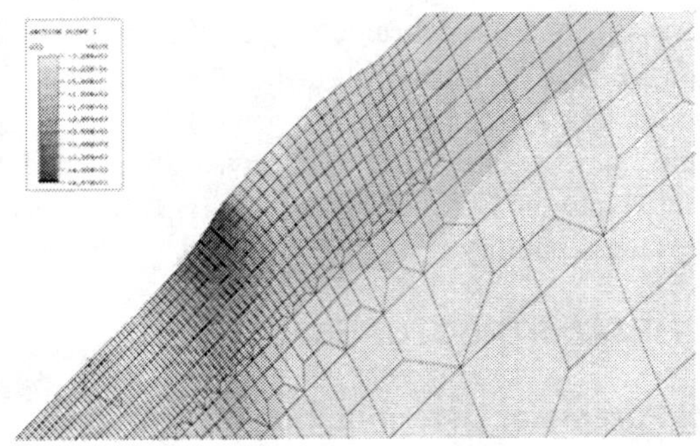

Figure 4: - Deformed shell and hoop stress in the crack tip region in the cylindrical shell under internal pressure and axial compressive load. The crack extension is equal to 5mm.

THIN FILM STRUCTURES

Problem Definition

In the present paper some preliminary results of modeling efforts on thin film adhesion tests are described. Specifically, focus is on the analysis of the pre-cracked-line scratch test, De Boer et al. 1996. Figure 5 depicts a schematic drawing of the test.

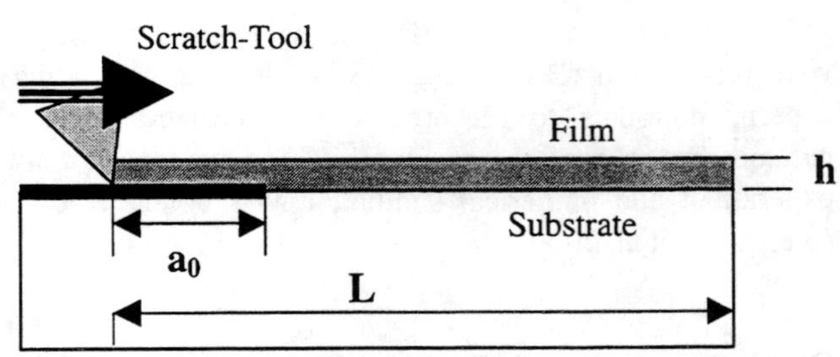

Figure 4 – Schematic drawing of the pre-cracked line scratch test.

A 2D finite element model for the configuration depicted in Figure 4 was developed. The model uses four node plane strain

A 2D finite element model for the configuration depicted in Figure 4 was developed. The model uses four node plane strain solid elements as well as cohesive zone elements along the substrate-film interface.

The material parameter used for the cohesive zone describing the interface were σ_{max}/E=0.0005 and h/δ=2000. A series of different geometries with initial crack lengths of a_0/L= 0.5, 0.375, 0.25, 0.1875, 0.125, 0.075 was investigated. Results from the crack growth-buckling analyses of these structures are depicted in Figure 5. The diagram depicts the computed results of load vs. film lift-off.

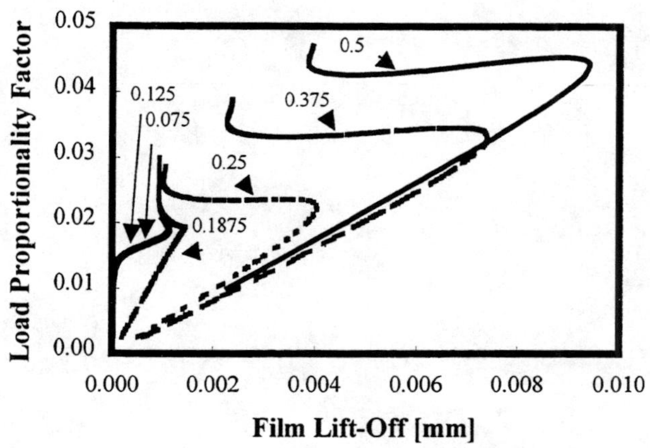

Figure 5:- Results of the computations of the pre-cracked line scratch test.

For initial crack lengths of a_0/L= 0.5, 0.375, 0.25 the behavior is predicted such that buckling of the film occurs prior to crack initiation. For initial crack lengths of a_0/L= 0.1875, 0.125, 0.075 and for the material parameters investigated the initiation of a mode II crack occurs and buckling occurs only later in the loading process.

SUMMARY

In the present paper first results of the application of a cohesive zone model to the analysis of structures that exhibit instability are presented. For the purpose of these analyses a cohesive model was extended to incorporate a full 3D kinematical relations.

For the thin sheet shell structures crack growth was predicted to be strongly influenced by the occurrence of the buckling event. During instability crack arrest and subsequent reinitiating occurred. The cohesive zone model is fully capable to predict such events.

For the pre-cracked line scratch test of the thin film configuration crack growth occurs under dominant mode II. The choice of crack length determines whether crack growth or buckling occur first. This transition is a function of the adhesive properties of the interface between substrate and film. This finding confirms that experiments including crack growth and instability are promising for the determination of adhesion properties.

ACKNOWLEDGMENT

This work was partially supported by National Computational Science Alliance under grant MSS990004N and utilized the NCSA SGI/CRAY Origin2000.

REFERENCES

Evans, A.G., Hutchinson, J.W., The thermomechanical integrity of thin films and multilayers, Acta metall. mater., 43 (1995) 2507-2530.

Nilson, K.-F., Giannakopoulos, A.E., A finite element analysis of configurational stability and finite growth of buckle driven delamination, J. Mech. Phys. Solids, 43 (1995) 1983-2021.

Choi, S.R., Hutchinson, J.W., Evans, A.G., Delamination of multilayer thermal barrier coatings, Mech. Materials, 31 (1999) 431-447.

De Boer, M.P., Gerberich, W.W., Microwedge indentation of the thin film fine line--I. Mechanics, Acta Mater., 44 (1996) 3169-3175.

Harris, C.E., Newman, J.C., Jr., Piascik, R.S., Starnes, J.H., Jr., Analytical methodology for predicting widespread fatigue damage onset in fuselage structures, J. Aircraft 35 (1998) 307-317.

Starnes, J.H., Rose, C.A., Buckling and stable tearing response of unstiffned aluminum shells with long cracks, AIAA paper No. 98-1991, 1998.

Nemeth, M.P., Starnes, J.H., The NASA monographs on shell stability design recommendations, NASA / TP-1998-206290.

Riks, E., Rankin, C.C., Brogan, F.A., The buckling behavior of a central crack in aplate under tension, Engng Fract. Mech., 43 (1992) 529-548.

Markstroem, K., Storakers, B., Buckling of cracked members under tension, Int. J. Solids Structures, 16 (1980) 217-229.

Rose, C.A., Young, R.D., Starnes, J.H., Nonlinear local bending response and bulging factors for longitudinal cracks in pressurized cylindrical shells, AIAA paper No. 99-1421.

Needleman, A., An analysis of decohesion along an imperfect interface, Int. J. Fract. 40 (1992) 1377-1397.

Gullerud, A.S., Dodds, R.H., Hampton, R.W., Dawicke, D.S., Three-dimensional modeling of ductile crack growth in thin sheet metals: computational aspects and validation, Eng. Fract. Mech., 63 (1999) 347-374.

Seshadri, B.R., Newman, J.C., Jr., Analyses of buckling and stable tearing in thin-sheet materials, ASTM STP 1332, T.L. Panontin et al., eds., American Society for Testing of Materials, Philadelphia, 1998.

Pandolfi, A., Guduru, P.R., Ortiz, M., Rosakis, A.J., Three dimensional cohesive-element analysis and experiments of dynamic fracture in C300 steel, Int. J. Solids Structures, 37 (2000) 3733-3760

HKS, Inc., ABAQUS, User's Manual (1998) Version 5.8, Pawtuket, RI.

Deng, X., Hutchinson, J.W., Approximate methods for analyzing the growth of large cracks in fatigue damaged aircraft sheet material and lap joints, Finite Elements in Analysis and Design 23 (1996) 101-114.

Siegmund, T., Brocks, T., Modeling Crack Growth in Thin Sheet Aluminum Alloys, Fatigue and Fracture Mechanics: 31st Volume, ASTM STP 1389, G.R. Halford, and J.P. Gallagher, Eds., in print, 2000.

Broek, D., Thomson, D. and Jeong, D.Y., "Testing and Analysis of Flat and Curved Panels with Multiple Cracks, FAA-NASA International Symposium on Advanced Structural Integrity Methods for Airframe Durability and Damage Tolerance, NASA Conference Publication 3274, Part I, September 1994, pp. 85-98.

PVP-Vol. 415, Recent Advances in Solids and Structures — 2000
ASME 2000

HIGHER ORDER SINGULARITIES AND THEIR ENERGETICS IN ELASTIC-PLASTIC FRACTURE

Seyoung Im
Department of Mechanical Engineering, KAIST, Science Town, TaeJon 305-701, S. Korea;
sim@kaist.ac.kr

Insu Jeon
Department of Mechanical Engineering, KAIST, Science Town, TaeJon 305-701, S. Korea;
jis@imhp.kaist.ac.kr

ABSTRACT

The higher order singularities[1] are systematically examined, and discussed are their complementarity relation with the nonsingular eigenfunctions and their relations to the configurational forces like *J*-integral and *M*-integral. By use of the so-called two state conservation laws[2] or interaction energy, originally proposed by Eshelby[3] and later treated by Chen and Shield[4], the intensities of the higher order singularities are calculated, and their roles in elastic-plastic fracture are investigated. Numerical examples are presented for illustration.

INTRODUCTION

The purpose of the present work is to review the complementarity relationships among the eigenvalues in an eigenfunction expansion for a generic isotropic wedge together with our recent applications to wedge and crack problems[2,5,6]. Furthermore, we demonstrate their application for characterizing elastic-plastic cracks via high order singularities and their complementary eigenpairs, which are nonsingular terms.

GOVERNING EQUATIONS AND *J*- AND *M*-INTEGRAL

For the plane problem, the *J*-integral and *M*-integral[2] may be written as:

$$J = \int_{\Gamma} (Wn_1 - t_i u_{i,1}) ds \tag{1}$$

$$M = \int_{\Gamma} (Wn_l - t_i u_{i,l}) x_l ds \tag{2}$$

where n_l is the component of unit outward normal on the contour Γ; W and t_i indicate the strain energy density and the traction component, given as $W = C_{ijkm}\varepsilon_{ij}\varepsilon_{km}/2$ and $t_i = \sigma_{ij}n_j$.

Consider two independent elastic states "A" and "B" for the plane strain problems. We suppose another elastic state "C", which is obtained by superposing the two equilibrium states "A" and "B". Then the path-independent integrals *J* and *M* are written as

$$J^C = J^A + J^B + J^{(A,B)} \tag{3}$$

$$M^C = M^A + M^B + M^{(A,B)} \tag{4}$$

where the superscripts "A", "B" and "C" indicate the aforementioned elastic states, and $J^{(A,B)}$ and $M^{(A,B)}$ are given as

$$M^{(A,B)} = \int_C [C_{ijkm}\varepsilon^A_{ij}\varepsilon^B_{km} n_n - (t^A_i \frac{\partial u^B_i}{\partial x_n} + t^B_i \frac{\partial u^A_i}{\partial x_n})] x_n ds \tag{5}$$

$$J^{(A,B)} = \int_C [C_{ijkm}\varepsilon^A_{ij}\varepsilon^B_{km} n_1 - (t^A_j \frac{\partial u^B_j}{\partial x_1} + t^B_j \frac{\partial u^A_j}{\partial x_1})] ds \tag{6}$$

The integrals $J^{(A,B)}$ and $M^{(A,B)}$ result from the mutual interaction between two elastic states "A" and "B". These are conservation integrals for two equilibrium states, since the area integral version of these contour integrals vanishes identically for the domains with no singularities

We briefly summarize the structure of the asymptotic solutions in the form of eigenfunction series for the two-dimensional wedge problem. The eigenfunction expansion for the stress and displacement components for the generic wedge may be written in the following power type function of $z = x_1 + ix_2$ and $\bar{z} = x_1 - ix_2$ [2]:

$$\sigma_{\alpha\beta}^{(m)} = Re[\sum_{\delta_n} \beta_n \sum_{k=1}^{2} \{C_{kn}^{(m)}(\Lambda_{\alpha\beta k} g_n'(z) + \Gamma_{\alpha\beta k} \bar{z} g_n''(z)) + C_{(k+2)n}^{(m)}(\bar{\Lambda}_{\alpha\beta k} g_n'(\bar{z}) + \bar{\Gamma}_{\alpha\beta k} z g_n''(\bar{z}))\}] \tag{7}$$

$$u_\alpha^{(m)} = \frac{1}{2\mu^{(m)}} Re[\sum_{\delta_n} \beta_n \sum_{k=1}^{2} \{C_{kn}^{(m)}(p_{\alpha k}^{(m)} g_n(z) + q_{\alpha k} \bar{z} g_n'(z)) + C_{(k+2)n}^{(m)}(\bar{p}_{\alpha k}^{(m)} g_n(\bar{z}) + \bar{q}_{\alpha k} z g_n'(\bar{z}))\}] \tag{8}$$

with $g_n'(z) = z^{\delta_n}$ *and* μ being the shear modulus; the non zero components of $\Lambda_{\alpha\beta k}$, $\Gamma_{\alpha\beta k}$, $p_{\alpha k}$ and $q_{\alpha k}$:

$$\begin{aligned} -\Lambda_{111} = \Lambda_{221} = i\Lambda_{121} = 1,\ \Lambda_{112} = \Lambda_{222} = 2, \\ \Gamma_{112} = -\Gamma_{222} = -i\Gamma_{122} = -1, p_{11}^{(m)} = -ip_{21}^{(m)} = -1, \\ p_{12}^{(m)} = ip_{22}^{(m)} = 3 - 4\nu^{(m)},\ q_{12} = -iq_{22} = -1 \end{aligned} \tag{9}$$

where δ_n is an eigenvalue and ν is Poisson's ratio; C_{kn}, short for $C_k(\delta_n)$, is the corresponding eigenvector; $\beta_n = \beta(\delta_n)$ represents the load parameter or the intensity of the elastic field associated with eigenvalue δ_n. Note that β_n is real for a real δ_n, but it is, in general, complex for a complex δ_n. For a complex δ_n, it is self-evident from the expression (7) that its conjugate $\bar{\delta}_n$ also belongs to the eigenvalues. For clarity, we assume that the imaginary part of complex δ_n is positive in the expression (7) since a complex eigenvalue δ_n and its conjugate $\bar{\delta}_n$ lead to the same eigenfunction. The superscript "*(m)*" indicates the *m*-th sector.

For an arbitrary eigenvalue δ_l in equations (7) and (8), we first define its complementary eigenvalue δ_l^c in the *M*-integral sense such that

$$\delta_l^c + \delta_l = -2 \tag{10}$$

It has been shown that δ_l^c is also an eigenvalue for a given wedge [2, 7]. For cracks around which the *J*-integral becomes path-independent, in a similar way we can define the complementary eigenvalue δ_l^c in the *J*-integral sense for an arbitrary eigenvalue δ_l :

$$\delta_l^c + \delta_l = -1 \tag{11}$$

Equation (10) can be utilized for finding the intensities of the singular terms in the eigenfunction expansion for the re-entrant vertices of thin films[2] and adhesive lap-joints[5]. Recently it is shown that equation (11) together with (10) is useful for decomposing local three dimensional crack tip field under mixed mode to obtain the individual stress intensities[6]. In addition to the recent results of ref. [2,5,6], in the present presentation we will show that equation (10) or (11) may be applied for computing the intensities of the high order singularities[1] and the nonsingular terms as well for elastic-plastic crack tips. This enables us to characterize the elastic-plastic cracks via these additional terms together with the inverse square root singularity. Specializing equation (7) for elastic-plastic cracks with plastic zone removed, we obtain the expressions:

$$\begin{aligned} \sigma_{ij} = \cdots + \beta_{-2} r^{-3/2} f_{ij}(\theta, -3/2) + \beta_{-1} r^{-1} f_{ij}(\theta, -1) \\ + (K_I / \sqrt{2\pi})r^{-1/2} f_{ij}(\theta, -1/2) + \beta_1 \delta_{1i} \delta_{1j} \\ + \beta_2 r^{1/2} f_{ij}(\theta, 1/2) + \beta_3 r f_{ij}(\theta, 1) + \cdots \end{aligned} \tag{12}$$

where K_I and β_1 represent the stress intensity factor and the *T*-stress, respectively. Utilizing the aforementioned two-state *J*- or *M*-integral with the aid of finite element analysis, we calculate the intensities β_j as well as K_I for elastic-plastic crack tip. The result shows that the contribution to *J*-integral comes from each complementary pair of eigenfunctions, defined by equation (11), in addition to the inverse square root singularity, whose complementary pair is itself or $\delta_0 = \delta_0^c = -1/2$.

For the elastic-plastic plane problem, the *J*-integral and *M*-integral can be written as

$$J = J_0 + \sum_{n=1}^{\infty} J^{(\delta_n, \delta_n^c)} \tag{13}$$

$$M = \sum_{n=0}^{\infty} M^{(\delta_n, \delta_n^c)} \tag{14}$$

where $J^{(\delta_n, \delta_n^c)}$ and $M^{(\delta_n, \delta_n^c)}$ are the two-state conservation integrals[2], which provide the method for calculating the intensities of higher order singularities. For the eigenvalue δ_n in plane strain crack problems, we have complementary eigenvalue $\delta_n^c = -1 - \delta_n$ in the *J*-integral sense and $\delta_n^c = -2 - \delta_n$ in the *M*-integral sense wherein $J^{(\delta_n, \delta_n^c)}$ and $M^{(\delta_n, \delta_n^c)}$ retain the path independence. $J_0 = J(-1/2, -1/2)$ is the classical expression of the *J*-integral for elastic crack problem calculated with $\delta_0 = -1/2$. The summation

$\sum_{n=1}^{\infty} J^{(\delta_n,\delta_n^c)}$ and $\sum_{n=0}^{\infty} M^{(\delta_n,\delta_n^c)}$ are the *J*-integral and the *M*-integral contribution due to each of the eigenvalues δ_n and their conjugate eigenvalues δ_n^c, respectively. J_0, related to the translation of $\delta_0 = -1/2$ singularity, means the extension of the crack, and the remains terms $\sum_{n=1}^{\infty} J^{(\delta_n,\delta_n^c)}$ are associated with translation of the higher order singularities. Furthermore $\sum_{n=0}^{\infty} M^{(\delta_n,\delta_n^c)}$ is identically zero in the absence of the plastic zone, but increases in proportion to the size of the plastic zone due to the increasing influence of the higher order singularities.

NUMERICAL RESULTS AND CONCLUSION

For a numerical example, we choose a homogeneous isotropic material with the properties: $E = 71GPa$, $\nu = 0.33$ and $\sigma_Y = 303MPa$. For the model of this study, we select single edge notched tension(SENT) panel(see Fig. 1) and the Mode I loading is applied on the boundary of the model(see Fig. 2). The calculation was processed with a power-law hardening material in a uniaxial stress-strain law that is the plastic part of the Ramberg-Osgood formula. We choose $m = 10$ and $\alpha = 1.0$ for numerical computation. The package code ABAQUS is used for the finite element solution.

Fig. 3 and 4 show that the computed J, J_0 and each value of $J^{(\delta_n,\delta_n^c)}$ as well as M and $M^{(\delta_n,\delta_n^c)}$ increase as the maximum plastic zone size increases. From these figures we may draw the conclusion that $J^{(1/2,-3/2)}$ and $M^{(-1/2,-3/2)}$ have the most dominant effect on $\sum_{n=1}^{\infty} J^{(\delta_n,\delta_n^c)}$ and $\sum_{n=0}^{\infty} M^{(\delta_n,\delta_n^c)}$, respectively.

REFERENCES

1. Hui, C. Y., and Ruina, A., 1995, "Why K ? High order singularities and small scale yielding," Int. J. Fracture, Vol. 72, pp. 97-120
2. Im, S., and Kim, K. S, 2000, "An application of two-state M-integral for computing the intensity of the singular near-tip field for a generic wedge," J. Mech. Phys. of Solids, Vol. 48, pp. 129-151.
3. Eshelby, R. D., 1956, *The Continuum Theory of Lattice Defects*, Solid State Physics, **III**, eds., Seitz, F. and Turnbull, D., Academic Press.
4. Chen, F. H. K., and Shield, R. T., 1977, "Conservation Laws in Elasticity of the J-integral type," Z. Angw. Math. Phys.(ZAMP), Vol. 28, pp. 1-22.
5. Lee, Y., Goo, N. S., and Im, S., 2000, "Analysis of adhesive lap joints with the aid of a conservation integral," in preparation.
6. Kim, Y., and Im, S., 2000, "Mode decomposition of three-dimensional mixed-mode cracks via two-state integrals," submitted to Int. J. Solids & Struct.
7. Leguillon, D., and Sanchez-Palencia, E., 1987, *Computation of Singular Solutions in Elliptic Problems and Elasticity*, Masson(or John Wiley & Sons), New York.

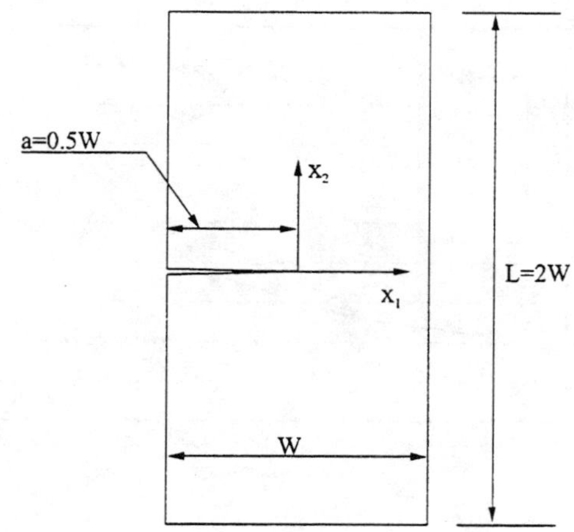

Fig. 1 The SENT panel under plane strain

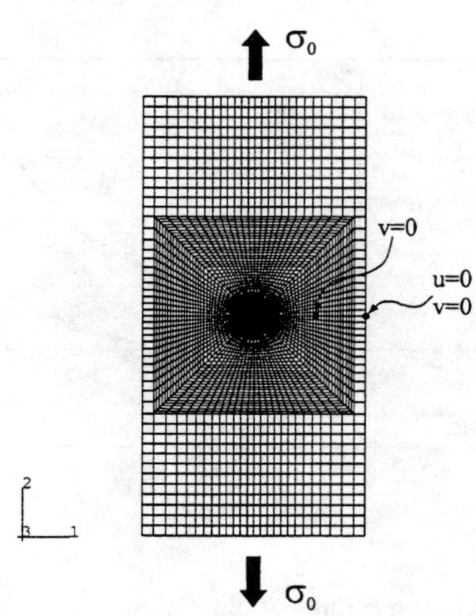

Fig. 2 F.E. mesh and boundary condition

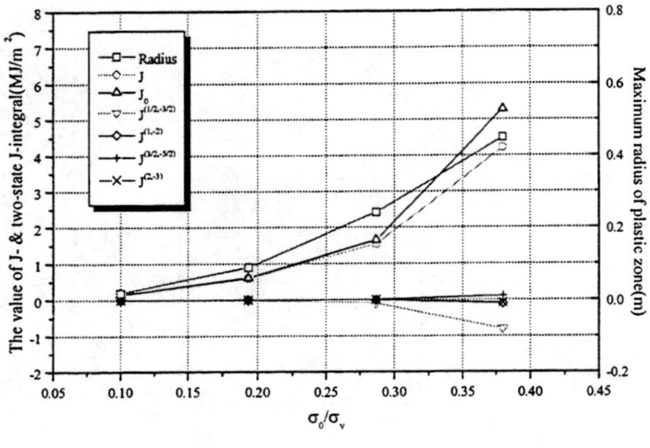

Fig. 3 Maximum radius of plastic zone and computed *J*-integral and two-state *J*-integrals versus applied loading

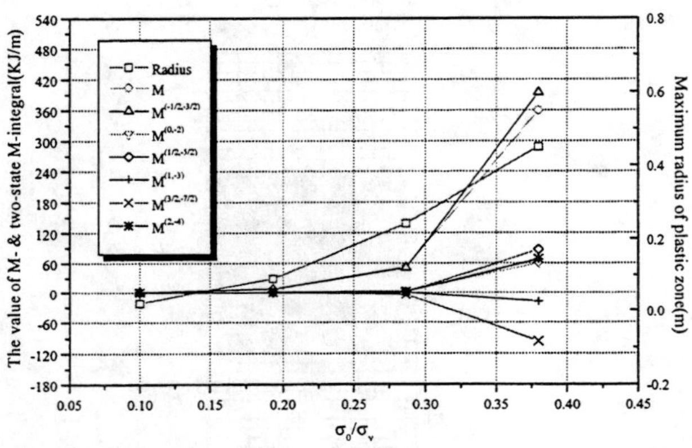

Fig. 4 Maximum radius of plastic zone and computed *M*- and two-state *M*-integrals versus applied loading

PVP-Vol. 415, Recent Advances in Solids and Structures — 2000
ASME 2000

MODELING AND VIBRATION FEEDBACK CONTROL OF ROTATING TAPERED BEAMS INCORPORATING ADAPTIVE CAPABILITIES

Sungsoo Na

Korea Polytechnic University
Automation Engineering Department
101 3-Ga, Shiwha-Gongdan, Shiheung-si
429-450 Keonggi-do, South Korea

Liviu Librescu *

Department of Engineering Science and Mechanics
Virginia Polytechnic Institute and State University
Blacksburg, Virginia, 24061-0219, USA
librescu@vt.edu

ABSTRACT

This paper addresses the problem of the modeling and vibration control of tapered rotating blades modeled as thin-walled beams and incorporating adaptive capabilities. The blade model incorporates non-classical features such as anisotropy, transverse shear, secondary warping and includes the centrifugal and Coriolis force fields.

For the non-adaptive system, a thorough validation and assessment of a number of nonclassical features including the taper characteristics is accomplished. The adaptive capabilities are provided by a system of piezoactuators bonded or embedded into the structure and spread over the entire span of the beam.

Based on the converse piezoelectric effect, the piezoactuators produce a localized strain field in response to a voltage and consequently, a change of the dynamic response characteristics is induced. A combined feedback control law relating the piezoelectrically induced transversal bending moment at the beam tip with the kinematical response quantities appropriately selected is used and the beneficial effects upon the closed-loop dynamic characteristics of the blade are highlighted.

INTRODUCTION

The study of the eigenvibration response of rotating beams is an important prerequisite in the design of helicopter, tilt rotor aircraft and turbomachinery. An important step toward the rational design of modern rotor blades and propelers consists of the development of analytical models that are capable to accurately predicting their dynamic behavior. Moreover, in order to enhance their dynamic behavior and avoid vibration-induced fatigue failure, new technologies have to be implemented. One of the ways to accomplish such goals consists of the incorporation into the host structure of adaptive materials able to respond actively to changing conditions. In contrast to traditional passive structures, in a structure featuring adaptive capabilities, the natural frequencies, damping and mode shapes can be tuned to avoid structural resonance and enhance dynamic response characteristics. In addition, due to the nature of *intelligent structures* that feature a highly distributed networks of sensors and actuators, more encompassing control schemes that are impractical when a small number of actuators are used, would be possible to be implemented. For helicopter and tilt rotor aircraft, the incorporation of adaptive materials technology for vibration control could result in significant increases in comfort, range, life, etc. In this sense, piezoelectric materials are excellent candidates for the roles of senors and actuators.

In this paper, a study of the in-plane (lagging) and transverse to the plane of rotation (flapping) free vibrations of rotating beams featuring nonuniform cross sections is addressed. Moreover, by incorportating the capability referred to as induced strain actuation a control of flapping vibration is carried out. The adaptive capability is achieved through the converse piezoelectric effect that consists of the generation of localized strains in response to an applied voltage. The induced strain field produces, in turn, an adaptive change in the dynamic response characteristics of the structure.

*Address all correspondence to this author.

Implementation of a control law relating the applied electric field to one of the mechanical quantities characterizing the blade response according to a prescribed functional relationship, results in an eigenvalue problem. Its solution consists of closed-loop eigenvalues which are functions of the applied voltage, i.e. of the feedback control gain.

Under consideration is a blade rotating with constant angular velocity and modelled as a thin/thick walled, single-cell untwisted beam.

Although of an evident practical importance, to the best of the authors' knowledge no such studies can be found in the specialized literature (see in this sense the comprehensive review papers, by Hodges (1990), Rosen (1991), Ewins and Henry (1992), and Kunz (1994)). To the best of the authors' knowledge the free vibration problem of tapered rotating beames was done within the solid beam model only. Some results addressing the issue can be found e.g. in the papers by Bazoune et al. (1999) and Khulief and Bazoune (1992). The results obtained herein constitute a generalization of the ones previously obtained in the papers by Librescu et al. (1994), Na and Librescu (1999), Song and Librescu (1997, 1999), Librescu et al. (1993 and 1997).

GENERAL CONSIDERATIONS

The case of straight untwisted nonuniform blades rotating with constant angular velocity is considered. The origin of the rotating axis system (x,y,z) is located at the blade root at an offset R_o of the rotation axis fixed in space. R_o denotes also the radius of the hub in which the blade is mounted and which rotates about its polar axis through the origin O.

The inertial system of coordinates (X,Y,Z) is assumed to be attached to the center of the hub O. By $(\mathbf{i}, \mathbf{j}, \mathbf{k})$ and $(\mathbf{I}, \mathbf{J}, \mathbf{K})$ we define the unit vectors associated with the coordinate systems (x,y,z) and (X,Y,Z), respectively. In addition, a local (surface) coordinate system (s,z,n) associated with the beam is considered. Its geometric configuration and the typical cross-section along with the associated system of coordinates are presented in Fig. 1.

Within the present work, it is assumed that the precone, presetting and sweep angles of the blade are zero. It is further assumed that the rotation takes place in the plane (X,Z) with the constant angular velocity $\Omega (\equiv \Omega\mathbf{J} = \Omega\mathbf{j})$, the spin axis being along the Y-axis.

The structural model corresponds to a thin/thick walled beam. In this context the case of a single-cell thin-walled beam of non-uniform closed-sections is considered, where the spanwise, z- coordinate axis coincides with a straight unspecified reference axis. It is assumed that the piezoactuator layers are distributed over the entire beam span and that the polarization is in their thickness direction.

The equations of rotating thin-walled beams are based upon the following statements (see Song and Librescu (1997, 1999)):

i) The original cross-section of the beam is preserved, ii) the transverse shear effects are taken into account, iii) the circumferential stress resultant N_{ss}(*i.e.*, the hoop stress resultant) is negligibly small when compared to the remaining ones, iv) doubly-tapered beam in both the horizontal and vertical planes are considered; v) the case of a bi-convex beam cross-section profile is adopted and vi) both the materials of the host structure and of the piezoactuators exhibit transversely-isotropic properties, the surface of isotropy being parallel at each point to the mid-surface of the beam.

By virtue of assumption c), the following linear distribution of the chord $c(\eta)$ and height $b(\eta)$ of the mid-line cross-section profiles along the wing span (see Librescu et al. [1994]), is considered

$$\begin{Bmatrix} c(\eta) \\ b(\eta) \end{Bmatrix} = [1-\eta\,(1-\sigma)] \begin{Bmatrix} c_R \\ b_R \end{Bmatrix}, \qquad (1a)$$

Herein $\sigma \equiv c_T/c_R (\equiv b_T/b_R)$ denotes the taper ratio, $\eta \equiv z/L$ is the dimensionless spanwise coordinate $\eta \in [0,1]$), where L denotes the wing semi-span, $c(\eta)$ and $b(\eta)$ denote the local wing chord and height, respectively, while superscripts R and T identify the wing characteristics at the root and tip cross-sections, respectively. Moreover, the radius of curvature of the circular arc associated with the midline contour at section η along the wing span is expressed as

$$R(\eta) = [1-\eta\,(1-\sigma)]R_R. \qquad (1b)$$

As result of the considered anisotropy of both the piezoactuators and of the structure, an exact decoupling of transverse bending (flapping - expressed in terms of variables v_o and θ_x), chordwise bending (lagging, u_o and θ_y), twist (Θ) and axial (w_o) motions is resulting. For the problem studied herein, the flapping and lagging motions will be considered only. However, since secondary warping induces transverse bending, this effect will also be incorporated. Finally, it is assumed that the electric field vector $\mathcal{E}_i$ is represented in terms of its component $\mathcal{E}_3$ in the n-direction only.

THE DYNAMIC EQUATIONS OF ADAPTIVE ROTATING CONTILEVERED BEAMS

In order to obtain the equations of motion of adaptive rotating beams and the associated boundary conditions, use is made of Hamilton's variational principle.

By virtue of this principle, of all displacements $v_i \equiv v_i(x,y,z;t)$ that satisfy the boundary conditions (BCs) $v_i = \underline{v}_i$ over Ω_V and fulfill also the condition $\delta v_i = 0$ at $t = t_0$ and $t = t_1$, where t_0 and t_1 are two arbitrary instants of time, the actual ones fulfill the variational equation:

$$\delta J = \int_{t_0}^{t_1} \left[\int_\tau \sigma_{ij}\delta\varepsilon_{ij}d\tau - \delta K - \int_{\Omega_\sigma} \underset{\sim}{s}_i\delta v_i d\Omega - \int_\tau \rho H_i \delta v_i d\tau \right] dt = 0 \quad (2a)$$

were

$$K = \frac{1}{2}\int_\tau \rho(\dot{\mathbf{R}}_i \cdot \dot{\mathbf{R}}_i)d\tau \quad (2b)$$

denotes the kinetic energy.

In these equations $d\tau(\equiv dndsdz)$ denotes the differential volume element; $\underset{\sim}{s}_i(\equiv \underset{\sim}{\sigma}_j n_j)$ denote the prescribed components of the stress vector on a surface element of the undeformed body characterized by the outward normal components n_i; H_i denote the components of the body forces; Ω_σ denotes the external area of the body over which the stresses are prescribed; ρ denotes the mass density; an undertilde sign identifies a prescribed quantity while δ denotes the variation sign. In Eqs. (12) the Einsteinian summation convention applies to repeated indices where Latin indices range from 1 to 3. In the same equations, $(v_1,v_2,v_3) \equiv (u,v,w), (x_1,x_2,x_3) \equiv (x,y,z)$ and the superposed dots denote time derivatives. In order to evaluate the quantities entering the energy functional, the position vector $\mathbf{R}(\equiv \mathbf{R}(x,y,z;t))$ of a deformed point of the beam is defined as:

$$\mathbf{R} = \mathbf{R}_o + \mathbf{r} + \Delta; \qquad \mathbf{R}_M = (R_o + L)\mathbf{k} + \Delta_o. \quad (3a,b)$$

In Eq. (2a), $\mathbf{r}(\equiv x\mathbf{i} + y\mathbf{j} + z\mathbf{k})$ denotes the position vector measured in the rotating axis system; $\mathbf{R}_o = R_o\ \mathbf{K}$ and $\Delta(\equiv u\mathbf{i} + v\mathbf{j} + w\mathbf{k})$ denotes the displacement vectors of the points of the blade. We will make also use of the equations expressing the time derivates of unit vectors, namely $(\dot{\mathbf{i}}, \dot{\mathbf{j}}, \dot{\mathbf{k}}) = (-\Omega\mathbf{k}, \mathbf{0}, \Omega\mathbf{i})$; of expressions of components R_i of the position vector of a deformed point of the blade and tip mass, as well as those of velocity V_i and acceleration a_i components. Their distribution throughout the beam is (see Song and Librescu, (1997, 1999)).

$$V_1 = \dot{u} + (R_o + z + w)\Omega; \quad V_2 = \dot{v}; \quad V_3 = \dot{w} - (x+u)\Omega, \quad (4a-c)$$

and

$$a_1 = \ddot{u} + \underset{\sim}{2\dot{w}\Omega} - \underset{\approx}{(x+u)\Omega^2}; \qquad a_2 = \ddot{v}; \quad (5a,b)$$

$$a_3 = \ddot{w} - \underset{\sim}{2\dot{u}\Omega} - \underset{\approx}{(R_0 + z + w)\Omega^2} \quad (5c)$$

respectively,

In Eqs. (4) and (5), as well as in the forthcoming ones, the terms underscored by a single and double undulated lines identify the Coriolis and centrifugal acceleration terms, respectively whereas.

In addition, u, v and w denote the components of the 3-D displacement vector defined in terms of 1-D generalized displacement measures as

$$\begin{aligned} u(x,y,z;t) &= \underline{\underline{u_o(z;t)}} - y\Theta(z;t), \\ v(x,y,z;t) &= \underline{v_o(z;t)} + x\Theta(z;t), \qquad (6a\text{-}c) \\ w(n,s,z;t) &= w_o(z;t) + \underline{\underline{x(s)\theta_y(z;t)}} + \underline{y(s)\theta_x(z;t)} - F_\omega(s)\Theta'(z;t) \\ &\quad + n\left[\frac{dy}{ds}\underline{\underline{\theta_y(z;t)}}\frac{dy}{ds}\underline{\theta_x(z;t)} - a(s)\Theta'(z;t)\right]. \end{aligned}$$

In these equations, the six kinematic variables $u_0(z;t)$, $w_0(z;t)$, $\theta(z;t)$ and $\Theta(z;t)$ represent three translations in the x, y, z directions and three rotations about the x, y, and z axis, respectively.

In Eq. (6) the terms underscored by one and two solid lines belong to the flapping and lagging motions, respectively. The remaining terms in these expressions are associated with the axial warping, and twist motions.

The first integral in Eq. (2a) should be considered in the sense

$$\int_\tau (\cdot)d\tau = \int_o^L \oint \sum_{k=1}^{r+p} \int_{h_{(k-1)}}^{h_{(k)}} (\cdot)dndsdz, \quad (7)$$

where $r+p$ denotes the total number of layers, while $\oint(\cdot)ds$ denotes the integral around the circumference of the mid-line cross-section of the beam. In the previously displayed equation L denotes the span of the beam.

THE GOVERNING SYSTEM

The governing equations of nonuniform rotating blades can be represented in terms of displacement quantities. This can be done by replacing in the equations of motion the stress-resultants and stress-couples in terms of displacement variables. These equations that can be found in the papers by Song and Librescu (1997, 1999), are not displayed here.

It should be stressed, however, that even by neglecting Coriolis' effects, for a general type of anisotropy of the layer materials (*i.e.* of the master structure, of the actuator patches or of both of them), the system of governing equations result in a complete coupled form.

However, for the type of anisotropy considered herein, (*i.e.* of transverse isotropy) an exact split of the governing system of equations and of the associated BCs into four uncoupled subsystems arises. In the sequence considered above, the subsystems of equations govern the coupling between the chordwise shear and flapwise bending, between chordwise shear and flapwise bending, as well as extension and twist motions, respectively. These subsystems of equations are labelled as system (A), (B), (C) and (D), respectively.

Since the analysis is confined here to rotating blades featuring lagging and flapping montions only, the associated governing equations and boundary conditions for cantilevered beams are given explicitely as:

Equations involving the coupling chordwise bending - chordwise shear (i.e. coupling (A)) are:

The Governing Equations:

$$\delta u_o: \quad [a_{44}(u_o'+\theta_y)]' + \Omega^2\left\{\left[\left[b_1\left[R_o(L-z)+\frac{1}{2}(L^2-z^2)\right]\right]\right]u_o'\right\}'$$
$$-\ \underline{2b_1\Omega\dot{w}_o'} + \underline{\underline{b_1\Omega^2 u_o}} - b_1\ddot{u}_o = 0. \quad \text{(8a)}$$

$$\delta\theta_y: \quad (a_{22}\theta_y')' - a_{44}(u_o'+\theta_y) - I_{yy}(\ddot{\theta}_y - \underline{\underline{\Omega^2\theta_y}}) = 0 \quad \text{(8b)}$$

and
the *Boundary Conditions:* At $z=0$:

$$u_o = \theta_y = 0, \quad (9a,b)$$

and at $z=L$:

$$\delta u_o: \quad a_{44}(u_o'+\theta_y) = 0 \quad \text{(9c)}$$
$$\delta\theta_y: \quad a_{22}\,\theta_y' = \tilde{\bar{M}}_y \quad \text{(9d)}$$

The equations involving the coupling flapwise shear-flapwise bending (i.e. coupling (B)) are:

$$\delta v_o: \quad [a_{55}(v_o'+\theta_x)]' + \Omega^2\{[b_1[R_o(L-z)+\frac{1}{2}(L^2-z^2)]]v_o'\}'$$
$$-b_1\ddot{v}_o + p_y = 0 \quad \text{(10a)}$$

$$\delta\theta_x: \quad (a_{33}\theta_x')' - a_{55}(v_o'+\theta_x) - I_{xx}(\ddot{\theta}_x - \underline{\underline{\Omega^2\theta_x}})$$
$$-\ \underline{2I_{xx}^{\circ}\Omega\dot{\Theta}} = 0 \quad \text{(10b)}$$

and the BCs: At $z=0$:

$$v_o = \theta_x = 0. \quad (11a,b)$$

and at $z=L$:

$$\delta v_o: \quad a_{55}(v_o'+\theta_x) = 0 \quad \text{(11c)}$$
$$\delta\theta_x: \quad a_{33}\theta_x' = \tilde{\bar{M}}_x \quad \text{(11d)}$$

Elimination of $a_{44}(u_o'+\theta_y)$ and $a_{55}(v_o'+\theta_x)$ in the group of equations (8), (8c), (9c), (10) and (11c) respectively, followed by consideration of $\theta_x \Rightarrow -v_o'$ and $\theta_y = -u_o'$, results in the non-shearable counterpart of the two groups of equations, (A) and (B). These are:

Classical counterpart of the equation group (A):

$$\delta u_o: (a_{22}u_o'')'' - \Omega^2\left\{\left[b_1\left[R_o(L-z)+\frac{1}{2}(L^2-z^2)\right]u_o'\right\}'$$
$$+\ \underline{2b_1\Omega\dot{w}_o'} - \underline{\underline{b_1\Omega^2 u_o}} + b_1\ddot{u}_o - [I_{yy}(\ddot{u}_o' - \underline{\underline{\Omega^2 u_o'}})]' = 0 \quad (12)$$

and the BCs: at $z=0$:

$$u_o = u_o' = 0, \quad (13a,b)$$

and at $z=L$:

$$\delta u_o: \quad [a_{22}u_o'']' - I_{yy}(\ddot{u}_o' - \underline{\underline{\Omega^2 u_o'}}) = 0,$$
$$\delta u_o': \quad a_{22}u_o'' = \tilde{\bar{M}}_y. \quad \text{(14a,b)}$$

Non-shearable counterpart of the equation group (B):

$$\delta v_o: \quad [a_{33}v_o'']'' - \Omega^2\{[b_1[R_o(L-z)+\frac{1}{2}(L^2-z^2)]v_o']\}'$$
$$-\ [I_{xx}(v_o' - \underline{\underline{\Omega^2 v_o'}})]' + b_1\ddot{v}_o = 0 \quad (15)$$

and the BCs: At $z=0$:

$$v_o = v_o' = 0. \quad (16a,b)$$

and at $z = L$:

$$\delta v_o: \quad [a_{33} v_o'']' - I_{xx}(\ddot{v}_o' - \underline{\underline{\Omega^2 v_o'}}) + I_{xx}^{\circ} \Omega \dot{\Theta} = 0,$$

$$\delta v_o': \quad a_{33} v_o'' = \hat{\tilde{M}}_x \tag{16c,d}$$

As is readily seen, the coupling between the lagging-extension and flapping-twist motions induced by the Coriolis effect, occurs also within the *classical* beam model.

Consistent with the number of four boundary conditions, the governing equations associated with lagging and flapping are each of fourth order. This feature is valid for both shear-deformable and non-shearable rotating beams.

It should be remarked that in the light of the actuator configuration, the piezoelectrically-induced stress-resultant and stress-couples are independent on the z-coordinate. As a result, in the governing equations their contribution is immaterial while in the BCs they intervene as non-homogeneous terms, only. This feature renders the control to be accomplished via the piezoelectrically induced *bending moments* at the beam tip (see Librescu et al. (1993, 1997)). However, due to the shape of the cross-section beam contour, only the control of flapping motion can be accommodated.

THE CONTROL LAW

In the previously displayed equations, due to the special distribution of piezoactuators, it was shown that the piezoelectrically induced moment $\hat{\tilde{M}}_x$ intervenes solely in the boundary conditions associated with the bending motion, prescribed at the beam tip and, hence, it plays the role of the *boundary moment control.*

The adaptive nature of the rotating beam is introduced by requiring the applied electric field $\mathcal{E}_3$ to be related to one of the mechanical quantities characterizing its dynamic response. As a result, a number of control laws can be implemented.

Within the adopted feedback control law the piezoelectrically induced bending moment at the blade tip, is expressed as $\hat{\tilde{M}}_x$,

$$\tilde{M}_x(L) = k_v \dot{\theta}_x(L) + k_a \ddot{\theta}_x(L) \tag{17}$$

Herein k_v and k_a denote the velocity and acceleration feedback gains, respectively.

NUMERICAL ILLUSTRATIONS AND DISCUSSION

The geometric characteristics of the structure are displayed in Fig. 1 while the properties of the PZT-4 piezoceramic can be found in the paper by Berlincourt et al. (1964).

For the free vibration problem, it is necessary to solve the closed-loop eigenvalue problem. To this end the unknown variables are represented in a generic form as

$$\mathcal{F}(z,t) = \overline{\mathcal{F}}(z) \exp(\lambda t). \tag{18}$$

Use of the representation (18) in Eqs. (8) and (9) associated with lagging motion, and in Eqs. (10) and (11), associated with the flapping, motion, results in two differential eigenvalue problems in terms of $\overline{u}_o(z)$ and $\overline{\theta}_y(z)$ on one hand, and in terms of $\overline{v}_o(z)$ and $\overline{\theta}_x(z)$, on the other hand. The discretization of the obtained differential eigenvalue problem in the spatial domain was done via the extended Galerkin method which is carried out directly in the Hamilton's principle, Eqs. (2). To this end $\overline{u}_o(z), \overline{v}_o(z), \overline{\theta}_x(z)$ and $\overline{\theta}_y(z)$ are expanded in series of trial functions satisfying the essential boundary condition. In the case $k_v \neq 0$, the solution of the algebraic eigenvalue problem yields the closed-loop eigenvalues

$$(\lambda_r, \overline{\lambda}_r) = \sigma_r \pm i\omega_{dr} \qquad r = 1, 2, \dots n. \tag{19}$$

which depend on the feedback control gain k_v where σ_r is a measure of the damping in the rth mode, while ω_{dr} is the *rth* frequency of damped oscillations. The damping factor in the *rth* mode results as

$$\zeta_r = -\frac{\sigma_r}{(\sigma_r^2 + \omega_{dr}^2)^{1/2}}. \tag{20}$$

When $k_a \neq 0, (\lambda_r, \overline{\lambda}_r) = \pm i\omega_r$ and, as a result, undamped free vibration results.

In Figs. 2 through 4 there are emphasized the effects of the taper parameter σ considered in conjunction with that of the blade rotational speed $\overline{\Omega}(\equiv \Omega b_1 L^4 / a_{33})$ and of the hub ratio $r_0(\equiv R_0/L)$ on dimensionless eigenfrequencies $\bar{\omega}_1(\equiv \omega_1 (b_1 L^4 / a_{33})^{1/2})$.

From Figs. 2 through 4 it becomes evident that as the beam taper increases (i.e when σ decreases), the first natural frequency increases whereas the second and the third ones decrease. However, in Fig. 3 it becomes apparent that with the increase of $\overline{\Omega}$ and decrease of σ there is a critical taper ratio for a beam rotating at relatively high speed for which the decay of the natural frequency with the decrease of σ stops, and a slight change of trend is manifested.

A general remark emerging from Figs. 2 through 4 is that the stiffening effect due to beam rotation contributes to the increase of natural frequencies for all taper ratios. Moreover, the hub radius has a beneficial effect of increasing the eigenfrequencies, specially at larger rotational speeds where the effect appears to be more prominent.

In Figs. 5 through 7 there is displayed the dependence of $\bar{\omega}_i(i = 1,2,3)$, in flapping vs. $\overline{\Omega}$ for two values of the taper ratio and for selected values of the hub ratio. These plots enforce the conclusions emerging from Figs. 2 through 4 that have been already outlined, related with the effects of σ, $\overline{\Omega}$ and $\bar{r}_0$ on the natural frequencies.

In Fig. 8 there are depicted the implications of the rotational speed considered in conjunction with those of the taper ratio and hub radius on the dimensionless first natural frequency in lagging. Since the frequencies in lagging are larger than their counterparts in flapping, the stiffening due to the centrifugal effect is weaker than that exerted on flapping frequencies. This trend clearly appears when comparing Fig. 8 with the flapping counterpart , Fig. 5. The same trend also emerges from Fig. 9 where the Campbell diagram was completed as to also include the effect of the beam taper-ratio.

As was already mentioned, the shape of the cross-section profile considered in conjunction with implemention of the control methodology enables one to control the flapwise motion only. With this in mind, Figs. 12 through 14 depict the variation of the first three closed-loop eigenfrequencies with the dimensionless velocity feedback gain for two values of the rotational speed $\overline{\Omega}$, for two values of the taper ratio and for $\eta(\equiv I_{xx}/b_1L^2) = 0.05$ and $\mu(\equiv a_{33}/L^2a_{55}) = 0.007647$. The results reveal the dramatic role played by the implementation of piezoelectric actuation upon the enhancement of eigenfrequencies. As concerns the implications of the taper, these are similar to those already considered in the case of the unactivated rotating beams.

In Fig. 15 the variation of the piezoelectrically induced damping of rotating beam for two values of the taper ratio, vs. the velocity feedback gain is presented. Whereas the effect of the taper ratio is similar to that featured in the case of non-rotating beams (see Na and Librescu (1999)), the increase of the rotational speed yields a decrease of the induced damping. This is attributed to the fact that the beam stiffened by the centrifugal effect that increases with the increase of $\overline{\Omega}$, is less prone to induce damping as compared to its less stiff beam counterpart, i.e. of that rotating at lower values of $\overline{\Omega}$.

A similar trend, but less pronounced than in the previous case is featured by ζ_2, (see Fig. 16). Figures 10 and 11 constitute the counterparts for the second and third frequence in flapping-lagging, of Fig. 9, where, in, addition the implications of transverse shear flexibility are emphasized. These reveal that: a) the effect of rotation is much more reduced for the second and third flap frequencies than for the first one; b) The effect of σ and of transverse shear are much stronger for the lagging frequencies (second and third), than for their flapping frequencies counterparts.

CONCLUSIONS

A dynamic structural model of rotating beam of non-uniform cross-sections was developed, and the effect of the taper ratio was assessed. Moreover, based upon the converse piezoelectric effect, a distributed actuator methodology aimed at controlling the flapping eigenvibration characteristics of rotating cantilevered thin-walled beams of non-uniform cross-sections was developed. This control is achieved through the piezoelectrically-induced flapwise bending moments at the tip of the beam.

The obtained results reveal that via this control capability it is possible to tune conveniently the eigenfrequencies of the system, and consequently to modify in a beneficial and predictable way the dynamic response characteristics of the structure.

It is believed that this control capability can play a major role in enhancing the dynamic response and flutter instability as well as in prolonging the fatigue life of the structural booms for space missions, helicopter rotor blades as well as of tilt rotor aircraft.

REFERENCES

Bazoune, A., Khulief, Y. A. and Stephen N. G., (1999), "Further Results for Model Characteristies of Rotating Tapered Timoshenko Beams," *Journal of Sound and Vibration*, 219, 1, 157-174.

Berlincourt, D.A., Curran, D.R., and Jaffe, H., (1964), "Piezoelectric and Piezomagnetic Materials and Their Function in Transducers," *Physical Acoustics–Principles and Methods*, (Eds. W.P. Mason), Vol. 1, Part A, Academic Press, New York and London, pp. 169–270.

Ewins, D.J. and Henry, R., (1992), "Structural Dynamic Characteristics of Individual Blades," *Vibration and Rotor Dynamics*, Von Karman Institute for Fluid Dynamics, Lecture Series 1992-06, pp. 14.1-14.27.

Hodges, D.H., (1990), "Review of Composite Rotor Blade Modeling," *AIAA Journal*, **28**, pp 561-565.

Khulief, Y. A. and Bazoune, A., (1992), "Frequencies of Rotating Tapered Timoshenko Beams with Different Boundary Conditions," *Computers & Structures*, 42, 781-795.

Kunz, D.L., (1994), "Survey and Comparison of Engineering Beam Theories for Helicopter Rotor Blades," *Journal of Aircraft* 31, 3, pp. 473-479.

Librescu, L., Song, O. and Rogers, C. A., (1993), "On Adaptive Vibrational Behavior of Cantilevered Structures Modelled as Composite Thin- Walled Beams," *International Journal of Engineering Science*, Vol. 31, No. 5, pp. 775-792.

Librescu, L., Meirovitch, L. and Song, O., (1994), "Vibration and Static Aeroelatic Instability of Nonuniform Thin-Walled Beam Composite Wings," *AIAA/ASME/ASCE/AHS/ASC 35th Structures, Structural Dynamics, and Materials Conference*, Paper AIAA-94-1491, Hilton Head, SC.

Librescu, L., Meirovitch, L. and Na, S. S., (1997), "Control of Cantilevers Vibration Via Structural Tailoring and Adaptive Materials," *AIAA Journal*, Vol. 35, No 8, pp. 1309-1315.

Na, S. S. and Librescu, L., (1999), "Dynamic Behavior of Aircraft Wings Modeled as Doubly-Tapered Composite Thin-Walled Beams," P.V.P. Vol. 398, Recent Advances in Solids and Structures, Ed. Y. W. Kwon, H. H. Chung, ASME 1999, pp. 59-68.

Rosen, A., (1991), "Structural and Dynamic Behavior of Pretwisted Rods and Beams," *Applied Mechanics Reviews*, 44, 12, Part 1, pp. 483–515.

Song, O. and Librescu, L., (1997), "Structural Modeling and Free Vibration Analysis of Rotating Composite Thin-Walled Beams," *Journal of the American Helicopter Society*, Vol. 42, No. 4, pp. 358-369.

Song, O. and Librescu, L., (1999), "Modeling and Dynamic Behavior of Rotating Blades Carrying a Tip Mass and Incorporating Adaptive Capabilities," *Acta Mechanica*, **134**, pp. 169-197.

FIGURE CAPTIONS

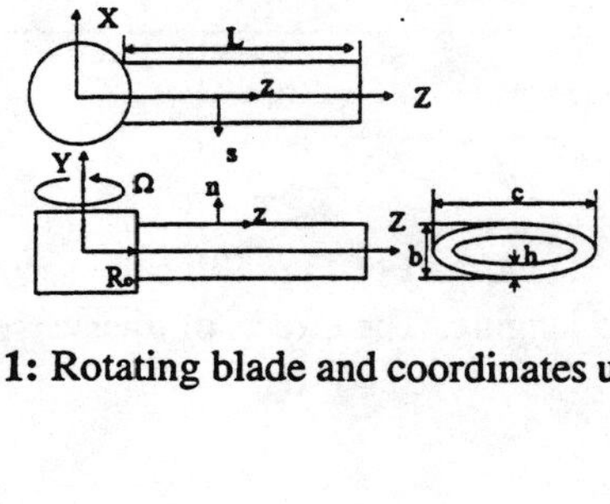

Fig. 1: Rotating blade and coordinates used.

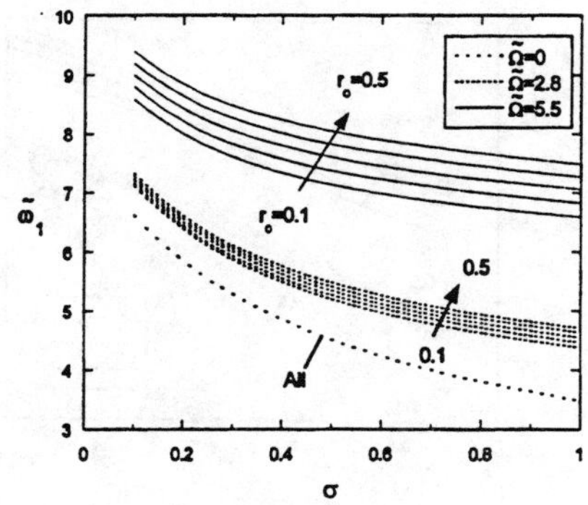

Fig. 2: Variation of the fundamental dimensionless flapping frequency with the taper ratio for selected values of the rotational speed $\overline{\Omega}$ and hub ratio r_0.

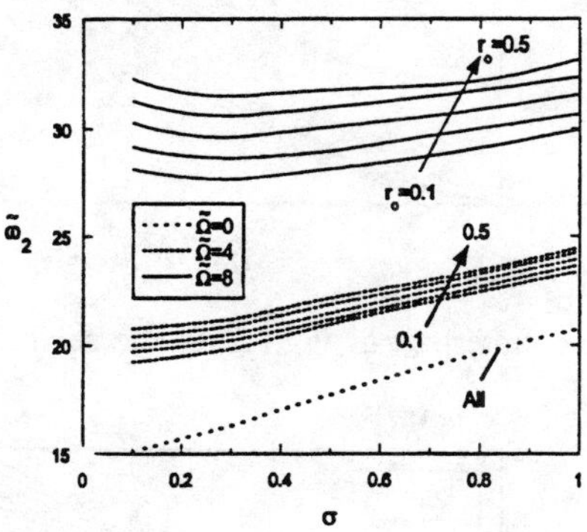

Fig. 3: The counterpart of Fig. 2 for the second flapping frequency

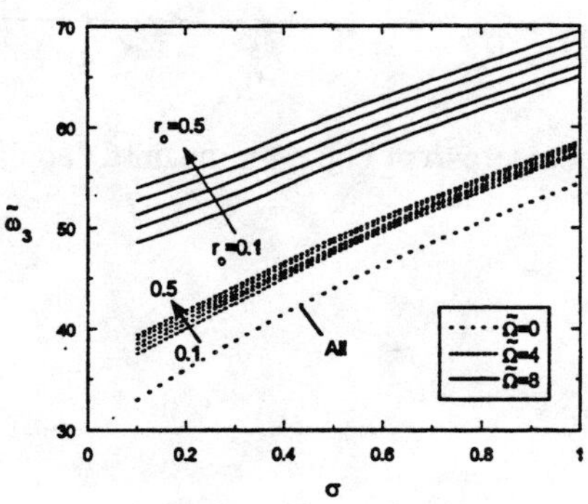

Fig. 4: The counterpart of Fig. 2 for the third flapping frequency.

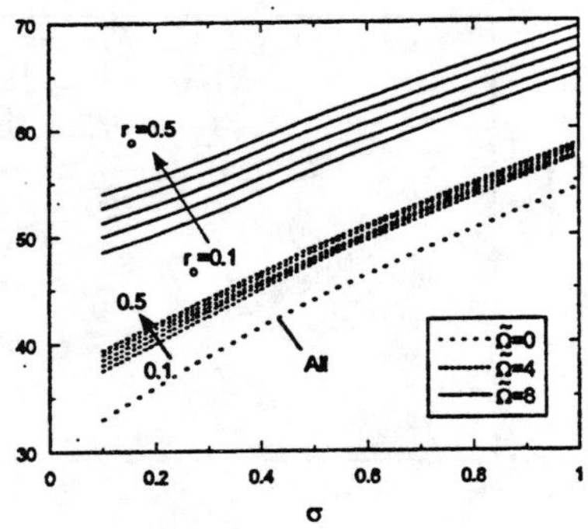

Fig. 5: Fundamental dimensionless flapping frequency vs. $\overline{\Omega}$ for selected values of the taper ratio and hub ratio.

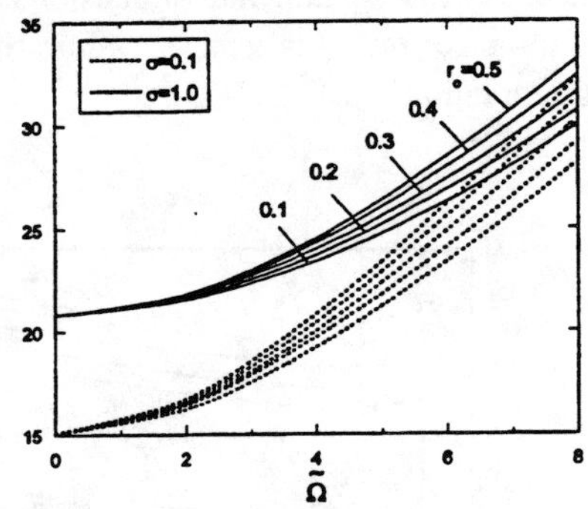

Fig. 6: The counterpart of Fig. 5 for the second flapping frequency.

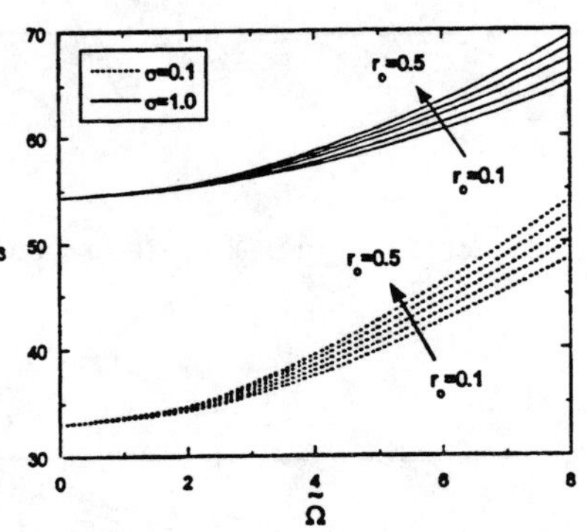

Fig. 7: The counterpart of Fig. 5 for the third flapping frequency.

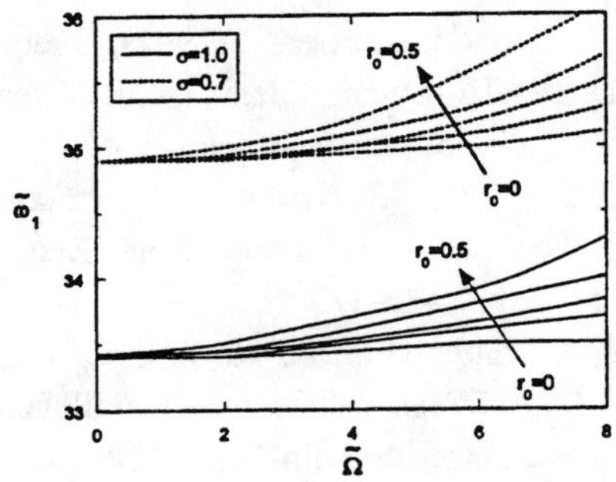

Fig. 8: Fundamental lagging frequency (dimensionless) vs. $\overline{\Omega}$ for selected values of the taper ratio and hub ratio, (classical beam model).

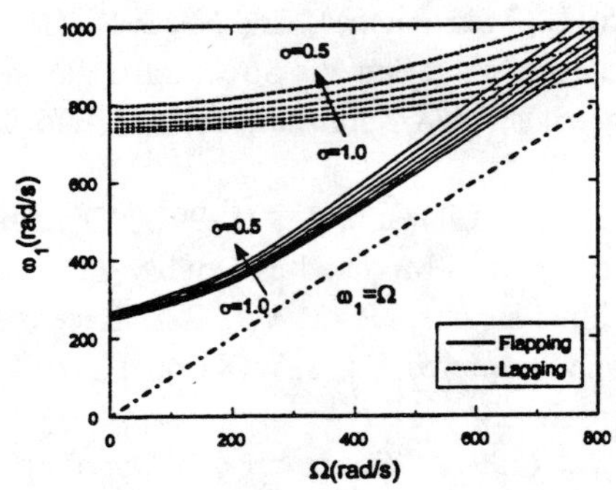

Fig. 9: Fundamental flapping and lagging frequencies vs. rotating speed for selected values of the taper ratio ($E/G' = 100, r_0 = 0.1$).

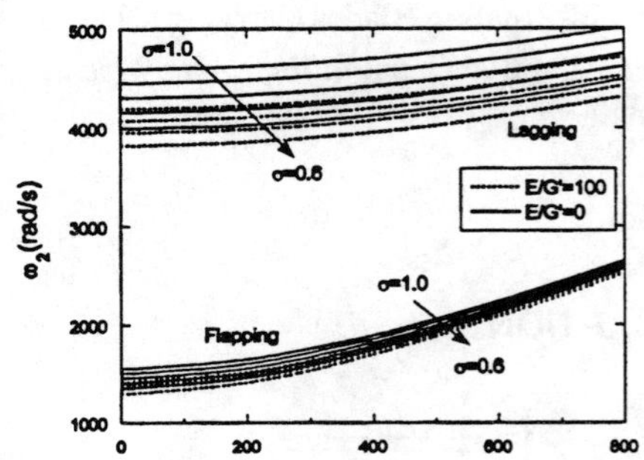

Fig. 10: The counterpart of Fig. 9 for the second frequencies in flapping and lagging. The effects of transverse shear are also emphasized.

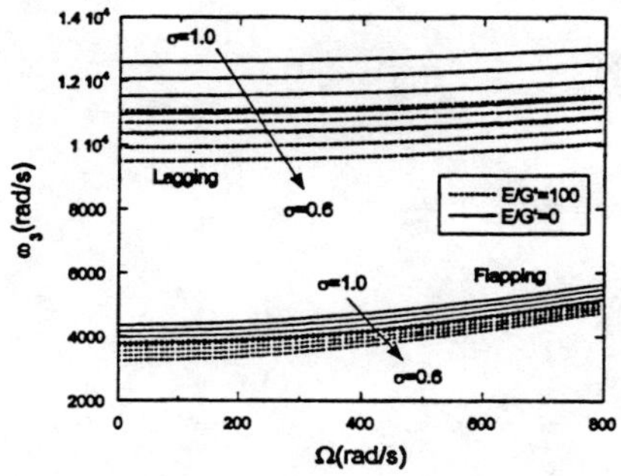

Fig. 11: The counterpart of Fig. 10 for the third frequencies in flapping and lagging.

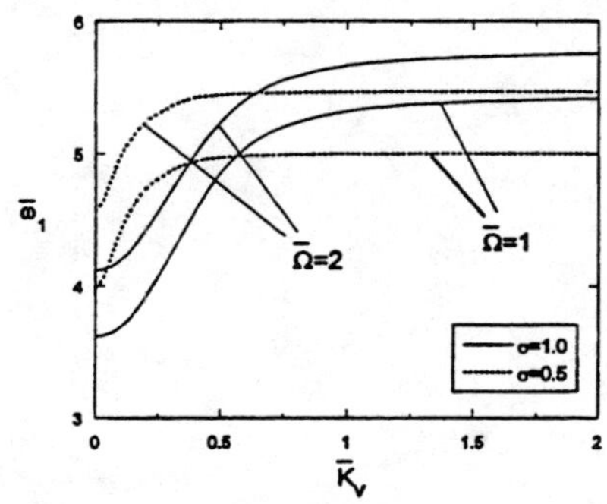

Fig. 12: Closed-loop fundamental flapping frequencies vs. $\overline{K}_v$ for two selected values of rotating speed and taper ratio (shearable beam model),

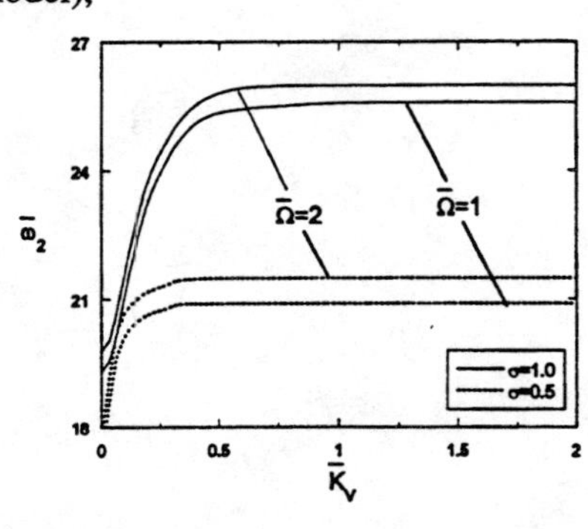

Fig. 13: The counterpart of Fig. 10 for the second closed-loop eigenfrequency.

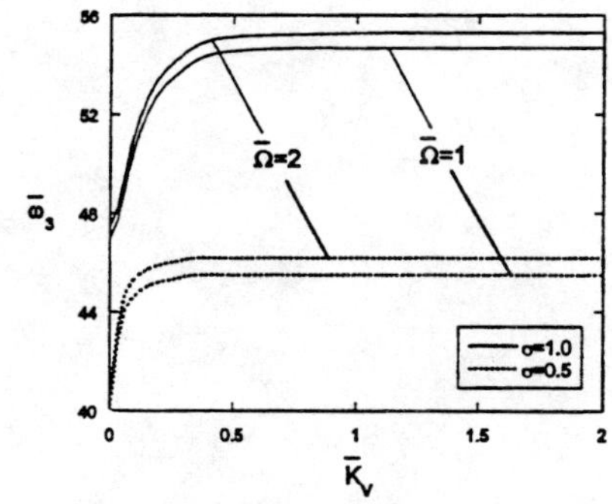

Fig. 14 The counterpart of Fig. 10 for the third closed-loop eigenfrequency.

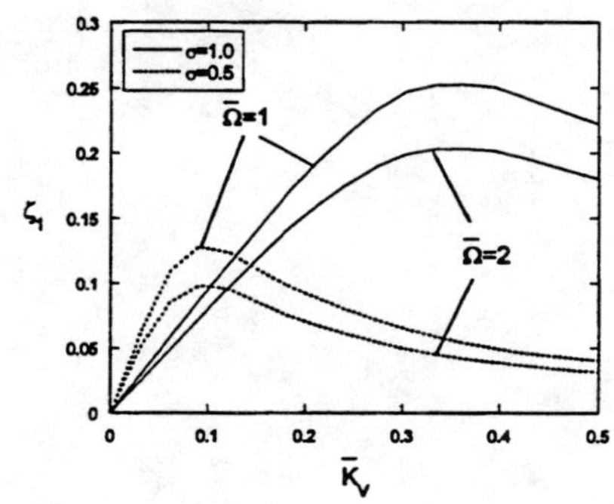

Fig. 15: Piezoelectrically induced damping ζ_1 in flapping vs. $\overline{K}_v$ for two values of the rotational speed and taper ratio ($r_0 = 0.1$).

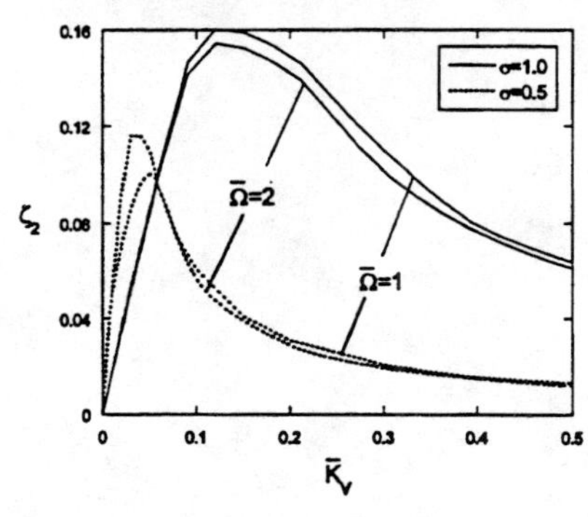

Fig. 16: The counterpart of FIg. 15 for ζ_2.

PVP-Vol. 415, Recent Advances in Solids and Structures — 2000
ASME 2000

FINITE ELEMENT MODELING FOR VIBRATION ANALYSIS OF CONSTRAINED LAYER DAMPING TREATED STRUCTURES

Yanchu Xu and Dennis Chen
Maxtor Corporation
510 Cottonwood Drive,
Milpitas, California 95035

ABSTRACT

A compound plate element is presented for modeling all layers of the damped structure into a single element. By associating deformations of the damping layer to those of base structure layer and constraining layer, the kinetic and potential energies of the damping layer as well as those of base and constraining layers can be derived. Then the element mass and stiffness matrices can be obtained for all layers as a whole. The newly derived element formulation results in significant simplification of constrained layer modeling and dramatic reduction of element density while maintaining the desired accuracy. The use of the element also allows direct and more accurate calculation of structural modal damping in modal analysis.

INTRODUCTION

To control the excessive noise and vibration, the application of damping treatment to the structure is becoming standard practice in many industries. In particular, the surface damping treatments associated with sheet metal structures are often used to solve a variety of noise and vibration problems. These applications normally involve bonding a relatively stiff thin member, also called constraining layer, to sheet metal structure with a soft visco-elastic material such that strain is induced in the adhesive during vibration. The flexural vibration analysis of such constrained layer damping treated structure has received considerable attention in the past 50 years [1].

Of those the analysis developed by Ross, Ungar, and Kerwin [2] has been the most widely used. This analysis has been developed for a three layer system and outlined the dominant design parameters for the case where all layers vibrate with the same sinusoidal spatial dependence. The outer layers are assumed to deform as Eular-Bernoulli beams and the adhesive is assumed to deform only in shear, which leads to a single fourth order beam equation where the equivalent complex bending stiffness depends on the properties of the three layers. The beauty of the fourth order equation is not only in that one can estimate the natural frequencies and loss factors of the laminate providing that one knows the complex properties of the visco-elastic material, but also in that it can be used in the inverse problem. By knowing the natural frequencies and modal loss factors of the laminate, one can estimate the complex material properties analytically using fourth order equation, which has been used for decades in the visco-elastic material complex property identification. The problem with the fourth order equation is that it is derived from simply supported-simply supported boundary conditions which is not always true in real world application. Also it neglects the thickness deformation of the visco-elastic layer which has been shown that it is not very accurate for flexible cores. To extend Ross, Ungar and Kerwin's analysis to beams with general boundary conditions in which sinusoidal spatial dependence cannot be assumed, Mead etc.[3] obtained a sixth order equation of motion. It is assumed that the beam's deflection is small and uniform across a section, the axial displacements are continuous, the base and constraining layers bend according to the Eular hypothesis, the damping layer deforms only in shear, and the longitudinal and rotary inertia effects are insignificant. The validity of the analysis is therefore limited to some upper range of core stiffness. Miles etc.[4] obtained a sixth order equation of motion by using Hamilton's principal. The assumptions ware equivalent to those of Mead except that relative transverse deflection is permitted between the outer layers and longitudinal inertia is included. The analysis neglects rotary inertia in all layers and shear deformation in the outer layers, which limits the applicability to the lower order, long wavelength modes. However, the errors introduced by these simplifications may be neglected when the wavelength is more than four times of the thickness of the thickest outer layer.

Though analytical methods are useful for predicting damping characteristics of some simple structures such as beams and plates, finite element approach remains to be the method of choice when complex physical systems have to be analyzed. Under the framework of finite element analysis, several different modeling techniques have been developed to deal with damped structures. A common approach to model multiple layer structure is to use shell element with equivalent cross sectional properties [5]. This approach has a drawback on boundary condition handling because it treats all layers as one layer so any boundary condition applies to all layers, which is normally not true for constrained layer damping treatment where the constrained layer usually has free boundary condition while the base structure might have any kind of boundary condition. To estimate damping loss factors associated with each mode, modal strain energy method was developed by using the real eigen-vectors obtained from finite element analysis of the corresponding undamped structure. In the modal strain energy method, the strain energy in each layer is calculated based on the modal shape, the dissipative energy is calculated proportional to the strain energy in the damping layer, and the modal loss factor is obtained by calculating the ratio of the dissipative energy to the total structural energy. There are some limitations, however, in the analysis of the damped structures using conventional finite element method. For achieving the best damping result, the thickness of the damping layer is normally very thin, which results in the dramatic increase of element density used in the finite element modeling to get a proper aspect ratio for accurate prediction. Modal strain energy method becomes quite inaccurate when the damping of the structure becomes high because it is always assumed that the structural modal shapes are real while actual modal shapes of the damped structures are quite complex. Xu etc. [6,7] derived a new finite element type, compound beam element, for sandwich beam by considering the complex modulus of the adhesive layer and the effects of both shear and thickness deformation in the adhesive layer.

In this paper, the modeling approach developed by Xu etc. is extended for finite element analysis of sandwich plate. All layers of the constrained layer damping treated structure are modeled into a single element, called compound plate element. By associating deformations of the damping layer to those of base structure and constraining layer, the kinetic and potential energies of the damping layer as well as those of base and constrain layers can be derived from deformations of outer layers. Then the element mass and stiffness matrices can be obtained for all layers as a whole. The newly derived element formulation results in significant simplification of constrain layer modeling and dramatic reduction of element density while maintaining the desired accuracy. The use of the element also allows direct and more accurate calculation of structural modal damping.

EQUATION OF MOTION OF THE STRUCTURE

For the vibration analysis of constrained layer damping treated structures, the equation of motion of the structure need to be derived. In this study, system equation of motion for the vibration analysis will be obtained using Hamilton's Principle. Figure 1 shows the sandwich plate geometry, which is composed of two layers of stiff base layers, normally metal, and a layer of soft core of adhesive. The deformations to be considered here will only include the bending and in-plane displacements of both outer layers. It is to be shown that the deformation of the middle, adhesive layer will be derived from displacements of both base structure layer and the constraining layer such that both shear and thickness deformations in the adhesive layer can be determined.

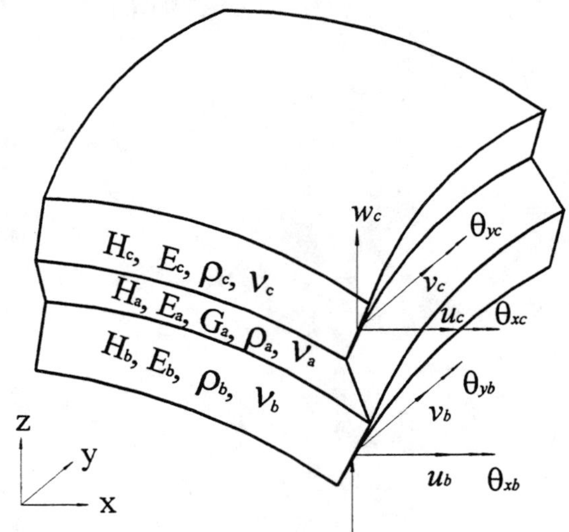

Figure 1. Geometry of Sandwich Plate

The bending deformation of the thin plate is to be described entirely by lateral displacement w of the middle plane of the plate using Timishenko's [8] classical plate. In-plane deformations are to be described by the middle plane displacement, u and v. Timishenko plate theory assumes that the plates are thin and the deflection small with certain approximations such as the linear variation of strains and stresses on lines normal to the plane of the plate. However, for an approximate satisfaction of slope continuity in finite element formulation, three displacement components are used to describe the bending deformation: the first actual displacement w in the z direction, the second a rotation about the y axis, θ_x, and the third a rotation about the x axis, θ_y. Clearly the slope of w and the rotation are identical. We have the following assumptions[9]: (1) Each layer is isotropic, linear elastic, and of constant thickness. (2) The plate thickness is very small compared to its length and width. (3) Stresses acting in the xy plane dominate the plate behavior, then stress components, σ_z, τ_{xz} and τ_{yz}, are assumed to be zero and an approximate state of plane stress is said to exist in the outer layers. (4) Negligible transverse normal strain ε_z, and transverse shear strains, γ_{xz} and γ_{yz}, constitutes a statement of non-deformable normals to the middle surface although there is an inherent, but commonly ignored, conflict with the assumption of zero transverse normal stress, σ_z. (5) Displacements, u, v, and w, are small compared to the thickness of the plate (generally, although not necessarily, indicative of small deflection theory). (6) Strains are small compared to unity (small strain theory). The small displacement approximation is important here for combining the bending and in-plane terms because it is assumed that the displacements prescribed for in-plane forces do not affect the bending deformation and *vice versa*. Hamilton's Principle states:

$$\delta \int_{t_1}^{t_2} (T - V + W)dt = 0 \tag{1}$$

where T is the kinetic energy, V is the potential energy, W is the work done by external forces, and δ is the variation operator. When the system under study is a continuum, T and V are given by integration over the volume of the body. In the following it is assumed that no external forces are acting on the body so that W is set to zero. Since we are primarily interested in the energy dissipation, or damping, by the visco-elastic adhesive layer, its dissipative abilities will be included in the model by considering the shear modulus, G_a, and Young's modulus, E_a, of the adhesive, to be complex.

The kinetic energy and potential energy for the three layers will be derived in the following discussion. For both base structure layer and constraining layer as shown in Fig. 1, the kinetic energy can be written as,

$$\begin{aligned} T_m = \frac{1}{2}\int_S \Bigg\{ & \rho_b H_b \left[\left(\frac{\partial w_b}{\partial t}\right)^2 + \left(\frac{\partial u_b}{\partial t}\right)^2 + \left(\frac{\partial v_b}{\partial t}\right)^2 \right] \\ & + \rho_b \frac{H_b^3}{12} \left[\left(\frac{\partial \theta_{xb}}{\partial t}\right)^2 + \left(\frac{\partial \theta_{yb}}{\partial t}\right)^2 \right] \\ & + \rho_c H_c \left[\left(\frac{\partial w_c}{\partial t}\right)^2 + \left(\frac{\partial u_c}{\partial t}\right)^2 + \left(\frac{\partial v_c}{\partial t}\right)^2 \right] \\ & + \rho_c \frac{H_c^3}{12} \left[\left(\frac{\partial \theta_{xc}}{\partial t}\right)^2 + \left(\frac{\partial \theta_{yc}}{\partial t}\right)^2 \right] \Bigg\} ds \end{aligned} \tag{2}$$

and the potential energy can be expressed as,

$$\begin{aligned} V_m = \frac{1}{2}\int_S \Bigg\{ & \frac{E_b H_b^3}{12(1-\nu_b^2)} \{\varepsilon'_{xb} \;\; \varepsilon'_{yb} \;\; \gamma'_{xyb}\} \begin{bmatrix} 1 & \nu_b & 0 \\ \nu_b & 1 & 0 \\ 0 & 0 & \frac{1-\nu_b}{2} \end{bmatrix} \begin{Bmatrix} \varepsilon'_{xb} \\ \varepsilon'_{yb} \\ \gamma'_{xyb} \end{Bmatrix} \\ & + + \frac{E_b H_b}{1-\nu_b^2} \{\varepsilon''_{xb} \;\; \varepsilon''_{yb} \;\; \gamma''_{xyb}\} \begin{bmatrix} 1 & \nu_b & 0 \\ \nu_b & 1 & 0 \\ 0 & 0 & \frac{1-\nu_b}{2} \end{bmatrix} \begin{Bmatrix} \varepsilon''_{xb} \\ \varepsilon''_{yb} \\ \gamma''_{xyb} \end{Bmatrix} \\ & + \frac{E_c H_c^3}{12(1-\nu_c^2)} \{\varepsilon'_{xc} \;\; \varepsilon'_{yc} \;\; \gamma'_{xyc}\} \begin{bmatrix} 1 & \nu_c & 0 \\ \nu_c & 1 & 0 \\ 0 & 0 & \frac{1-\nu_c}{2} \end{bmatrix} \begin{Bmatrix} \varepsilon'_{xc} \\ \varepsilon'_{yc} \\ \gamma'_{xyc} \end{Bmatrix} \\ & + \frac{E_c H_c}{1-\nu_c^2} \{\varepsilon''_{xc} \;\; \varepsilon''_{yc} \;\; \gamma''_{xyc}\} \begin{bmatrix} 1 & \nu_c & 0 \\ \nu_c & 1 & 0 \\ 0 & 0 & \frac{1-\nu_c}{2} \end{bmatrix} \begin{Bmatrix} \varepsilon''_{xc} \\ \varepsilon''_{yc} \\ \gamma''_{xyc} \end{Bmatrix} \Bigg\} ds \end{aligned} \tag{3}$$

where, the strain due to bending deformation can be expressed as,

$$\begin{Bmatrix} \varepsilon'_{xb} \\ \varepsilon'_{yb} \\ \gamma'_{xyb} \end{Bmatrix} = \begin{Bmatrix} \frac{\partial^2 w_b}{\partial x^2} \\ \frac{\partial^2 w_b}{\partial y^2} \\ -2\frac{\partial^2 w_b}{\partial x \partial y} \end{Bmatrix}, \quad \begin{Bmatrix} \varepsilon'_{xc} \\ \varepsilon'_{yc} \\ \gamma'_{xyc} \end{Bmatrix} = \begin{Bmatrix} \frac{\partial^2 w_c}{\partial x^2} \\ \frac{\partial^2 w_c}{\partial y^2} \\ -2\frac{\partial^2 w_c}{\partial x \partial y} \end{Bmatrix}, \tag{4}$$

and the strain due to in-plane deformation,

$$\begin{Bmatrix} \varepsilon''_{xb} \\ \varepsilon''_{yb} \\ \gamma''_{xyb} \end{Bmatrix} = \begin{Bmatrix} \frac{\partial u_b}{\partial x} \\ \frac{\partial v_b}{\partial y} \\ \frac{\partial u_b}{\partial y} + \frac{\partial v_b}{\partial x} \end{Bmatrix}, \quad \begin{Bmatrix} \varepsilon''_{xc} \\ \varepsilon''_{yc} \\ \gamma''_{xyc} \end{Bmatrix} = \begin{Bmatrix} \frac{\partial u_c}{\partial x} \\ \frac{\partial v_c}{\partial y} \\ \frac{\partial u_c}{\partial y} + \frac{\partial v_c}{\partial x} \end{Bmatrix} \tag{5}$$

The small displacement assumptions are used here such that the bending and in-plane terms in the potential energy can be uncoupled.

The kinetic energy and potential energy in the damping layer will be derived from the displacements of both outer layers in the following. It is assumed that there exists perfect adhesion between damping layer and both base structure layer and constraining layer and hence, no slippage at the interfaces between the adhesive and outer layers. It is also assumed that all displacement components in the adhesive layer vary linearly through the thickness of the adhesive, such as the transverse and in-plane displacements, as well as extensional and shear strains, etc. Therefore, we can express the transversal vibration velocity in the damping layer as,

$$\frac{\partial w(\zeta)}{\partial t} = \frac{\partial w_b}{\partial t} + \frac{\zeta\left(\frac{\partial w_c}{\partial t} - \frac{\partial w_b}{\partial t}\right)}{H_a} \tag{6}$$

where ζ is the transverse distance from the bottom of the adhesive layer, and H_a is the nominal adhesive thickness. The kinetic energy due to the transversal displacement of the adhesive layer can be obtained by the following integration,

$$\frac{1}{2}\int \rho_a H_a \left\{ \int_0^{H_a} \left[\frac{\partial w_b}{\partial t} + \frac{\zeta\left(\frac{\partial w_c}{\partial t} - \frac{\partial w_b}{\partial t}\right)}{H_a} \right]^2 d\zeta \right\} ds$$
$$= \frac{1}{2}\int \frac{\rho_a H_a}{6}\left[\left(\frac{\partial w_b}{\partial t}\right)^2 + \left(\frac{\partial w_c}{\partial t}\right)^2 + \left(\frac{\partial w_b}{\partial t} + \frac{\partial w_c}{\partial t}\right)^2\right] ds \tag{7}$$

This integration will be used in all similar terms in the following derivations for both kinetic and potential energies of the adhesive layer. Thus the kinetic energy in the adhesive layer can be expressed as,

$$\begin{aligned} T_a = \frac{1}{2}\int \frac{\rho_a H_a}{6} \Bigg\{ & \left(\frac{\partial w_b}{\partial t}\right)^2 + \left(\frac{\partial w_c}{\partial t}\right)^2 + \left(\frac{\partial w_b}{\partial t} + \frac{\partial w_c}{\partial t}\right)^2 \\ & + \left(\frac{\partial u_b}{\partial t} + \frac{\partial \theta_{yb}}{\partial t}\frac{H_b}{2}\right)^2 + \left(\frac{\partial u_c}{\partial t} - \frac{\partial \theta_{yc}}{\partial t}\frac{H_c}{2}\right)^2 \\ & + \left(\frac{\partial u_b}{\partial t} + \frac{\partial \theta_{yb}}{\partial t}\frac{H_b}{2} + \frac{\partial u_c}{\partial t} - \frac{\partial \theta_{yc}}{\partial t}\frac{H_c}{2}\right)^2 \\ & + \left(\frac{\partial v_b}{\partial t} - \frac{\partial \theta_{xb}}{\partial t}\frac{H_b}{2}\right)^2 + \left(\frac{\partial v_c}{\partial t} + \frac{\partial \theta_{xc}}{\partial t}\frac{H_c}{2}\right)^2 \\ & + \left(\frac{\partial v_b}{\partial t} - \frac{\partial \theta_{xb}}{\partial t}\frac{H_b}{2} + \frac{\partial v_c}{\partial t} + \frac{\partial \theta_{xc}}{\partial t}\frac{H_c}{2}\right)^2 \Bigg\} ds \end{aligned} \tag{8}$$

The potential energy in the adhesive layer comes from two major contributions. One of the contributions to the potential energy is due to the extensional deformations of the adhesive. The extensional deformation in the thickness direction can be considered as plane strain, and the in-plane extensions as plane stresses, therefore, the potential energy in adhesive layer due to extensional deformation can be obtained as,

$$\begin{aligned} V_a^e = \frac{1}{2}\int E_a H_a \Bigg\{ & \frac{1-\nu_a}{1-\nu_a - 2\nu_a^2}\varepsilon_z^2 \\ & + \frac{1}{6\left(1-\nu_a^2\right)}\Big[\left(\varepsilon_{xb}^2 + \varepsilon_{xc}^2 + \left(\varepsilon_{xb} + \varepsilon_{xc}\right)^2\right) \\ & + \left(\varepsilon_{yb}^2 + \varepsilon_{yc}^2 + \left(\varepsilon_{yb} + \varepsilon_{yc}\right)^2\right) \\ & + 4\nu_a\left(\varepsilon_{xb}\varepsilon_{yb} + \varepsilon_{xc}\varepsilon_{yc}\right) \\ & + 2\nu_a\left(\varepsilon_{xb}\varepsilon_{yc} + \varepsilon_{xc}\varepsilon_{yb}\right)\Big]\Bigg\} ds \end{aligned} \tag{9}$$

where,

$$\begin{aligned} \varepsilon_z &= \frac{w_c - w_b}{H_a} \\ \varepsilon_{xb} &= \frac{\partial u_b}{\partial x} + \frac{\partial \theta_{yb}}{\partial x}\frac{H_b}{2} \\ \varepsilon_{xc} &= \left(\frac{\partial u_c}{\partial x} - \frac{\partial \theta_{yc}}{\partial x}\frac{H_c}{2}\right) \\ \varepsilon_{yb} &= \left(\frac{\partial v_b}{\partial y} - \frac{\partial \theta_{xb}}{\partial y}\frac{H_b}{2}\right) \\ \varepsilon_{yc} &= \left(\frac{\partial v_c}{\partial y} + \frac{\partial \theta_{xc}}{\partial y}\frac{H_c}{2}\right) \end{aligned} \tag{10}$$

Bending strain in the adhesive layer is neglected here. This is due to the fact that the adhesive layer is usually very thin and the adhesive materials used in damping applications have modulus that are several orders of magnitude lower than those of the outer layers. Therefore, the contribution due to bending strain of the adhesive to the total energy is negligible. It is also assumed that the all displacement components in the adhesive layer vary linearly through the thickness of the adhesive, so the derivations follow equation (6) and (7).

The remaining effect that contributes to the potential energy is the shear in the adhesive layer. Like above, it is assumed that the shear strain varies linearly through the thickness of the adhesive. The potential energy due to adhesive shear strain therefore equals to,

$$V_a^s = \frac{1}{2}\int_s G_a H_a \left\{ \left[\frac{\gamma_{xyb}^2 + \gamma_{xyc}^2}{6} + \frac{(\gamma_{xyb} + \gamma_{xyc})^2}{6} \right] + \left[\frac{\gamma_{yzb}^2 + \gamma_{yzc}^2}{6} + \frac{(\gamma_{yzb} + \gamma_{yzc})^2}{6} \right] + \left[\frac{\gamma_{zxb}^2 + \gamma_{zxc}^2}{6} + \frac{(\gamma_{zxb} + \gamma_{zxc})^2}{6} \right] \right\} ds \tag{11}$$

From Fig. 1, it can be seen that the shear strain of the adhesive layer at the interfaces with both base structure and the constraining layer, γ_b and γ_c, may be related to displacement components of both base plate and constrained plate by,

$$\begin{aligned}
\gamma_{xyb} &= \frac{\partial u_b}{\partial y} + \frac{\partial v_b}{\partial x} - 2\frac{\partial^2 w_b}{\partial x \partial y} \\
\gamma_{xyc} &= \frac{\partial u_c}{\partial y} + \frac{\partial v_c}{\partial x} - 2\frac{\partial^2 w_c}{\partial x \partial y} \\
\gamma_{yzb} &= \theta_{xb}\left(1 + \frac{H_b}{2H_a}\right) + \theta_{xc}\frac{H_c}{2H_a} - \frac{v_b - v_c}{H_a} \\
\gamma_{yzc} &= \theta_{xc}\left(1 + \frac{H_c}{2H_a}\right) + \theta_{xb}\frac{H_b}{2H_a} - \frac{v_b - v_c}{H_a} \\
\gamma_{zxb} &= \theta_{yb}\left(1 + \frac{H_b}{2H_a}\right) + \theta_{yc}\frac{H_c}{2H_a} + \frac{u_b - u_c}{H_a} \\
\gamma_{zxc} &= \theta_{yc}\left(1 + \frac{H_c}{2H_a}\right) + \theta_{yb}\frac{H_b}{2H_a} + \frac{u_b - u_c}{H_a}
\end{aligned} \tag{12}$$

where, as shown in Fig. 1, H_b, H_a and H_c are the thickness of base plate layer, adhesive layer and constraining layer, respectively. It should be noted that above equations are valid only for small deformations of the plate.

FORMULATION OF COMPOUND PLATE ELEMENT

The constrained layer damping treated structure as shown in Fig. 1 can be discretized into a series of elements, called compound plate element in this paper, which consists of three layers. To solve the problem using finite element method, mass matrix and stiffness matrix for the compound plate element need to be derived. In the following, the equation of motion for the vibration of a sandwich plate obtained above will be used for the derivation of matrices. Follow the trivial approach of finite element modeling, The displacement components of nodal point i will be defined the same as those in Fig. 1 and expressed as,

$$\begin{aligned}
\{d_i\} &= \{d_{bi}^T \; d_{ci}^T\}^T \\
&= \{w_{bi} \; \theta_{xbi} \; \theta_{ybi} \; u_{bi} \; v_{bi} \; w_{ci} \; \theta_{xci} \; \theta_{yci} \; u_{ci} \; v_{ci}\}^T
\end{aligned} \tag{13}$$

For the compound plate element of m nodal points, the state vector of the element can be defined as,

$$\{a\} = \{d_1 \; d_2 \; \ldots \; d_m\}^T \tag{14}$$

where, the number of nodes, m, depends on the element type chosen. The total number of displacement components of the element will be $n=10xm$.

The displacements within each element have to be uniquely defined by these state vectors. Following the general practice in finite element analysis [9,10], a set of shape functions can be defined for both bending and 'in-plane' displacements, respectively. For the bending deformation, the shape function needs to be defined to ensure that deformation components for bending of the plate not only remains continuous but also does not 'kink' between elements. By properly choosing shape functions for bending deformation, the displacement components related to bending can be expressed as,

$$\begin{aligned}
w_b &= [B_1(x,y)]\{a\} = [B_1]\{a\} \\
\theta_{xb} &= \left[\frac{\partial B_1(x,y)}{\partial y}\right]\{a\} = [B_2(x,y)]\{a\} = [B_2]\{a\} \\
\theta_{yb} &= \left[-\frac{\partial B_1(x,y)}{\partial x}\right]\{a\} = [B_3(x,y)]\{a\} = [B_3]\{a\} \\
w_c &= [B_4(x,y)]\{a\} = [B_4]\{a\} \\
\theta_{xc} &= \left[\frac{\partial B_4(x,y)}{\partial y}\right]\{a\} = [B_5(x,y)]\{a\} = [B_5]\{a\} \\
\theta_{yc} &= \left[-\frac{\partial B_4(x,y)}{\partial x}\right]\{a\} = [B_6(x,y)]\{a\} = [B_6]\{a\}
\end{aligned} \tag{15}$$

Also, by properly choosing shape functions for the in-plane deformation, which needs to ensure in-plane deformation components of the plate to remain continuous, the related displacement components can be expressed as,

$$u_b = [B_7(x,y)]\{a\} = [B_7]\{a\}$$
$$v_b = [B_8(x,y)]\{a\} = [B_8]\{a\}$$
$$u_c = [B_9(x,y)]\{a\} = [B_9]\{a\}$$
$$v_c = [B_{10}(x,y)]\{a\} = [B_{10}]\{a\}$$

(16)

Thus, displacement derivatives for kinetic and potential energies used in Eq. (3) through (12) can be derived from shape function and expressed as,

$$-\frac{\partial\theta_{yb}}{\partial y} = \frac{\partial\theta_{xb}}{\partial x} = \frac{\partial^2 w_b}{\partial x\partial y} = \left[\frac{\partial B_1^2(x,y)}{\partial x\partial y}\right]\{a\} = [B_{11}(x,y)]\{a\} = [B_{11}]\{a\}$$
$$-\frac{\partial\theta_{yb}}{\partial x} = \frac{\partial^2 w_b}{\partial x^2} = \left[\frac{\partial B_1^2(x,y)}{\partial x^2}\right]\{a\} = [B_{12}(x,y)]\{a\} = [B_{12}]\{a\}$$
$$\frac{\partial\theta_{xb}}{\partial y} = \frac{\partial^2 w_b}{\partial y^2} = \left[\frac{\partial B_1^2(x,y)}{\partial y^2}\right]\{a\} = [B_{13}(x,y)]\{a\} = [B_{13}]\{a\}$$
$$-\frac{\partial\theta_{yc}}{\partial y} = \frac{\partial\theta_{xc}}{\partial x} = \frac{\partial^2 w_c}{\partial x\partial y} = \left[\frac{\partial B_4^2(x,y)}{\partial x\partial y}\right]\{a\} = [B_{14}(x,y)]\{a\} = [B_{14}]\{a\}$$
$$-\frac{\partial\theta_{yc}}{\partial x} = \frac{\partial^2 w_c}{\partial x^2} = \left[\frac{\partial B_4^2(x,y)}{\partial x^2}\right]\{a\} = [B_{15}(x,y)]\{a\} = [B_{15}]\{a\}$$
$$\frac{\partial\theta_{xc}}{\partial y} = \frac{\partial^2 w_c}{\partial y^2} = \left[\frac{\partial B_4^2(x,y)}{\partial y^2}\right]\{a\} = [B_{16}(x,y)]\{a\} = [B_{16}]\{a\}$$

(17)

And those related to 'in-plane' displacements can be expressed as,

$$\frac{\partial u_b}{\partial x} = \left[\frac{\partial B_7(x,y)}{\partial x}\right]\{a\} = [B_{17}(x,y)]\{a\} = [B_{17}]\{a\}$$
$$\frac{\partial u_b}{\partial y} = \left[\frac{\partial B_7(x,y)}{\partial y}\right]\{a\} = [B_{18}(x,y)]\{a\} = [B_{18}]\{a\}$$
$$\frac{\partial v_b}{\partial x} = \left[\frac{\partial B_8(x,y)}{\partial x}\right]\{a\} = [B_{19}(x,y)]\{a\} = [B_{19}]\{a\}$$
$$\frac{\partial v_b}{\partial y} = \left[\frac{\partial B_8(x,y)}{\partial y}\right]\{a\} = [B_{20}(x,y)]\{a\} = [B_{20}]\{a\}$$
$$\frac{\partial u_c}{\partial x} = \left[\frac{\partial B_9(x,y)}{\partial x}\right]\{a\} = [B_{21}(x,y)]\{a\} = [B_{21}]\{a\}$$
$$\frac{\partial u_c}{\partial y} = \left[\frac{\partial B_9(x,y)}{\partial y}\right]\{a\} = [B_{22}(x,y)]\{a\} = [B_{22}]\{a\}$$
$$\frac{\partial v_c}{\partial x} = \left[\frac{\partial B_{10}(x,y)}{\partial x}\right]\{a\} = [B_{23}(x,y)]\{a\} = [B_{23}]\{a\}$$
$$\frac{\partial v_c}{\partial y} = \left[\frac{\partial B_{10}(x,y)}{\partial y}\right]\{a\} = [B_{24}(x,y)]\{a\} = [B_{24}]\{a\}$$

(18)

Then, for the element in consideration, the kinetic energy and potential energy can be expressed in matrix format,

$$T = T_m + T_a = \frac{1}{2}\frac{\partial\{a^T\}}{\partial t}[M_e]\frac{\partial\{a\}}{\partial t}$$
$$V = V_m + V_a^e + V_a^s = \frac{1}{2}\{a^T\}[K_e]\{a\}$$

(19)

where, $[M_e]$ and $[K_e]$ are mass matrix and stiffness matrix of the compound plate element, respectively, and,

$$[M_e] = [M_e^m] + [M_e^a]$$
$$[K_e] = [K_e^m] + [K_e^{ae}] + [K_e^{as}]$$

(20)

$[M_e^m]$ is the contribution from the mass of metal layers (both base plate and constraining plate), and $[M_e^a]$ the adhesive layer, to the element mass matrix. $[K_e^m]$, $[K_e^{ae}]$, and $[K_e^{as}]$ are the contributions to the element stiffness matrix from the metal layers, elastic deformation of adhesive layer, and shear deformation of the adhesive layer, respectively. All these matrices can be obtained by the following integration over the area of the plate element, e.g.,

$$\begin{aligned}[M_e^m] = \int_s \Big\{ & \rho_b H_b [B_1^T B_1 + B_7^T B_7 + B_8^T B_8] \\ & + \rho_b \frac{H_b^3}{12}[B_2^T B_2 + B_3^T B_3] \\ & + \rho_c H_c [B_4^T B_4 + B_9^T B_9 + B_{10}^T B_{10}] \\ & + \rho_c \frac{H_c^3}{12}[B_5^T B_5 + B_6^T B_6]\Big\} ds\end{aligned}$$

(21)

All other matrices, $[M_e^a]$, $[K_e^m]$, $[K_e^{ae}]$, and $[K_e^{as}]$ can be obtained in the similar way, which will be listed in Appendix A.

CASE STUDY AND COMPARISON WITH CONVENTIONAL METHOD

To assess the effectiveness of the finite element modeling technique presented in this paper in dealing with the damping treated structures, a rectangular sandwich plate of *254 mm* by *304.8 mm* is used as a case example. The base plate layer and the constraining plate layer are assumed to be identical steel plate of thickness *1.52 mm* with Young's modulus of $E=0.207x10^{12}$ N/m^2, poison ratio of $\nu=0.3$, and density of $\rho=7.846x10^3$ kg/m^3. The viscoelastic layer is a *0.05 mm* soft core with shear storage modulus of $G=0.14x10^6$ N/m^2, poison ratio of

ν=0.45, and density of ρ=3.158x10^3 kg/m^3. The base plate layer is

Although the mechanical properties of viscoelastic materials, shear modulus G and loss factor η, are both temperature and frequency dependent, the case example used here is intended to demonstrate the modeling effectiveness for both convergence speed and direct computation of modal loss factor. In real world application, an iterative scheme needs to be used for each mode.

In the case study, the modeling technique presented in this paper and conventional 3-D element in commercial package are compared with each other. For the sandwich plate given above, the convergence comparison results are given in Tab. 1. When the sandwich plate is modeled by the present method, 4 by 5, 8 by 10, and 16 by 20 rectangular elements, with analysis DOF's of 300, 990, and 3570, respectively, are used to study the convergence speed. The results shown in Tab. 1 indicate that analysis model with DOF of 990 is already convergent with error less than one percent for up to mode 7. When conventional 3-D solid elements are used, 10x10, 30x30, and 60x60 element densities which result in 1452, 11532, and 44650 DOF's, respectively, are analyzed. The results shown in Tab. 1 indicate that analysis model with DOF of 11532 is convergent with similar error range. Table 2 shows the natural frequency and modal

clamped on one of the longer sides.

loss factor comparison between the present modeling technique and the conventional method. The numbers of DOF used are 990 and 11532, respectively, for two different methods. The strain energy approximation method is used to estimate the modal loss factors for the conventional method while complex modulus is used in the present method to directly calculate the modal loss factors. When material has different material loss factor, conventional method will result in the same natural frequencies because the modulus used in the model is only the storage modulus, and modal loss factors are proportional to the material loss factor. When the present method is used, the results in Tab. 2 show that natural frequencies increase with the material loss factor and modal loss factors are not proportional to the material loss factors. It is obvious because the conventional method does not consider the contribution of imaginary modulus to the structural stiffness and nor the natural frequencies. The modal loss factor estimation using strain energy method is not actuate either because it assumes the real modal shape while the actual modal shapes are quiet complex.

Table 1. Convergence Comparison

	DOF used in Compound Plate Element			DOF used in Traditional 3-D Element		
Mode	**300**	**990**	**3570**	**1452**	**11532**	**44652**
1	21.30(4.84)	20.45(0.66)	20.32	20.76(2.12)	20.38(0.24)	20.33
2	50.81(2.87)	49.61(0.42)	49.40	52.93(11.2)	47.76(0.30)	47.62
3	129.68(2.95)	126.59(0.50)	125.96	134.27(7.77)	125.03(0.35)	124.59
4	141.12(1.96)	138.73(0.23)	138.41	153.44(11.7)	137.78(0.33)	137.33
5	179.46(2.53)	175.71(0.39)	175.03	213.29(24.9)	171.70(0.58)	170.71
6	283.00(0.11)	282.34(-.12)	282.68	344.33(26.1)	275.91(1.04)	273.07
7	285.61(0.48)	284.36(0.04)	284.23	381.48(35.3)	283.90(0.68)	281.97

Note: Values in () indicates error percentage toward converged solution

Table 2. Natural frequency and modal loss factor comparison

	Using Compound Beam Element						Using Traditional 3-D Element					
	η=0.125		η=0.5		η=2.0		η=0.125		η=0.5		η=2.0	
Mode	f_n	ζ_n	f_n	ζ_n	f_n	ζ_n	f_n	ζ_n	f_n	ζ_n	f_n	ζ_n
1	20.57	.0188	20.78	.0827	21.40	.2131	20.35(-1.08)	.0152(-19.7)	20.35(-2.09)	.0609(-26.3)	20.35(-4.93)	.2437(14.35)
2	49.41	.0297	49.84	.1185	54.41	.3924	47.67(-3.52)	.0244(-17.7)	47.67(-4.36)	.0978(-17.5)	47.67(-12.4)	.3910(-0.34)
3	126.27	.0254	127.03	.1004	137.17	.3220	124.76(-1.20)	.0235(-7.62)	124.76(-1.79)	.0938(-6.63)	124.76(-9.05)	.3751(16.51)
4	138.55	.0226	139.00	.0866	145.11	.3185	137.51(-0.75)	.0192(-15.0)	137.51(-1.07)	.0768(-11.3)	137.51(-5.24)	.3073(-3.52)
5	175.31	.0207	175.87	.0830	183.88	.3001	171.06(-2.42)	.0152(-26.7)	171.06(-2.73)	.0607(-27.0)	171.06(-6.97)	.2427(-19.1)
6	282.37	.0185	282.96	.0739	289.32	.2043	273.97(-2.97)	.0105(-43.2)	273.97(-3.18)	.0421(-43.1)	273.97(-5.30)	.1682(-17.7)
7	284.31	.0133	284.62	.0528	291.20	.2795	282.69(-0.57)	.0109(-17.6)	282.69(-0.68)	.0438(-17.1)	282.69(-2.92)	.1751(-37.4)

Note: Values in () indicates the percentage of inaccuracy of traditional method

CONCLUSIONS

A compound plate element is developed by modeling all layers of the sandwich plate into a single element and associating the effects of both shear and thickness deformation in the adhesive layer to the deformation of outer layers. The use of the element results in significant simplification for finite element modeling of sandwich plate and dramatic reduction of element density used while maintaining the desired accuracy. The modeling can easily take complex modulus of the adhesive material into consideration for accurate prediction of structural natural frequencies and modal loss factors. The use of the element will find very useful in design optimization for the structure with constrained layer damping treatment where extensive function analysis is required.

REFERENCES

1. Nashif, A. D., Jones, D. I. G., Henderson, J. P., 1985, *Vibration Damping*, John Wiley & Sons, New York
2. Ross, D., Ungar, E. E. and Kerwin, E. M. Jr., 1959, "Damping of Flexural Vibrations by Means of Viscoelastic Laminates", *Structural Damping*, ASME, New York
3. Mead, D. J., Markus, S., 1969, "The Forced Vibration of a Three-Layer, Damped Sandwich Beam with Arbitrary Boundary Conditions," *Journal of Sound and Vibration*, Vol. 10(2), pp. 163-175
4. Miles, R. N., and Reinhall, P. G., 1986, "An analytical model for the vibration of laminated beams including the effects of both shear and thickness deformation in the adhesive layer", *Journal of Vibration, Acoustics, Stress, and Reliability in Design*, Vol. 108, pp. 56-64
5. ANSYS Theory Manual for Revision 5.5.
6. Xu, Y., and Nashif, A. D., 1996, "Measurement, Analysis and Modeling of the Dynamic Properties of Materials", *Sound and Vibration, Materials Reference Issue*, July
7. Xu, Y., and Chen, D., 2000, "Finite Element Modeling for the Flexural Vibration of Damped Sandwich Beams Considering Complex Modulus of the Adhesive Layer", *Proceedings of 7th SPIE Symposium on Smart Structures and Materials*, Long Beach, CA
8. Timishenko, S. and Woinowsky-Krieger, S., 1959, *Theory of Plates and Shells*, McGraw-Hill, 2nd ed.
9. Zienkiewicz, O. C., 1977, *The Finite Element Method*, McGraw-Hill, 3rd ed.
10. Cook, R. D., 1981, *Concepts and Applications of Finite Element Analysis*, John Wiley & Sons, 2nd ed.

APPENDIX A: DERIVATION OF ELEMENT MATRICES

Similar to Eq. (21), the rest of element mass and stiffness matrices are given below. The contribution from the mass of the adhesive layer to the element mass matrix,

$$
\begin{aligned}
\left[M_e^a\right] = \int_s \Bigg\{ \frac{\rho_a H_a}{6} \Bigg[& B_1^T B_1 + B_4^T B_4 + (B_1 + B_4)^T (B_1 + B_4) \\
& + \left(B_7 + \frac{H_b}{2} B_3\right)^T \left(B_7 + \frac{H_b}{2} B_3\right) \\
& + \left(B_9 - \frac{H_c}{2} B_6\right)^T \left(B_9 - \frac{H_c}{2} B_6\right) \\
& + \left(B_7 + \frac{H_b}{2} B_3 + B_9 - \frac{H_c}{2} B_6\right)^T \\
& \quad \left(B_7 + \frac{H_b}{2} B_3 + B_9 - \frac{H_c}{2} B_6\right) \\
& + \left(B_8 - \frac{H_b}{2} B_2\right)^T \left(B_8 - \frac{H_b}{2} B_2\right) \\
& + \left(B_{10} + \frac{H_c}{2} B_5\right)^T \left(B_{10} + \frac{H_c}{2} B_5\right) \\
& + \left(B_8 - \frac{H_b}{2} B_2 + B_{10} + \frac{H_c}{2} B_5\right)^T \\
& \quad \left(B_8 - \frac{H_b}{2} B_2 + B_{10} + \frac{H_c}{2} B_5\right) \Bigg] \Bigg\} ds
\end{aligned}
\tag{A.1}
$$

The contribution to the element stiffness matrix from both base layer and constraining layer,

$$
\begin{aligned}
\left[K_e^m\right] = \int_s \Bigg\{ & \frac{E_b H_b^3}{12\left(1 - \nu_b^2\right)} \left\{ B_{12}^T \quad B_{13}^T \quad -2B_{11}^T \right\} \begin{bmatrix} 1 & \nu_b & 0 \\ \nu_b & 1 & 0 \\ 0 & 0 & \frac{1-\nu_b}{2} \end{bmatrix} \begin{Bmatrix} B_{12} \\ B_{13} \\ -2B_{11} \end{Bmatrix} \\
& + \frac{E_b H_b}{1 - \nu_b^2} \left\{ B_{17}^T \quad B_{20}^T \quad (B_{18} + B_{19})^T \right\} \begin{bmatrix} 1 & \nu_b & 0 \\ \nu_b & 1 & 0 \\ 0 & 0 & \frac{1-\nu_b}{2} \end{bmatrix} \begin{Bmatrix} B_{17} \\ B_{20} \\ B_{18} + B_{19} \end{Bmatrix} \\
& + \frac{E_c H_c^3}{12\left(1 - \nu_c^2\right)} \left\{ B_{15}^T \quad B_{16}^T \quad -2B_{14}^T \right\} \begin{bmatrix} 1 & \nu_c & 0 \\ \nu_c & 1 & 0 \\ 0 & 0 & \frac{1-\nu_c}{2} \end{bmatrix} \begin{Bmatrix} B_{15} \\ B_{16} \\ -2B_{14} \end{Bmatrix} \\
& + \frac{E_c H_c}{1 - \nu_c^2} \left\{ B_{21}^T \quad B_{24}^T \quad (B_{22} + B_{23})^T \right\} \begin{bmatrix} 1 & \nu_c & 0 \\ \nu_c & 1 & 0 \\ 0 & 0 & \frac{1-\nu_c}{2} \end{bmatrix} \begin{Bmatrix} B_{21} \\ B_{24} \\ B_{22} + B_{23} \end{Bmatrix} \Bigg\} ds
\end{aligned}
\tag{A.2}
$$

The contribution to the element stiffness matrix from extensional deformation of adhesive layer,

$$[K_e^{as}] = \int_s \left\{ \frac{G_a H_a}{6} \left[B_{30}^T B_{30} + B_{31}^T B_{31} + (B_{30} + B_{31})^T (B_{30} + B_{31}) \right.\right.$$
$$+ B_{32}^T B_{32} + B_{33}^T B_{33} + (B_{32} + B_{33})^T (B_{32} + B_{33})$$
$$\left.\left. + B_{34}^T B_{34} + B_{35}^T B_{35} + (B_{34} + B_{35})^T (B_{34} + B_{35}) \right] \right\} ds \qquad \text{(A.3)}$$

where,

$$[B_{25}] = \frac{B_4 - B_1}{H_a}$$
$$[B_{26}] = B_{17} - \frac{H_b}{2} B_{12}$$
$$[B_{27}] = B_{21} + \frac{H_c}{2} B_{15}$$
$$[B_{28}] = B_{20} - \frac{H_b}{2} B_{13}$$
$$[B_{29}] = B_{24} + \frac{H_c}{2} B_{16} \qquad \text{(A.4)}$$

The contribution to the element stiffness matrix from shear deformation of adhesive layer,

$$[K_e^{ae}] = \int_s E_a H_a \left\{ \frac{1 - \nu_a}{1 - \nu_a - 2\nu_a^2} B_{25}^T B_{25} \right.$$
$$[B_{29}] = B_{24} + \frac{c}{2} B_{16}$$
$$+ \frac{1}{6(1 - \nu_a^2)} \left[B_{26}^T B_{26} + B_{27}^T B_{27} \right.$$
$$+ (B_{26} + B_{27})^T (B_{26} + B_{27})$$
$$+ B_{28}^T B_{28} + B_{29}^T B_{29}$$
$$+ (B_{28} + B_{29})^T (B_{28} + B_{29})$$
$$+ 2\nu_a (B_{26}^T B_{28} + B_{27}^T B_{29})$$
$$+ \nu_a (B_{26}^T B_{29} + B_{27}^T B_{28})$$
$$+ 2\nu_a (B_{28}^T B_{26} + B_{29}^T B_{27})$$
$$\left.\left. + \nu_a (B_{29}^T B_{26} + B_{28}^T B_{27}) \right] \right\} ds \qquad \text{(A.5)}$$

where,

$$[B_{30}] = [B_{18} + B_{19} - 2B_{11}]$$
$$[B_{31}] = [B_{22} + B_{23} - 2B_{14}]$$
$$[B_{32}] = \left[\left(1 + \frac{H_b}{2H_a}\right) B_2 + \frac{H_c}{2H_a} B_5 - \frac{1}{H_a}(B_8 - B_{10}) \right]$$
$$[B_{33}] = \left[\left(1 + \frac{H_c}{2H_a}\right) B_5 + \frac{H_b}{2H_a} B_2 - \frac{1}{H_a}(B_8 - B_{10}) \right]$$
$$[B_{34}] = \left[\left(1 + \frac{H_b}{2H_a}\right) B_3 + \frac{H_c}{2H_a} B_6 + \frac{1}{H_a}(B_7 - B_9) \right]$$
$$[B_{35}] = \left[\left(1 + \frac{H_c}{2H_a}\right) B_6 + \frac{H_b}{2H_a} B_3 + \frac{1}{H_a}(B_7 - B_9) \right] \qquad \text{(A.6)}$$

FREE-FREE VIBRATION EXTRACTION FROM AVAILABLE VIBRATION DATA

Thomas T. Yi, Ph.D.
Technology Center, CNH Global N.V.
7 S. 600 County Line Rd, Burr Ridge, IL 60521
tong.yi@cnh.com

ABSTRACT

This paper presents a procedure for identifying the free-free vibration data of a structure from the available vibration data of the same structure with boundary conditions. For a structure in a mechanical system, depending upon the dynamic formulation used, we may need a set of free-free modal data or a set of constrained modal data. If a finite element model of the structure is available, its vibration data can be obtained analytically under any desired boundary conditions. However, if a finite element model is not available, the vibration data may be determined experimentally. Since experimentally measured vibration data are obtained for a structure supported by some form of boundary conditions, we may need to determine its free-free vibration data. The aim of this study is to extract, based on experimentally obtained vibration data, the necessary free-free frequencies and the associated modes for structures to be used in dynamic formulations. The available vibration data may be obtained for a structure supported either by springs or by fixed boundary conditions. Furthermore, the available modes may be either a complete set; i.e., as many modes as the number of degrees of freedom of the associated FE model is available, or it can be an incomplete set.

1 INTRODUCTION

The dynamic simulation of a system requires the solution of the equations of motion with known initial conditions (Nikravesh, 1988). Since a structure in a mechanical system is modeled through the use of finite element method, incorporating its nodal stiffness and mass matrices in the equations of motion leads to a large number of degrees-of-freedom (Ambrosio and Nikravesh, 1992). To overcome this problem, the modal superposition technique is commonly used to reduce the number of coordinates of the structure (Song and Haug, 1980; Sunada and Dubowsky, 1981; Shabana and Wehage, 1983). The necessary modal data of a structure can be extracted from the finite element model. This process is always possible as long as we have an analytical model which has the nodal mass and stiffness matrices as known quantities. In the case of an experimental approach, since these matrices are not explicitly available, the problem becomes more complex. However, it is possible to identify the nodal mass and stiffness matrices from a set of experimentally obtained vibration data (Yi, 1996; Yi and Nikravesh, 1996 and 1997).

The boundary conditions for a structure, for under which the vibration measurements are obtained, may be different from the boundary conditions of that structure in its system setup. From the measured vibration data, we must obtain the necessary modal data for the dynamic modeling of the structure in its mechanical system environment. Therefore, for the utilization of the experimental vibration data, some structural identification problems need to be solved.

Most structural identification problems mainly deal with obtaining the mass and stiffness matrices based on an analytical model as reviewed in reference (Yi, 1996). Some studies in structural identification intend to update or improve the mass and stiffness matrices which are constructed from a low-resolution finite element model of the structure (Caesar and Peter, 1987; Foster and Mottershead, 1990). The updating processes use an incomplete set of vibration test data. Berman and Flannelly (1971) presented a method for computing the effects of mass and stiffness changes on natural frequencies and normal modes. Their purpose was to also modify an analytical model identifying modal parameters by using measured vibration data.

An attempt to determine the vibration data of unconstrained structures, such as rockets and airplanes in flight, can be found in reference (Przemieniecki, 1968). In this reference the difficulties of determining the free-free modes of a structure are explained, and an experimental approach for identifying the free-free modal data using experimentally obtained constrained data is proposed. The main drawback of this formulation is the need for a full set of experimentally obtained vibration data which is normally nonpractical. Furthermore, the proposed method requires the determination of some of the elements of the stiffness matrix by a series of static tests on the constrained structure.

In order to overcome the drawbacks of the existing methods, some analytical approaches are presented in this paper. The essence of this research is to obtain the necessary modal data from the available data without performing additional experiments. These methods utilize either a complete set or an incomplete set of measured vibration data to determine the necessary modal data for a structure in a mechanical system.

1.1 Problem statement

The equations of motion for a structure within a system can be described either in terms of the nodal or the modal coordinates. If the modal form of the equations of motion is used, then the following issues must be answered:

(a) do the equations of motion require the mode shapes in the free-free form or in the constrained form?
(b) are the modal data obtained from a finite element model or from laboratory or field experimentation?
(c) are all (or only some) of the modal data available?

If a finite element model representation of the structure is available, then its modal data can be extracted either in free-free or in constrained form. However, a structure in a system environment has moving boundary conditions--the structure is not free-free nor it has nonmoving boundary conditions in the standard finite element sense. This issue has been discussed by many researchers during the past decade and various formulations have been suggested (Shabana, 1997). Some formulations require the mode shapes to be obtained in a free-free state and some allow non-moving boundary conditions when the mode shapes are computed. Regardless of how the boundary conditions are described, some of these formulations can produce correct results only if all the modes are extracted and used in the model. If an incomplete set of modes is used, then the computed response can be far from being correct (Craig, 1981). This problem has led to the introduction of a variety of corrective techniques such as static mode correction (Yoo and Haug, 1986).

These problems become even more complex if the modal data can only be obtained experimentally and not from a finite element model. When the vibration data is obtained experimentally, only a few mode shapes associated with lower frequencies are obtained. Furthermore, if the dynamic formulation requires the mode shapes to be obtained in a free-free state, it is obvious that in a laboratory setup, due to the presence of gravity, such experiments cannot be conducted properly. In a laboratory, the experiment is performed on a structure that is supported at some points either by springs or by kinematic joints. The measured response represents the vibration characteristics of the constrained structure that is different from the actual structure with moving boundary conditions.

2 STIFFNESS MATRIX FROM VIBRATION DATA

In the finite element sense, assume that a structure has a total of n degrees of freedom (DoF). For the structure, we assume a nodal mass matrix $\mathbf{M}$, a nodal stiffness matrix $\mathbf{K}$, a modal mass matrix $\overline{\mathbf{M}}$, and a modal stiffness matrix $\overline{\mathbf{K}}$. The eigenvalue and modal matrices respectively are represented as:

$$\Lambda = diag[\lambda_1, \ \lambda_2, \ \cdots, \ \lambda_n] \tag{1}$$

$$\Psi = [\Psi_1, \ \Psi_2, \ \cdots, \ \Psi_n] \tag{2}$$

where λ_i and Ψ_i are the ith eigenvalue and the corresponding mode shape, respectively. We observe the following identities:

$$\overline{\mathbf{M}} = \mathbf{\Psi}^T \mathbf{M} \mathbf{\Psi} \tag{3}$$

$$\overline{\mathbf{K}} = \mathbf{\Psi}^T \mathbf{K} \mathbf{\Psi} \tag{4}$$

Based on references (Berman and Flannelly, 1971; Yi, 1996), we may consider two cases where a complete or an incomplete set of modal data is available:

(a) If a complete set of modal data is available, then we compute the complete stiffness matrix $\mathbf{K}$ as:

$$\mathbf{K} = \sum_{i=1}^{n} \frac{\lambda_i}{\overline{\mathrm{m}}_i} \left[\mathbf{M} \Psi_i \Psi_i^T \mathbf{M} \right] \tag{5}$$

where $\overline{\mathrm{m}}_i$ is the ith modal mass.

(b) If an incomplete set of modal data is available, then we compute the incomplete stiffness matrix $\mathbf{K}^k$ as:

$$\mathbf{K}^k = \sum_{i=1}^{m} \frac{\lambda_i}{\overline{\mathrm{m}}_i} \left[\mathbf{M} \Psi_i \Psi_i^T \mathbf{M} \right] \tag{6}$$

where the super-script "k" denotes an entity associated with the "kept" modes. For a structure with a truncated set of m modal data, where $m < n$, the incomplete stiffness matrix can be obtained from this equation. This computed incomplete stiffness matrix is highly singular. A complete stiffness matrix can be described as the summation of an incomplete stiffness matrix $\mathbf{K}^k$ due to the "kept" modes and a residual stiffness matrix $\mathbf{K}^d$ due to the unavailable or "deleted" modes (Craig, 1981); i.e.,

$$\mathbf{K} = \mathbf{K}^k + \mathbf{K}^d \tag{7}$$

The deleted modes in most cases are the higher modes. The residual stiffness matrix can be expressed as:

$$\mathbf{K}^d = \sum_{i=m+1}^{n} \frac{\lambda_i}{\bar{m}_i}\left[\mathbf{M}\Psi_i\Psi_i^T\mathbf{M}\right] \quad (8)$$

Based upon this partitioning, a set of elastic modes can also be separated into the kept modes Ψ^k and the deleted modes Ψ^d. On the basis of Eq. 6, where an incomplete set of modal data is available, we are only able to determine the incomplete stiffness matrix.

Note that in order to compute the stiffness matrix using Eq. 5 or 6, we assume that the nodal mass matrix is available. However, if the mass matrix is not available, then we must identify the mass distribution in advance. A procedure for identification of the mass matrix can be found in references (Yi and Nikravesh, 1996 and 1997).

In the next section we show how the free-free modes can be extracted from the constrained modes. we will consider two forms of constrained modal data--the structure is supported by springs or the structure is supported by kinematic joints. The available modal data can be either a complete set or an incomplete set.

3 FREE-FREE MODES FROM SPRING SUPPORTED MODES

In order to obtain the free-free modes from the modal data of a spring-supported structure, we assume that the supporting springs do not provide any kinematic constraints. Furthermore, we assume that the spring attachment points on the structure are also considered as measurement nodes. Nodes or measurement points are points on the structure, or its finite element representation, where the mode shapes are determined. Thus, the spring support points appear as finite element nodes in the model. Assuming that we know the characteristics of the springs, we can explicitly construct that portion of the stiffness matrix associated with the springs. We will refer to the combination of the structure and the supporting springs as the "combined system".

Before computation of the stiffness matrix, we first eliminate the mass effect due to the spring masses. The lumped mass matrix $\mathbf{M}$ of a free-free structure is given by

$$\mathbf{M}=\mathbf{M}_{(c)} - \mathbf{M}_{(s)} \quad (9)$$

where $\mathbf{M}_{(c)}$ is the mass matrix of the combined system, and $\mathbf{M}_{(s)}$ is the mass matrix associated with the springs.

We will consider two cases where a complete set or an incomplete set of constrained modal data is available. In each case, we will show how the modal data of the free-free structure can be extracted from its constrained modal data.

3.1 Complete Set of Spring Supported Modes

When a full set of constrained modal data is available, we use Eq. 5 to obtain the stiffness matrix $\mathbf{K}_{(c)}$. This computed stiffness matrix is the combination of the structure stiffness matrix $\mathbf{K}$ and the spring stiffness matrix $\mathbf{K}_{(s)}$. Therefore, the corrected stiffness matrix $\mathbf{K}$ for the free-free structure can be described as

$$\mathbf{K}=\mathbf{K}_{(c)} - \mathbf{K}_{(s)} \quad (10)$$

Here we realize that by subtracting the spring stiffness matrix from the definite stiffness matrix of the constrained structure, the resultant stiffness matrix becomes a semi-definite matrix. From the corrected mass matrix and stiffness matrix of Eqs. 9 and 10, we obtain the corrected eigenvalues and the corresponding mode shapes for the free-free structure. Note that for a well-defined problem, six of the modes will be rigid body modes.

3.2 Incomplete Set of Spring Supported Modes

Using an incomplete set of modal data, we can identify a partial (or an incomplete) stiffness matrix from Eq. 6. To completely eliminate the effects of the support springs, we must correct the mass and stiffness matrices from Eqs. 9 and 10. However, the stiffness matrix correction of incomplete models cannot be directly performed. A reasonable approximation to the stiffness matrix for an incomplete model can be made as (Yi, 1996)

$$\mathbf{K}^k = \mathbf{K}^k_{(c)} - \mathbf{K}^k_{(s)} \quad (11)$$

The term $\mathbf{K}^k_{(c)}$ can be computed from Eq. 6 from an incomplete set of vibration data. Since the stiffness matrix associated with the support springs is not orthogonal with respect to the measured modes, we can recompute the incomplete portion of the spring stiffness matrix as

$$\mathbf{K}^k_{(s)} = \sum_{i=1}^{m}\sum_{j=1}^{m} \frac{\left(\Psi_i^T \mathbf{K}_{(s)} \Psi_j\right)}{\bar{m}_i \bar{m}_j}\left[\mathbf{M}_{(c)}\Psi_i\Psi_j^T\mathbf{M}_{(c)}\right] \quad (12)$$

After computing Eqs. 9 and 11, we may construct the eigenvalue equation for the incomplete model as

$$\left[\mathbf{K}^k - \lambda\mathbf{M}\right]\mathbf{\Psi}=\mathbf{0} \quad (13)$$

From this equation, we can obtain the corrected natural frequencies and eigenmodes for the free-free structure. Note that from Eq. 13 we obtain the same number of eigenmodes as the available number of constrained modes. Therefore, Eq. 13 generates as many zero eigenvalues as the number of deleted modes including six rigid-body modes.

3.3 Alternative Method In Modal Space

In this section we present another method to obtain the free-free modes from the spring-supported modes. We can use this method with either a complete set or an incomplete set of constrained modal data. Based on the assumptions made in the previous sections, we perform the mode identification in the free-free modal space (I) and the constrained modal space (II).

We assume that the constrained modal vectors form the basis for modal space (II). Thus, we denote the eigenvalues and the corresponding modal matrices as $\mathbf{\Lambda}_{\mathrm{II}}$ and $\mathbf{\Psi}_{\mathrm{II}}$, respectively. Using these matrices, the generalized modal mass $\overline{\mathbf{M}}_{\mathrm{II}}^{(c)}$ and stiffness $\overline{\mathbf{K}}_{\mathrm{II}}^{(c)}$ in the current modal space (II) are given by

$$\overline{\mathbf{M}}_{\mathrm{II}}^{(c)} = \mathbf{\Psi}_{\mathrm{II}}^{T} \mathbf{M}_{(c)} \mathbf{\Psi}_{\mathrm{II}} \tag{14}$$

$$\overline{\mathbf{K}}_{\mathrm{II}}^{(c)} = \mathbf{\Psi}_{\mathrm{II}}^{T} \mathbf{K}_{(c)} \mathbf{\Psi}_{\mathrm{II}} = \overline{\mathbf{M}}_{\mathrm{II}}^{(c)} \mathbf{\Lambda}_{\mathrm{II}} \tag{15}$$

As stated earlier, the mass contribution and the stiffness entries of the support-springs are known quantities. Therefore, we can compute the portion of the generalized modal mass matrix $\overline{\mathbf{M}}_{\mathrm{II}}^{(s)}$ and stiffness matrix $\overline{\mathbf{K}}_{\mathrm{II}}^{(s)}$ associated with the springs as

$$\overline{\mathbf{M}}_{\mathrm{II}}^{(s)} = \mathbf{\Psi}_{\mathrm{II}}^{T} \mathbf{M}_{(s)} \mathbf{\Psi}_{\mathrm{II}} \tag{16}$$

$$\overline{\mathbf{K}}_{\mathrm{II}}^{(s)} = \mathbf{\Psi}_{\mathrm{II}}^{T} \mathbf{K}_{(s)} \mathbf{\Psi}_{\mathrm{II}} \tag{17}$$

In Eqs. 16 and 17, the eigenvectors are not orthogonal with respect to the spring mass and stiffness matrices. By eliminating these modal mass and stiffness matrices, we can obtain a new set of modal mass and stiffness matrices in the modal space (II) as

$$\overline{\mathbf{M}}_{\mathrm{II}} = \overline{\mathbf{M}}_{\mathrm{II}}^{(c)} - \overline{\mathbf{M}}_{\mathrm{II}}^{(s)} \tag{18}$$

$$\overline{\mathbf{K}}_{\mathrm{II}} = \overline{\mathbf{K}}_{\mathrm{II}}^{(c)} - \overline{\mathbf{K}}_{\mathrm{II}}^{(s)} \tag{19}$$

In these equations, $\overline{\mathbf{M}}_{\mathrm{II}}$ and $\overline{\mathbf{K}}_{\mathrm{II}}$ are the corrected modal mass and stiffness matrices in the current modal space (II). Whether we have a complete set or an incomplete set of modal data, it is always possible to compute Eqs. 18 and 19. From these matrices we can extract a set of modal data--eigenvalue matrix $\mathbf{\Lambda}_{\mathrm{I}}$ in the free-free space (I) and the corresponding intermediate modal matrix $\mathbf{\Psi}_{\mathrm{I/II}}$. Here note that the modal matrix $\mathbf{\Psi}_{\mathrm{I/II}}$ is a transformation to the free-free modal space (I) from the constrained modal space (II).

Since we want to obtain a transformation between the physical space and the free-free modal space (I), the modal matrix is identified as

$$\mathbf{\Psi}_{\mathrm{I}} = \mathbf{\Psi}_{\mathrm{II}} \mathbf{\Psi}_{\mathrm{I/II}} \tag{20}$$

where $\mathbf{\Psi}_{\mathrm{I}}$ is the modal matrix of the free-free structure. Furthermore, if we have a complete set of modal data from this relation, we can get an inverse relationship as

$$\mathbf{\Psi}_{\mathrm{I/II}} \mathbf{\Psi}_{\mathrm{I/II}}^{-1} = \mathbf{I} \tag{21}$$

By applying this relation to Eq. 20, a constrained modal matrix $\mathbf{\Psi}_{\mathrm{II}}$ is obtained as

$$\mathbf{\Psi}_{\mathrm{II}} = \mathbf{\Psi}_{\mathrm{I}} \mathbf{\Psi}_{\mathrm{I/II}}^{-1} \tag{22}$$

This inverse relationship is useful when we have a complete set of modal data.

Example 1: In this section we provide an example to validate and to further describe the proposed methodologies. The data for the example is derived from a finite element model and not from actual experiments. Assume that a structure in the form of a block is supported by four tri-axial springs, as shown in Fig. 1. Based on the developed process, we would like to perform the free-free mode identification. The three-dimensional finite element model consists of 42 nodes and 24 elements. The material of the block has a Young's Modulus of 207 GPa, a Poisson's ratio of 0.292, and a mass density of 7650 Kg/m^3. The stiffness components of the tri-axial support-springs are identically assigned to be 6 KN/m. We assume that each spring attachment point has an additional mass of 2 Kg due to the support-springs. The model has a mass of 458 Kg and an inertia tensor $\mathbf{J} = diag[63.7 \quad 165.0 \quad 228.0]$.

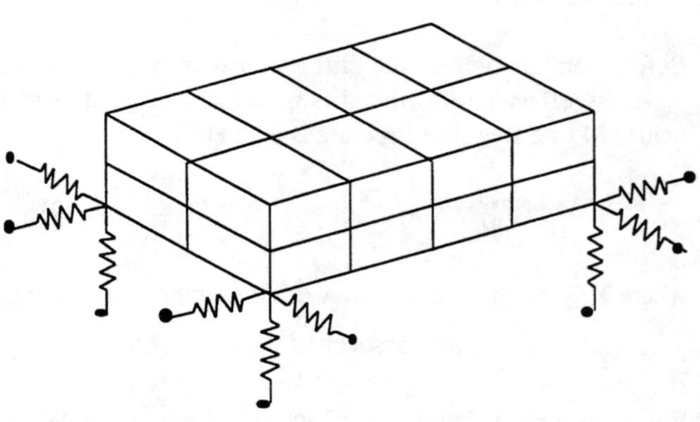

Figure 1. A FE model of a block structure supported by springs.

The modal analysis of the finite element model was conducted by ABAQUS (1996). The first twenty modal frequencies of the computed modal data are listed in column 2 of Table 1. Note that since the stiffness of the support springs are selected to be softer than the stiffness of the flexible body, the first six frequencies are very close to the reference rigid-body modes.

The free-free mode identification for the complete and three incomplete models are formulated on the basis of Eq. 13. The identified results are summarized in Table 1. Column 3 gives the results if we consider all 90 modes (complete set). The results for considering an incomplete set of modal data with 20, 12, or 9 modes are given in columns 4, 5, and 6, respectively. Column 7 contains the identified results from the alternative modal-space method with 12 modes. In order to verify the accuracy of the identified modes, the free-free modes are also extracted directly from the finite element model of the block structure, as shown in column 8. It can be observed that all the models give good approximation of the free-free modal data by eliminating the effect of the support-springs from the complete or the incomplete set of constrained modal data.

Table 1. Identified frequencies of complete and incomplete models

Mode No.	Spring Supported FE Model (Hz)	Free-Free Complete Model (Hz)	Free-Fre Incomplete Models (Hz)			ID in Modal Space (Hz)	Free-Free FE Model (Hz)
		m=90	*m=20*	*m=12*	*m=9*	*m = 12*	
01	1.152	0.008	0.000	0.000	0.000	-	0.020
02	1.153	0.004	0.004	0.004	0.004	-	0.007
03	1.153	0.006	0.006	0.006	0.006	-	0.004
04	1.153	0.006	0.006	0.006	0.007	-	0.000
05	1.385	0.007	0.007	0.007	0.007	-	0.049
06	1.854	0.008	0.008	0.009	0.009	-	0.024
07	16.712	16.997	17.021	17.021	17.031	17.010	17.010
08	109.160	111.640	111.804	111.821	111.820	111.630	111.632
09	184.290	188.497	188.802	188.802	188.802	188.492	188.493
10	231.750	236.066	236.346	236.510		236.061	236.064
11	271.370	274.418	274.620	274.620		274.409	274.408
12	421.520	425.673	426.056	426.422		425.670	425.669
13	422.640	431.108	431.968				431.103
14	464.990	473.560	474.082				473.554
15	470.620	475.854	476.202				475.850
16	490.450	491.838	491.931				491.830
17	506.980	513.102	513.712				513.106
18	545.690	555.951	557.266				555.950
19	600.180	610.467	611.234				610.472
20	616.360	626.427	627.134				626.425
...	...	...					...
col.	**=> 2**	**3**	**4**	**5**	**6**	**7**	**8**

4 FREE-FREE MODES FROM CONSTRAINED MODES

To obtain the free-free modes of a structure from the modal data of the same structure supported by kinematic joints, it is assumed that the kinematic joints are standard fixed boundary conditions such as spherical joints and roller supports. Using a complete set of constrained modal data with known mass distribution, we derive the necessary formulations to identify the free-free modes for the unconstrained structure.

4.1 Mathematical Approach

In order to utilize the available modal data of a structure supported by kinematic joints, the general form of eigenvalue equations for a free-free structure can be derived in terms of its modal quantities. Without considering the rigid-body motion of a free-free structure, the undamped free vibration equation for the structure is written as

$$\mathbf{M}\ddot{\mathbf{u}}+\mathbf{K}\mathbf{u} = \mathbf{0} \qquad (23)$$

In this equation, we assume that the displacement $\mathbf{u}$ can be divided into two displacement vectors, $\mathbf{u}_f$ and $\mathbf{u}_c$. Vector $\mathbf{u}_c$ contains the nodal displacements of those DoF that will be constrained by the boundary conditions later on, and $\mathbf{u}_f$ contains the nodal displacements which will remain free. Based on these displacement vectors, the nodal mass and stiffness matrices are also partitioned, and as a result, Eq. 23 is written as

$$\begin{bmatrix} \mathbf{M}_{ff} & \mathbf{M}_{fc} \\ \mathbf{M}_{cf} & \mathbf{M}_{cc} \end{bmatrix} \begin{Bmatrix} \ddot{\mathbf{u}}_f \\ \ddot{\mathbf{u}}_c \end{Bmatrix} + \begin{bmatrix} \mathbf{K}_{ff} & \mathbf{K}_{fc} \\ \mathbf{K}_{cf} & \mathbf{K}_{cc} \end{bmatrix} \begin{Bmatrix} \mathbf{u}_f \\ \mathbf{u}_c \end{Bmatrix} = \begin{Bmatrix} 0 \\ 0 \end{Bmatrix} \quad (24)$$

For the same structure, if we constrain the DoF associated with $\mathbf{u}_c$, since $\mathbf{u}_c = \ddot{\mathbf{u}}_c = \mathbf{0}$, the rows and columns associated with the constrained degrees of freedom can be eliminated to get

$$\mathbf{M}_{ff}\ddot{\mathbf{u}}_f + \mathbf{K}_{ff}\mathbf{u}_f = \mathbf{0} \quad (25)$$

This equation represents a positive definite system if all rigid body DoF are properly eliminated.

At this point we consider Eq. 24 for the free-free structure. The first row from Eq. 24 is written as

$$\mathbf{M}_{ff}\ddot{\mathbf{u}}_f + \mathbf{M}_{fc}\ddot{\mathbf{u}}_c + \mathbf{K}_{ff}\mathbf{u}_f + \mathbf{K}_{fc}\mathbf{u}_c = \mathbf{0} \quad (26)$$

Since Eq. 24 represents an unforced free vibration, the harmonic solution can be described as

$$\mathbf{u}_f = \mathbf{\Psi}_f e^{i\sqrt{\lambda}t} \quad (27)$$

$$\mathbf{u}_c = \mathbf{\Psi}_c e^{i\sqrt{\lambda}t} \quad (28)$$

where $\mathbf{\Psi}_f$ and $\mathbf{\Psi}_c$ are column matrices of the amplitudes of displacements $\mathbf{u}_f$ and $\mathbf{u}_c$, respectively, λ is an eigenvalue of the free-free structure, and t is the time. Using Eq. 28 and its second time derivative, Eq. 26 becomes

$$\mathbf{M}_{ff}\ddot{\mathbf{u}}_f + \mathbf{K}_{ff}\mathbf{u}_f = (\lambda\mathbf{M}_{fc} - \mathbf{K}_{fc})\mathbf{\Psi}_c e^{i\sqrt{\lambda}t} \quad (29)$$

In this equation, we observe that the right-side forcing term represents a harmonic loading. Therefore, using Eq. 27 and its second time derivative, and canceling the common terms yield

$$(\lambda\mathbf{M}_{ff} - \mathbf{K}_{ff})\mathbf{\Psi}_f = (\lambda\mathbf{M}_{fc} - \mathbf{K}_{fc})\mathbf{\Psi}_c \quad (30)$$

Now consider the second equation in Eq. 24. The second row in Eq. 24 is rewritten as

$$\mathbf{M}_{cf}\ddot{\mathbf{u}}_f + \mathbf{M}_{cc}\ddot{\mathbf{u}}_c + \mathbf{K}_{cf}\mathbf{u}_f + \mathbf{K}_{cc}\mathbf{u}_c = \mathbf{0} \quad (31)$$

By applying the defined displacements of Eqs. 27 and 28, and their second time derivatives, and canceling the common terms, we obtain

$$(\mathbf{K}_{cf} - \lambda\mathbf{M}_{cf})\mathbf{\Psi}_f + (\mathbf{K}_{cc} - \lambda\mathbf{M}_{cc})\mathbf{\Psi}_c = \mathbf{0} \quad (32)$$

Equating Eqs. 30 and 32, and rearranging the terms yields

$$\left(\begin{bmatrix} \mathbf{K}_{ff} & \mathbf{K}_{fc} \\ \mathbf{K}_{cf} & \mathbf{K}_{cc} \end{bmatrix} - \lambda \begin{bmatrix} \mathbf{M}_{ff} & \mathbf{M}_{fc} \\ \mathbf{M}_{cf} & \mathbf{M}_{cc} \end{bmatrix} \right) \begin{Bmatrix} \mathbf{\Psi}_f \\ \mathbf{\Psi}_c \end{Bmatrix} = \begin{Bmatrix} 0 \\ 0 \end{Bmatrix} \quad (33)$$

4.2 Stiffness Matrix and Free-Free Modes

This section presents an analytical method to determine the full nodal stiffness matrix by using the modal data of a kinematically constrained structure. In this approach we assume that the nodal mass distribution of the structure is known. After identifying the stiffness matrix, the free-free modes of the structure are extracted from the nodal mass and stiffness matrices.

If we apply the standard boundary conditions to Eq. 33, we can extract an eigenvalue equation for a constrained structure as

$$(\mathbf{K}_{ff} - \hat{\lambda}\mathbf{M}_{ff})\hat{\mathbf{\Psi}}_f = \mathbf{0} \quad (34)$$

where

$$\hat{\Lambda} = diag\left[\hat{\lambda}_1 \quad \hat{\lambda}_2 \quad \hat{\lambda}_3 \quad \cdots \quad \hat{\lambda}_n\right] \quad (35)$$

$$\hat{\Psi} = \left[\hat{\Psi}_1 \quad \hat{\Psi}_2 \quad \hat{\Psi}_3 \quad \cdots \quad \hat{\Psi}_n\right] \quad (36)$$

are the eigenvalues and the associated eigenmode matrix, respectively. In this study we assume that the constrained modal data are known quantities. With known modal data and known mass matrix $\mathbf{M}_{ff}$, the stiffness matrix $\mathbf{K}_{ff}$ is computed from Eq. 5 as

$$\mathbf{K}_{ff} = \sum_{i=1}^{n} \frac{\hat{\lambda}_i}{\hat{\mathrm{m}}_i}\left[\mathbf{M}_{ff}\hat{\mathbf{\Psi}}_i\hat{\mathbf{\Psi}}_i^T\mathbf{M}_{ff}\right] \quad (37)$$

Using this representation, Eq. 33 becomes

$$\left(\begin{bmatrix} \sum_{i=1}^{n} \frac{\hat{\lambda}_i}{\hat{\bar{m}}_i} \mathbf{M}_{ff} \hat{\Psi}_i \hat{\Psi}_i^T \mathbf{M}_{ff} & \mathbf{K}_{fc} \\ \mathbf{K}_{cf} & \mathbf{K}_{cc} \end{bmatrix} - \lambda \begin{bmatrix} \mathbf{M}_{ff} & 0 \\ 0 & \mathbf{M}_{cc} \end{bmatrix}\right) \begin{Bmatrix} \Psi_f \\ \Psi_c \end{Bmatrix} = 0 \tag{38}$$

Note that Eq. 38 uses a lumped mass formulation. The second line of this equation is associated with the constrained DoF. By eliminating the rows and columns associated with the constrained DoF, we can obtain a positive definite matrix which gives all the elastic eigenmodes.

In order to utilize Eq. 38, we first must recover $\mathbf{K}_{ff}$ from Eq. 37. We also need to determine the sub-stiffness matrices $\mathbf{K}_{fc}$, $\mathbf{K}_{cf}$, $\mathbf{K}_{cc}$, which must be identified either from experimental tests or by analytical methods. Since it is not practical to determine the sub-stiffness matrices experimentally, it will be useful if an analytical method is developed to use the existing modal data.

It is possible to directly construct the rigid-body modes without computing the characteristic equations (Yi, 1996). The rigid-body modes are mutually orthogonal to the physical mass distribution. Moreover, the semi-definite stiffness matrix and the rigid body modes are also mutually orthogonal. Utilizing the orthogonality and symmetric conditions, it might be possible to obtain the stiffness matrix, explicitly.

The rigid-body modal matrix Ψ_o is decomposed as

$$\Psi_o = \begin{bmatrix} \bar{\mathbf{I}} & \bar{\mathbf{s}} \end{bmatrix} = \begin{bmatrix} \bar{\mathbf{I}}_f & \bar{\mathbf{s}}_f \\ \bar{\mathbf{I}}_c & \bar{\mathbf{s}}_c \end{bmatrix} \tag{39}$$

where $\bar{\mathbf{I}}$ is an *nx3* matrix containing *3*x*3* identity matrices, and $\bar{\mathbf{s}}$ is an *n*x*3* matrix containing *3*x*3* skew matrices of nodal position vectors. Here, subscripts f and c represent the unconstrained and constrained nodal DoF, respectively. According to the defined nodal DoF, it is required that the rigid-body modal matrix to be adjusted. Since the semi-definite stiffness matrix and the rigid-body modes must be mutually orthogonal, we have

$$\mathbf{K}\Psi_o = \begin{bmatrix} \mathbf{K}_{ff} & \mathbf{K}_{fc} \\ \mathbf{K}_{cf} & \mathbf{K}_{cc} \end{bmatrix} \begin{bmatrix} \bar{\mathbf{I}}_f & \bar{\mathbf{s}}_f \\ \bar{\mathbf{I}}_c & \bar{\mathbf{s}}_c \end{bmatrix} = 0 \tag{40}$$

This equation provides the following four matrix equations as

$$\mathbf{K}_{fc}\bar{\mathbf{I}}_c = -\mathbf{K}_{ff}\bar{\mathbf{I}}_f \tag{41a}$$

$$\mathbf{K}_{fc}\bar{\mathbf{s}}_c = -\mathbf{K}_{ff}\bar{\mathbf{s}}_f \tag{41b}$$

$$\mathbf{K}_{cc}\bar{\mathbf{I}}_c = -\mathbf{K}_{cf}\bar{\mathbf{I}}_f \tag{41c}$$

and

$$\mathbf{K}_{cc}\bar{\mathbf{s}}_c = -\mathbf{K}_{cf}\bar{\mathbf{s}}_f \tag{41d}$$

Using Eqs. 41a through 41d, it is possible to determine the unknown entries in the sub-stiffness matrices. For more explicit expressions, a semi-definite stiffness matrix is expressed as

$$\mathbf{K} = \begin{bmatrix} \mathbf{K}_{ff} & \mathbf{K}_{fc(rr)} & \mathbf{K}_{fc(r\omega)} \\ \mathbf{K}_{fc(rr)}^T & \mathbf{K}_{cc(rr)} & \mathbf{K}_{cc(r\omega)} \\ \mathbf{K}_{fc(r\omega)}^T & \mathbf{K}_{cc(r\omega)}^T & \mathbf{K}_{cc(\omega\omega)} \end{bmatrix} \tag{42}$$

where the subscripts r and ω stand for rigid-body translational and rotational DoF, respectively. The rigid-body modal matrix associated with the constrained nodal DoF can be decomposed as

$$\bar{\mathbf{I}}_c = \begin{bmatrix} \mathbf{I} \\ \bar{\mathbf{0}} \end{bmatrix} \tag{43}$$

$$\bar{\mathbf{s}}_c = \begin{bmatrix} \bar{\mathbf{s}}_{c(r)} \\ \bar{\mathbf{I}} \end{bmatrix} \tag{44}$$

Using Eqs. 42 through 44, Eqs. 41 become

$$\mathbf{K}_{fc(rr)} = -\mathbf{K}_{ff}\bar{\mathbf{I}}_f \tag{45a}$$

$$\mathbf{K}_{fc(r\omega)} = -\mathbf{K}_{ff}\bar{\mathbf{s}}_f - \mathbf{K}_{fc(rr)}\bar{\mathbf{s}}_{c(r)} \tag{45b}$$

$$\begin{bmatrix} \mathbf{K}_{cc(rr)} \\ \mathbf{K}_{cc(r\omega)}^T \end{bmatrix} = -\mathbf{K}_{fc}^T\bar{\mathbf{I}}_f \tag{45c}$$

and

$$\mathbf{K}_{cc(\omega\omega)} = -\mathbf{K}_{fc}^T\bar{\mathbf{s}}_f - \mathbf{K}_{cc(r\omega)}^T\bar{\mathbf{s}}_{c(r)} \tag{45d}$$

From Eq. 37 we can compute $\mathbf{K}_{ff}$ and from Eq. 39 we can explicitly construct the rigid-body modal matrix. Using these computed quantities, it is possible to determine all of the unknown entries in the sub-matrices $\mathbf{K}_{fc}$ and $\mathbf{K}_{cc}$ from Eqs. 45.

After determining the full stiffness matrix, we can extract the eigenvalues and the associated modes from a conventional eigen-solver. It should be noted that in this section, we have assumed that the constrained structure is a deterministic one; i.e., a structure that is not over constrained. If the structure is over constrained, then it will not possible to recover the complete stiffness matrix of the corresponding free-free structure. We may lose some accuracy in identifying the free-free modes for such a structure. The following two examples will show the developed process to determine the free-free modes from the constrained modal data.

Example 2: Consider the constrained system shown in Fig. 2, containing two degrees of freedom. In this example, two moving bodies are interconnected by linear elastic springs, and the third body is fixed to the ground. For simplicity, the mass m and stiffness k are assumed to be of unit magnitude.

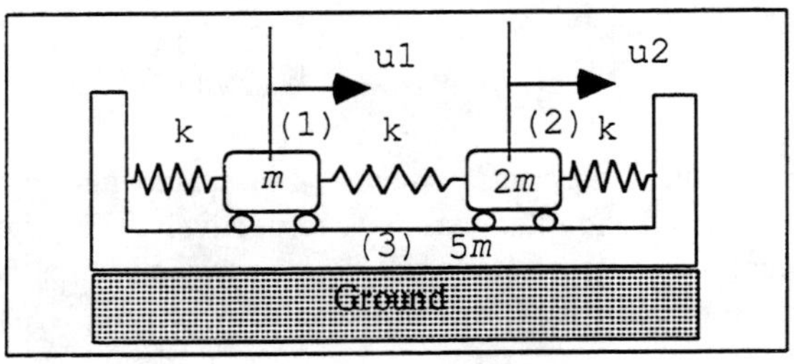

Figure 2. A constrained system.

Assume that the modal data for this constrained system is available as

$$\hat{\Lambda} = \begin{bmatrix} 0.634 & \\ & 2.366 \end{bmatrix} \text{ and } \hat{\Psi} = \begin{bmatrix} 0.7321 & 1.0000 \\ 1.0000 & -0.3660 \end{bmatrix}$$

Using this data, we want to obtain the modal data for the free-free system, where the third body is allowed to translate with respect to the ground. The solution steps are as follows:
For the constrained system, the mass matrix is $\mathbf{M}_{ff} = \begin{bmatrix} 1.0 & \\ & 2.0 \end{bmatrix}$. The modal mass matrix is computed as

$\hat{\bar{\mathbf{M}}} = \hat{\Psi}^T \mathbf{M}_{ff} \hat{\Psi} = \begin{bmatrix} 2.536 & \\ & 1.268 \end{bmatrix}$. From Eq. 37, the stiffness $\mathbf{K}_{ff}$ is determined as

$$\mathbf{K}_{ff} = \sum_{i=1}^{n} \frac{\hat{\lambda}_i}{\hat{\bar{m}}_i} \left[\mathbf{M}_{ff} \hat{\Psi}_i \hat{\Psi}_i^T \mathbf{M}_{ff} \right] = \begin{bmatrix} 1.9998 & -0.9998 \\ -0.9998 & 1.9998 \end{bmatrix}$$

Since there is no rotational DoF in this particular example, the rigid-body modes are given as

$$\Psi_o = \bar{\mathbf{I}} = \begin{bmatrix} \bar{\mathbf{I}}_f \\ \bar{\mathbf{I}}_c \end{bmatrix} \text{ where } \bar{\mathbf{I}}_f = \begin{bmatrix} 1 \\ 1 \end{bmatrix} \text{ and } \bar{\mathbf{I}}_c = [1]$$

From Eq. 45a, the sub-stiffness matrix $\mathbf{K}_{fc}$ is computed as

$$\mathbf{K}_{fc} = -\mathbf{K}_{ff} \bar{\mathbf{I}}_f = -\begin{bmatrix} 1.9998 & -0.9998 \\ -0.9998 & 1.9998 \end{bmatrix} \begin{bmatrix} 1 \\ 1 \end{bmatrix} = \begin{bmatrix} -1 \\ -1 \end{bmatrix}$$

From Eq. 45c, the sub-stiffness matrix $\mathbf{K}_c$ is determined as

$$\mathbf{K}_{cc} = -\mathbf{K}_{fc}^T \bar{\mathbf{I}}_f = -[-1 \quad -1] \begin{bmatrix} 1 \\ 1 \end{bmatrix} = [2]$$

Using the identified sub-matrices, the full stiffness and mass matrices are found to be

$$\mathbf{K} = \begin{bmatrix} 1.9998 & -0.9998 & -1 \\ -0.9998 & 1.9998 & -1 \\ -1 & -1 & 2 \end{bmatrix} \text{ and } \mathbf{M} = \begin{bmatrix} 1 & 0 & 0 \\ 0 & 2 & 0 \\ 0 & 0 & 5 \end{bmatrix}$$

Using any conventional eigen-solver, these stiffness and mass matrices yield the free-free modal data as

$$\Lambda = \begin{bmatrix} 0. & & \\ & 1.0 & \\ & & 2.4 \end{bmatrix} \text{ and } \Psi = \begin{bmatrix} 1.0 & -0.5 & -0.067 \\ 1.0 & 0.5 & 1.000 \\ 1.0 & 1.0 & -0.333 \end{bmatrix}$$

Example 3: As another example, consider a simple 3-D block finite element model consisting of eight nodes as shown in Fig. 3. The structure is constrained by kinematic joints such as a fixed spherical joint and three roller joints.

As shown in Table 2, the spherical joint at node (1) eliminates three degrees of freedom, and the roller joints at node (2), (3), and (4) each eliminates one DoF. We assume that the modal data for the constrained structure is available, and from that, we will identify the free-free modes when the kinematic joints are removed.

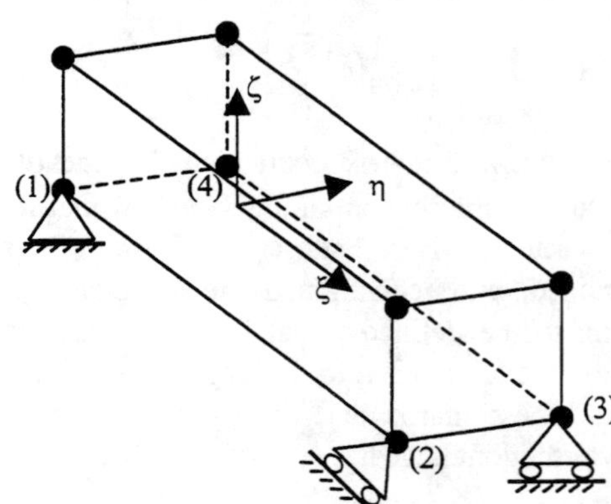

Figure 3. A structure constrained by kinematic joints.

Table 2. The constrained nodal degrees of freedom.

Node No.	Constrained ξ	Degrees of η	Freedom ζ
(1)	x	x	x
(2)	-	x	-
(3)	-	-	x
(4)	-	-	x

Note: x stands for the constrained DoF.

For the constrained system the modal data are obtained by the finite element program, ABAQUS (1996). The computed frequencies are listed in the second column of Table 3. Following the procedure explained in section 4.2, the identified frequencies for the corresponding free-free structure are summarized in the third column of Table 3. For validation purposes, the free-free modal data are also computed directly from the free-free structure, and the corresponding frequencies are listed in the fourth column of Table 3. It can be observed that the identified free-free frequencies are in agreement with the results obtained directly from the finite element model. Note that the constrained finite element model in this example is a deterministic structure; i.e., the boundary conditions have eliminated only 6 DoF. Therefore, it was possible to obtain all of the free-free modes without losing any information.

Table 3. Identified frequencies for free-free structure.

Mode No.	Constrained FE Model by Fixed-Constraints (Hz)	Free-Free Identification Results (Hz)	Direct Extraction from FE Model (Hz)
01	-	0.000	0.011
02	-	0.000	0.011
03	-	0.000	0.011
04	-	0.000	0.000
05	-	0.000	0.000
06	-	0.000	0.000
07	22.580	94.023	94.026
08	36.293	94.028	94.026
09	54.861	99.110	99.112
10	76.297	99.112	99.112
11	96.224	99.115	99.112
12	96.219	121.384	121.386
13	97.596	121.386	121.386
14	106.990	121.390	121.386
15	121.060	162.849	162.857
16	132.070	162.857	162.857
17	150.520	162.862	162.857
18	166.700	188.040	188.051
19	183.620	230.307	230.315
20	201.250	230.309	230.315
21	205.820	230.317	230.315
22	221.190	230.320	230.315
23	226.200	230.327	230.315
24	357.420	405.891	405.888
col.	==> 2	3	4

5. CONCLUSIONS

Most existing methods for generating structure models are based on the finite element method. These methods are generally restricted to bodies composed of relatively simple elements such as beam or shell. Nevertheless, the finite element method leads to computationally inefficient formulations due to the large number of degrees of freedom. The use of modal data allows for coordinate reduction by using the mode superposition technique and a small set of generalized coordinates. Another advantage is that the experimentally obtained modal data can be used as an alternative to having a finite element model.

Experimental vibration data are obtained for structure supported by springs or by fixed boundary conditions. However, the dynamic formulation may require the use of modal data for free-free structures. Therefore, in this paper general methodologies are developed to identify the necessary free-free modes from the available vibration data for flexible system dynamics. Since only a limited number of mode shapes and frequencies can be obtained experimentally, the proposed formulations can work for either complete or incomplete sets of modal data. Several simple examples show that the formulations can be used effectively and accurately.

REFERENCES

Nikravesh, P. E., 1988, *Computer Aided Analysis of Mechanical Systems*, Prentice-Hall.

Ambrosio, J. and Nikravesh, P. E., 1992, *Elaso-Plastic Deformation in Multibody Dynamics*, Nonlinear Dynamics 3, pp. 85-104.

Song, J. O. and Haug, E. J., 1980, *Dynamic Analysis of Planar Flexible Mechanisms, Computer Methods in Applied Mechanics and Engineering*, Vol. 24, pp. 359-381.

Sunada, W. and Dubowsky, S., 1981, *The Application of the Finite Element Method to the Dynamic Analysis of Flexible Spatial and Co-Planar Linkage Systems*, ASME J. of Mechanical Design, Vol. 103, pp. 643-651.

Shabana, A. and Wehage, R. A., 1983, *Coordinate Reduction Techniques for Transient Analysis of Spatial Substructures with Large Angular Rotations*, Journal of Structural Mechnics 11 (3), pp. 401-431.

Yi, T. Y., 1996, *Structural Identification Based On Vibration Data For Flexible and Rigid Multibody System Dynamics*, Ph.D. Dissertation, AME, University of Arizona, Tucson, AZ 85721.

Yi, T. Y. and Nikravesh, P. E., 1996, *Quasi-Finite Element Method*, Technical Report CAEL-96-2, AME, University of Arizona.

Yi, T. Y. and Nikravesh, P. E., 1997, *Prediction of Modal Characteristics Due To Design Change*, Proc. of DETC97/VIB-4190 ASME Design Engineering.

Caesar, B. and Peter, J., 1987, *Direct Update of Dynamic Mathematical Models from Modal Test Data*, AIAA Journal, Vol. 25, pp. 1494-1499.

Foster, C. D. and Mottershead, J. E., 1990, *A Method for Improving Finite Element Models by Using Experimental*

Data: Application Monitoring, Int. J. Mech. Vol. 32, No. 3, pp.191-203.

Berman, A. and Flannelly, W. G., 1971, *Theory of Incomplete Models of Dynamic Structures,* AIAA Journal, Vol. 9 pp. 1481-1487.

Przemieniecki, J. S., 1968, *Theory of Matrix Analysis*, McGraw-Hill.

Shabana, Ahmed A., 1997, *Flexible Multibody Dynamics: Review of Past and Recent Developments*, Multibody System Dynamics, Vol. 1 No. 2, pp. 189-222, Kluwer Academic Publishers, Netherlands.

Craig, R. R., 1981, *Structural Dynamics: An Introduction to Computer Methods*, John Wiley & Sons.

Yoo, W. S. and Haug, E. J., 1986, *Dynamics of Flexible Mechanical Systems Using Vibration and Static Correction Modes*, J. of Mechanisms, Transmissions, and Automation in Design, Vol. 108, pp. 315-322.

Hibbit, Karlson, and Sorensen, Inc., 1996, *ABAQUS Theory and User's Manual.*

PVP-Vol. 415, Recent Advances in Solids and Structures — 2000
ASME 2000

BALLISTIC IMPACT ANALYSIS OF THE HUMAN THORAX WITH A PROTECTIVE BODY ARMOR

Y. W. Kwon, J. E. Jolly, and T. A. Hughes
Mechanical Engineering Department
Naval Postgraduate School
Monterey, California

ABSTRACT

The biomechanical response of a finite element model of the human thorax and a protective body armor system was studied under impact loading from a projectile. The objective of the study was to create a viable finite element model of the human thorax. The model was validated by comparing the results of tests of body armor systems conducted on cadavers to results obtained from finite element analysis. A parametric study was undertaken to determine the essential components of the model. The results from this investigation determined that the path of force propagation from a body armor system to the thorax upon bullet impact is directly through the vest to the sternum and then through the skeleton to the rest of the body. Thus, any parameters that affect the components in this pathway were essential to the model. This included the muscles, their geometries, material properties, and viscosity, as well as the Young's modulus of the sternochondral cartilage and the bones themselves.

INTRODUCTION

Since the early twentieth century, the United States government has conducted research in the field of protective body armor systems. Initial experiments with silk and steel proved either to provide an inadequate level of protection or were too unwieldy as a body armor system for soldiers. Recent advances in the field of material science have lead to the advent of such polymers as Kevlar and Spectra, and ceramics, such as boron carbide, that are incorporated into modern body armor systems. Although these bulletproof vests are resistant to bullet penetrations, the forces that are imparted to the body are still significant and potentially lethal.

The objective of research described in this thesis is to create a viable three-dimensional finite element model of the human thorax (including ribs, sternum, vertebral column, cartilages, and major musculature) that would predict the response of a bullet impact on a body armor system. Validation of the model was conducted by comparing the finite element analysis results to those obtained during bullet impact analyses of body armor systems on cadavers. Upon validation, the model was subjected to a parametric study to determine its essential components.

With the model developed, it is possible to study the effects of different body armor systems, different projectiles with various velocities, and the thoracic dimensions of individuals, on the response of the thorax. This allows for realistic experimentation on new body armor systems for the United States government without the time and cost involved in cadaveric studies.

To date, there is little existing research into the dynamics of bullet impacts to the thorax. The predominant areas of study have been into the creation and validation of a human thorax finite element model and into analysis of the thorax reaction to blunt impacts in automobile collisions. A majority of the research has been concerned with static analyses to validate human thorax finite element models. Very few studies have examined dynamic effects on the thorax. The present research project breaks new ground in the use of the finite element method for investigations into the biomechanical response of the thorax to dynamic loading.

Research into a mathematical model of the thorax began with the studies of Roberts and Chen in 1970 [1]. Utilizing gross geometric data and approximate cross-sectional properties, they were able to develop an elastostatic finite element model of the thorax. Placing the sternum of the model under various static loading cases, they obtained reasonable data for sternal displacements, thereby validating their model. Due to the success of their model, it has been incorporated into most of the models utilized in later studies of the thorax. It was incorporated into the model developed for this study.

Andriacchi, Schultz, Belytschko, and Galante [2] used the model developed by Roberts and Chen in a study of the mechanics of the human skeleton. In order to study the interactions between the rib cage and the spine, their model further refined the existing model. In their study, they were able to examine the bending responses of the spine, the lateral stability of the spine, and the underlying mechanisms in scoliotic deformities. However, their analyses relied on static loading cases on the thorax and were not very useful in the development of a valid dynamic model.

Along with the development of the geometry of a three-dimensional thorax model, research was done into the material properties of the thorax. Due to the individual nature of these properties, these studies compiled results for a large population. In work by Yoganandan and Pintar, the mechanical properties of the seventh and eighth rib pairs were determined for 30 cadavers [3]. Utilizing three-point bending techniques on isolated ribs, the researchers were able to determine the cross-sectional area, moment of inertia, failure load, deflection, and Young's modulus. However, it is believed that further investigation into the material properties of the other rib pairs as well as the costal and articular cartilage is necessary due to the individual nature of these properties.

The only existing cadaveric study on the biomechanical response of the thorax to bullet impacts was performed by DeMaio, et al at the Armed Forces Institute of Pathology [4]. In their studies, cadavers were protected with a bulletproof vest incorporating either soft armor alone or soft armor plus one of two variations on a ceramic plate. These cadavers were then struck with either a NATO 9mm round or a NATO 7.62mm M80 ball round fired from a distance of 50 feet. Various parameters were recorded including the accelerations of the sternum, spine, and carina, and the left and right ventricular pressures. Post-shot autopsies were performed to judge injury and assess survivability. The model utilized in the present study was created in an attempt to obtain similar results from finite element analysis for these parameters, dependent upon the projectile and body armor used. As such, the boundary conditions and loads applied in our study were derived from the test parameters used at AFIP.

MODELING AND ANALYSIS

A. Human Thoracic Body Model

The human thorax presents a myriad of modeling problems for the researcher. Along with the complex geometries of the various ribs themselves, there are multiple connections, such as the articular cartilage between the rib pairs and the vertebrae that must be accurately modeled to create a viable model. In addition, the material properties of the ribs are not constant along their entire length. This adds an additional level of complexity to the model. Another fundamental concern is the wide variation that may take place between individuals. Thus, the correct material properties for a 20-year-old male are unlikely to match those of a 70-year-old female. The model created for this study is depicted in Figure 1 below.

Ribs

The modeling of the ribs was an area of some difficulty due to their complex shape and material properties. The size and shape of the ribs correspond to that of a smaller, female build. Each rib is represented by thirteen beam elements. A larger number would have been prohibitively costly in terms of computation time. Each element is given a specified cross-sectional area dependent upon its distance from the spine. This enables the correct size, shape, and angle of the head, tuberosity, midauxillary line junction, costrachondral joint, and sternochondral junction to be modeled. In addition to the geometry of the ribs, each element had to have its material property specified. The ribs were assumed to be composed of only compact bone and were modeled as a linear elastic material. The Young's Modulus chosen was 2.83 GPa; a density of 1000 kg/m^3 and a Poisson's ration of 0.2 were used.

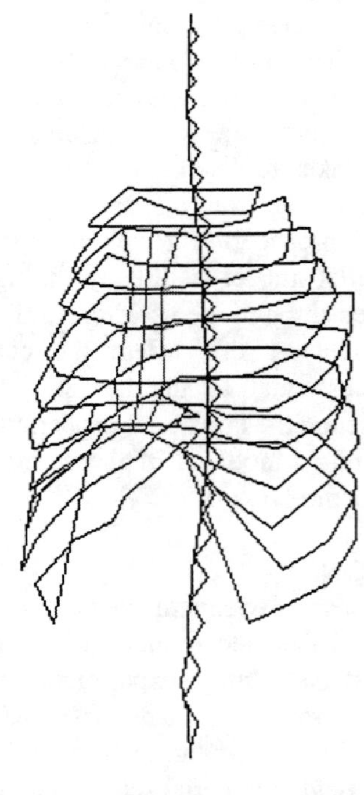

Figure 1. Plot of Sternum, Ribs, Spine, and Cartilage

Spine

Due to the nature of this study, the complexities of the spine were not fundamental to the research being conducted.

As such, the spine was modeled by a series of beam elements, based on previous experiments. [See Ref. 5] Each modeled vertebra consisted of two beam elements, with an intervening beam element between adjacent vertebra to model the intervertebral disc. The facet joints were modeled by connecting two beam elements, one from the midpoint of each adjacent vertebra. Figure 2 below depicts how the spine was modeled. The vertebra and facet joints were modeled with a Young's Modulus of 12.13 GPa, a density of 1000 kg/m^3, and a Poisson's ration of 0.2. The intervertebral discs were given the following material properties: Young's Modulus: 1.5 GPa, Density: 100 kg/m^3, and Poisson's ration: 0.2.

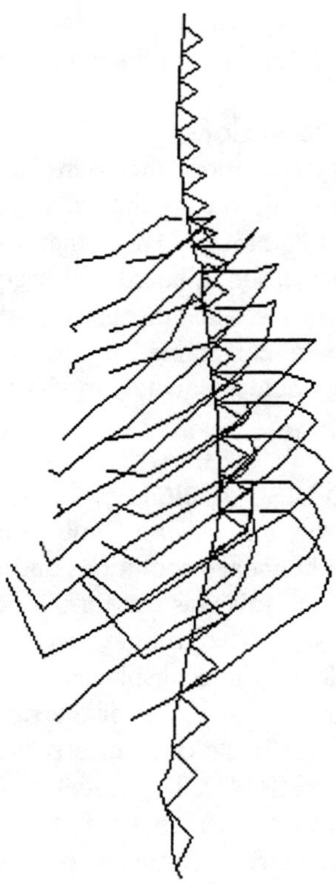

Figure 2. Medial View of the Ribs and Spine

Sternum

The sternum, as seen in Figure 3 below, was modeled with a series of twelve thin shell elements. Each element was given a specified thickness of 0.25 inches (0.63 cm) to create the three-dimensional model. The material properties for the sternum were derived from the compact bone properties of rib 4. As such, the sternum's Young's Modulus was 12.13 GPa, density was 1000 kg/m^3, and Poisson's ration was 0.2.

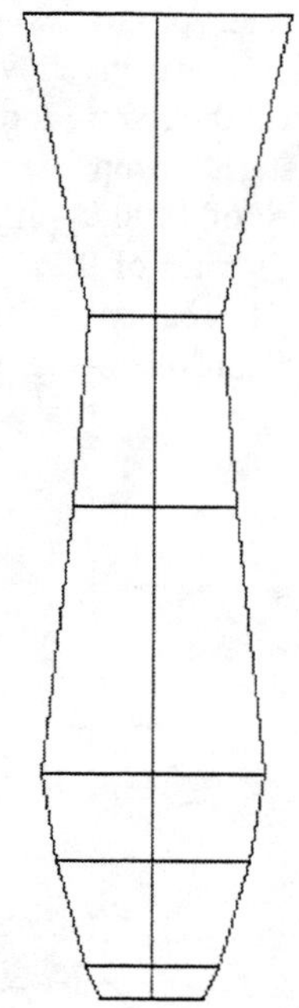

Figure 3: Sternum Shape in Thorax Model

Muscle

The modeling of the muscle was done with 24 solid elements. The elements varied in thickness from 0.25 inches (0.63 cm) along the central, anterior edge of the sternum to a maximum of 2.35 inches (5.97 cm) at the most distal position of rib #4. This accurately models the correct curvature of the muscle along the anterior side of the thorax. The material properties of muscle were modeled with a viscoelastic material type. The material property values utilized were a density of 950 kg/m^3, a bulk modulus of 5 x 10^5 Pa, a short-time shear modulus of 1 x 10^5 Pa, a long-time shear modulus of 1 x 10^4 Pa, and a decay constant of 1000. The equation used to calculate the shear relaxation behavior is as follows:

$$G(t) = G_\infty + (G_0 - G_\infty)e^{-\beta t}$$

G(t) = Shear Modulus
G_0 = Short-Time Shear Modulus
G_∞ = Long-Time Shear Modulus
β = Decay Constant

Cartilage

The cartilaginous connections of the thorax were modeled with beam elements, as depicted by the darker lines in Fig. 4 below. A single beam element was used to model the sternochondral joint whereas two beam elements were utilized to model the articular cartilage between the rib and the vertebrae. The material properties of the cartilage at the sternochondral junction were given values of 64.7 MPa for Young's Modulus, 1000 kg/m^3 for density, and 0.2 for Poisson's ratio. Material properties for the chondracostal cartilage were a density of 1000 kg/m^3, a Young's Modulus of 5 MPa, and a Poisson's ratio of 0.2. Additionally, a series of vertical cartilaginous connections at the inferior edge of the sternum, modeled as beam elements, were used to create the subcostal angle.

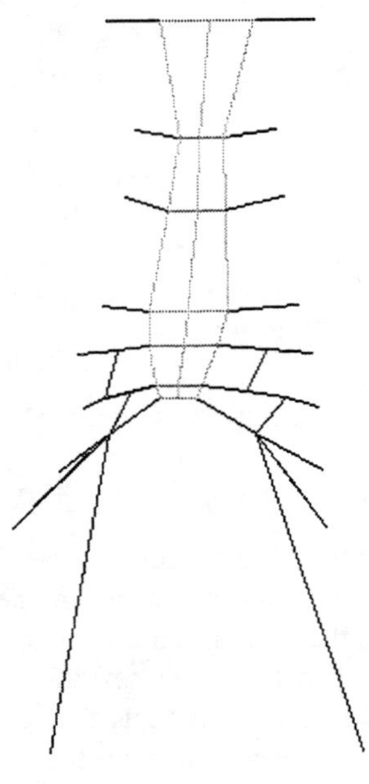

Figure 4. Articular Cartilage Connecting Sternum and Ribs

B. Body Armor

The body armor itself consisted of a plate of CBA (concealable body armor) composite material and sheet of Kevlar. The material properties of Kevlar were assumed to be a density of 1440 kg/m^3, a Young's Modulus of 1 GPa, and a Poisson's ratio of 0.2. The CBA plate was given a density of 2500 kg/m^3, a Young's Modulus of 448 GPa, and a Poisson's ratio of 0.2. In one investigation that was undertaken, a plate of CBA composite material was modeled with 126 thick shell elements. The thickness of this plate was set as 0.5 inches (0.51 cm). This plate was directly attached to a sheet of Kevlar, modeled by 126 thick shell elements as well and a thickness of 0.25 inches (0.63 cm). This sheet was modeled such that it projected directly across from the most anterior points of the muscle tissue. This created a gap between the muscle and the Kevlar above the sternum, which is an accurate representation of the actual placement of the vest when worn. In another investigation, the CBA plate was removed to simulate the usage of just soft armor.

C. Projectile

The projectile was given the correct mass and shape and was attached directly to the most anterior body armor component. This was either the CBA plate or the Kevlar sheet, depending upon the test being run. The projectile was modeled as a single solid element and given the dimensions and material properties of either a NATO 7.62 mm Ball M80 round or a NATO 9 mm full metal jacket round. In order to model the motion of the projectile, it was given an initial velocity. This imparted a momentum of the bullet to the body armor. The initial velocity of the projectile was determined experimentally during test firings [4]. These numbers were 2900 f/s for the 7.62mm round and 1500 f/s for the 9mm round.

D. Boundary Conditions

The boundary conditions that were chosen were based on the experimental set-up during the cadaveric studies at Armed Forces Institute of Pathology [4]. In their experiment, cadavers were laid flat on a wooden board and secured to it via wires. To approximate this set-up, the most posterior nodes of the thoracic model were constrained to prevent translations. It is believed that this would adequately model the experimental set-up.

RESULTS AND DISCUSSION

After construction of the finite element model and imposition of the boundary conditions and initial velocities for the projectile, an analysis was run for a model time duration of two milliseconds. The output were obtained for the acceleration, velocity, and displacement in the anterior to posterior direction for the center of the sternum as well as the T7 vertebra. These locations correspond to those used by DeMaio, et al at AFIP [4]. The model was analyzed with two variations. The first was that of the 7.62mm projectile impacting on a vest constructed of Kevlar and a CBA (concealable body armor) plate. The second was the impact of a 9mm projectile into a body wearing only the Kevlar vest. For purposes of validation, the acceleration data were compared between the numerical and experimental results. In order to set a standard for comparisons, the characteristics that were focused upon were the times and magnitudes of the first peak and trough, and the trend of the parameter at the end of the analysis period.

A. CBA Plate, Kevlar Vest, and NATO 7.62mm ball M80 Round

This test case utilized the NATO 7.62mm M80 ball round as the projectile, with an impact velocity of 966 m/s. This velocity is consistent with those obtained by DeMaio, et al from a chronometer during their test firings [4]. The vest consisted of a layer of Kevlar and a CBA plate. The time duration of the model analysis time was five milliseconds.

Figures 5 and 6 display the results that were obtained from the numerical analysis and the experimental test. They compare sternum and spinal accelerations in the anterior to posterior direction. This direction is chosen because it is the direction of the projectile upon impact.

The match between the experimental sternal acceleration data and the computer model was excellent for this case, as illustrated in Figure . The magnitude of the initial peak is of the right order, the location of the first peak and troughs are at the right time, and the long-term behaviors are similar. The only discrepancy occurs at approximately one millisecond. Postmortem examinations revealed that massive sternal fractures had occurred in this case. These fractures would significantly reduce the stiffness of the sternum and allow for greater accelerations, hence a greater magnitude of the trough, than those predicted by the numerical analysis. This is because there are no failure modes included in the numerical model allowing for bone fractures. Otherwise, there was an excellent correlation between the two cases. In this particular case, the analysis was run for five milliseconds. However, due to the long computer processing time necessary, as well as the consideration that the most important responses occurred in the first two milliseconds, all subsequent analyses were run for a shorter model time of two milliseconds.

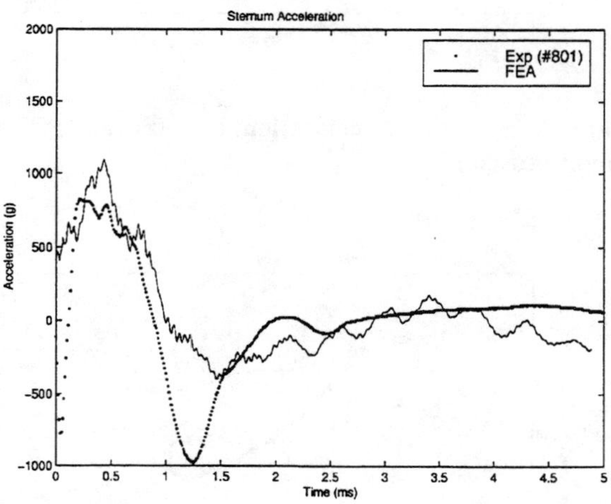

Figure 5. Sternum Acceleration (CBA & Kevlar)

DeMaio, et al [4] estimated about 30 millimeters of the maximum displacement at the sternum from experimental observation even though there was no direct measurement of the displacement. Due to the computer time required to run an analysis for this length of time, the period of investigation was limited to five milliseconds with the final displacement to be extrapolated from the resulting data. As the velocity begins to approach zero at the end of the five-millisecond period, the displacement will obtain a steady-state value. Extrapolating from the available result leads to an estimation of between 25 and 30 millimeters of displacement, corresponding to the results obtained by DeMaio [4].

The comparison between the spinal acceleration results from the computer model and the experimental data also correlated well, as depicted in Fig. 6. The magnitudes are approximately the same and the points of inflection are located at similar times. The discrepancy between the magnitudes lends credence to the idea that the model is too stiff. Additionally, it is important to note that the correlation is excellent until one millisecond. It is believed that the separation after this point is due to the massive sternal fractures that occurred and the resulting higher accelerations in the sternum than those predicted by the computer simulation. Because the forces imparted from the impact on the anterior face of the thorax must propagate through the connective and soft tissues as well as the skeleton to affect the spinal acceleration, these fractures in the sternum have a direct consequence on the acceleration of the spine.

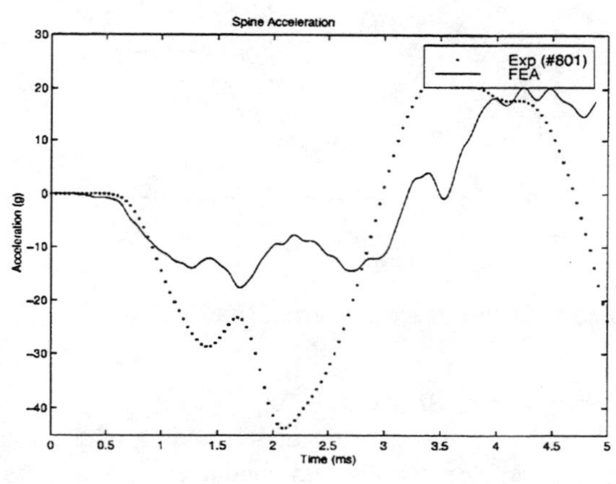

Figure 6. Spinal Acceleration (CBA & Kevlar)

B. Kevlar Vest and NATO 9mm Round

This test case utilized the NATO 9mm full metal jacket round as the projectile, with an impact velocity of 500 m/s. The armor vest consisted of only a layer of Kevlar. The analysis time duration was two milliseconds.

The correlation between the results from the computer simulation and the experimental work [4] were high for this case as well, as seen in Fig. 7 below. In the figure, note that the locations for the first and second peak match exactly. Unfortunately, the magnitudes of these peaks were clipped in the experimental data, yet extrapolation from the available data suggests that the magnitudes would match very well. It is believed that the method by which the accelerometer was mounted accounts for the discrepancy noted at approximately

0.9 milliseconds in Fig. 7. Securing the accelerometer to the posterior side of the sternum with a length of string allows for accurate measurements of acceleration as the sternum moves in the posterior direction, but stretching of the string under anterior acceleration of the sternum may create the noted discrepancy. Additionally, the massive sternal fractures noted in the previous validation case are not noted in post-mortem report for this test case. As such, the magnitude of the experimental response at the first trough does not exceed that predicted by the numerical analysis. However, the long-term behavior of the numerical predictions and those obtained experimentally match well, with both approaching zero.

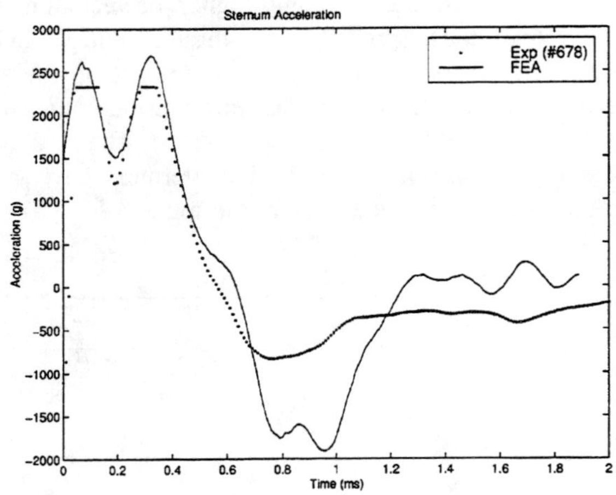

Figure 7. Sternum Acceleration (Kevlar only)

C. Parametric Study

In order to gain a better understanding of the constructed model, a parametric study was conducted. Parameters such as the Young's Modulus of the sternum, muscle, Kevlar, CBA (concealable body armor) plate, sternal cartilage, and intervertebral disc were changed individually to assess their importance to the overall behavior of the model through comparisons with the Kevlar and CBA plate case presented above. Through this study, the muscles, the Young's Modulus of the sternum articular cartilage and of the skeleton in general, the contact elements and gap between the vest and the sternum, the size of the thorax, and the densities of the CBA and Kevlar were determined to be the most important factors in determining the correct behavior of the model.

The following graphs are those that were obtained after changing one of the factors listed above. Note that in all subsequent figures, the three items graphed are the experimental results obtained by DeMaio, et al [4], the results from the first validation case (CBA plate, Kevlar vest, and NATO 7.62mm round), and the results returned with one of the parameters changed.

The most important element in the behavior of the model was the inclusion of muscle. Removing the muscle from the model had serious ramifications to the predictive power of the biomechanical response. It is believed that the muscles provide much of the damping and viscous effects that are essential to reduce the high forces that are imparted to the body upon impact. (Recall that the muscles were modeled as a viscoelastic material.) Additionally, muscle damping suppresses the high frequency oscillations that would otherwise dominate the dynamic response following impact. The graphs below, Figs. 8 and 9, depict the results that were obtained for the sternum and spinal accelerations with the muscle removed from the Kevlar and CBA plate case.

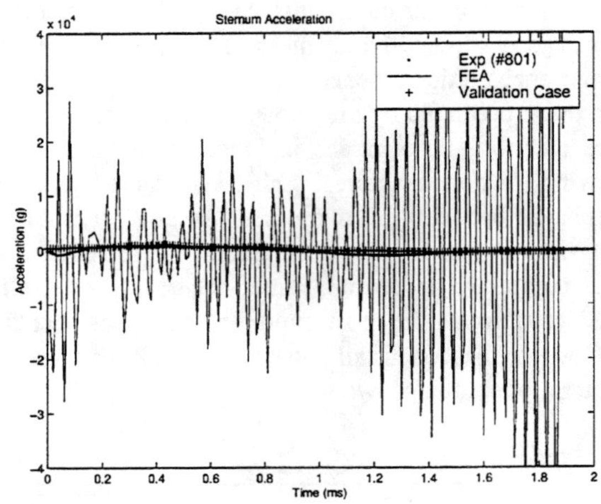

Figure 8. Sternum Acceleration (Kevlar and CBA plate without muscle)

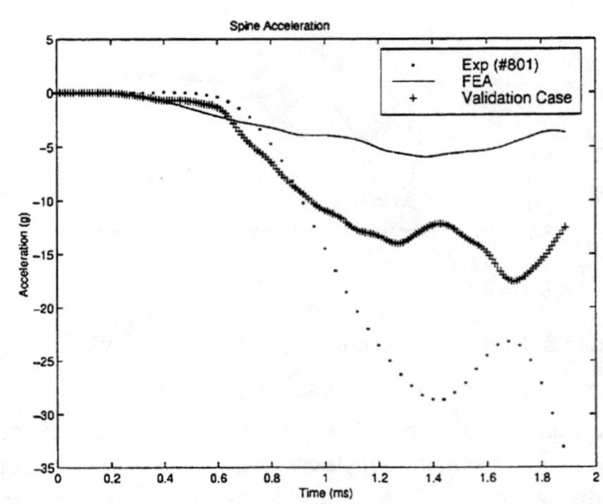

Figure 9. Spinal Acceleration (Kevlar and CBA plate without muscle)

The Young's Modulus of the muscle was also varied to determine its effect on the biomechanical behavior of the thorax model. In the following case, the Young's Modulus was increased by a factor of five from its original value. This increase in the modulus leads to an increase in the speed of sound calculated for the model.

The increase in the speed of sound results in an increase in frequency response for the model as illustrated below in Figs. 10 and 11. Thus, the locations for the first peak and trough in the sternum acceleration data are not the equivalent. Additionally, the change in the Young's Modulus also increases the magnitude of the first peak predicted by the computer model. There are no notable changes noticed in the behavior of the spine with this change in muscle Young's Modulus.

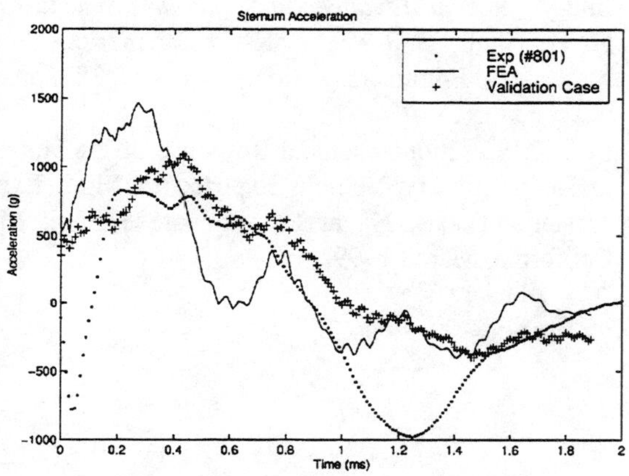

Figure 10. Sternum Acceleration (Kevlar and CBA plate with muscle Young's Modulus increased by five times)

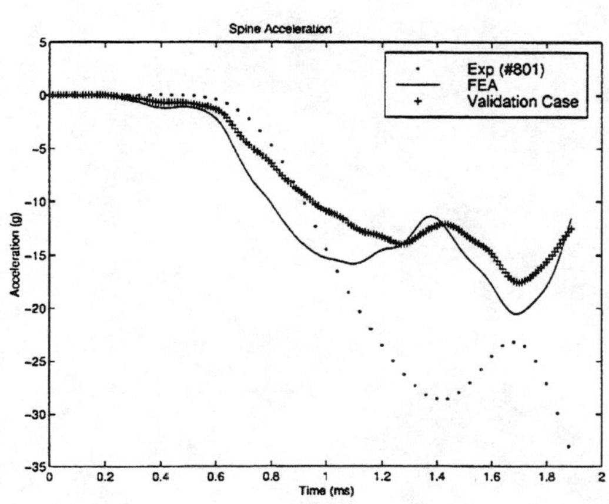

Figure 11. Spinal Acceleration (Kevlar and CBA plate with muscle Young's Modulus increased by five times)

A significant response was also noted for an increase in the Young's Modulus of the sternal articular cartilage. Due to the direct contact between the vest and the sternum via the muscle, a ten-fold increase in the Young's Modulus of the articular cartilage does not have any effect on the acceleration of the sternum. However, because of the connection of the sternum to the ribs through the articular cartilage, the change in the Young's Modulus does affect the response of the spine as illustrated in Fig. 12 below. Note that the point of inflection has advanced from approximately 0.7 milliseconds to 0.2 milliseconds. Additionally, the change in Young's Modulus introduced several new local maxima and minima in the predicted acceleration of the spine.

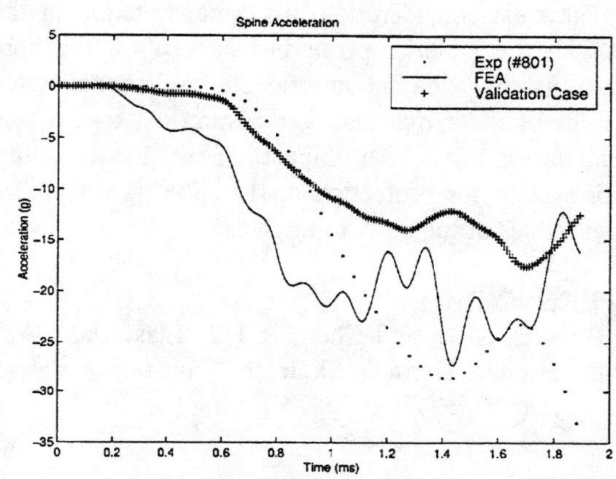

Figure 12. Spinal Acceleration (Kevlar and CBA plate with sternal cartilage Young's Modulus increased by ten times)

CONCLUSIONS

The research described in this study was undertaken to investigate the construction of a finite element model to model the effects of impact loads to a human thorax wearing a body armor system. The overriding goal for this study was to obtain a viable model of the human thorax that could be used to adequately predict the accelerations experienced by a body subjected to impact by a projectile, namely a bullet. After construction of the model, two studies were undertaken. The first was to prove the viability of the model. The results demonstrated that the model adequately matched the available experimental data for both load cases imposed. As such the model was deemed to be viable, and a parametric study was then undertaken to determine the critical components of the model. This analysis determined that the damping properties of the muscles were the most important element in the model. The inclusion of the musculature into the finite element model provided for the good correlation between the numerical results and the experimental data.

Additional elements were deemed significant in establishing the correct model of the thorax, including the

Young's Moduli of the sternum articular cartilage and the bones and the initial gap between the vest and the sternum. The parametric study demonstrated that the method of stress propagation from the body armor system to the sternum was direct, via the layer of muscle over the sternum and consequently through the ribs and sternochondral cartilage to the rest of the body. Therefore, variations in the stiffness of these elements, namely alterations in the Young's Modulus, resulted in significantly different responses in both the sternum and the spine. Additionally, this method of stress propagation was dependent upon the initial gap between the sternum and the bulletproof vest. Thus, removal of this space resulted in a different method of force transference, leading to a different predicted behavior.

The final consideration in the construction of this model was with the material properties chosen for the body armor system itself. The parametric study demonstrated that the densities of these materials are paramount to the behavior of the human thorax under impact. Thus, use of another body armor system for protection could have significant effects on the responses of the body to impacts.

REFERENCES

1. Roberts, S. B. and Chen, P. H., "Elastostatic Analysis of the Human Thoracic Skeleton," *Journal of Biomechanics*, v. 3, pp. 527-545, 1970.

2. Andriacchi, T., Shultz, A., Belytschko, T., and Galante, J., "A Model for the Studies of the Mechanical Interaction between the Human Spine and Rib Cage," *Journal of Biomechanics*, v. 10, pp. 497-507, 1974.

3. Yogananda, N. and Pintar, F., "Biomechanics of Human Thoracic Ribs," *Journal of Biomechanical Engineering*, v. 120, pp. 100-104, Feb 1998.

4. DeMaio, M., Parks, S., et al., "Biomechanical Evaluation of Chest Body Armor," Unpublished Report, Armed Forces Institute of Pathology (AFIP).

5. Kwon, Y. W. and King, Q. M., "Biomechanical Study of Ballistic Impact on Helmets: Injury of Head and Neck Injury", *Recent Advances in Solids and Structures* (ed. by H. H. Chung and Y. W. Kwon), American Society of Mechanical Engineers (ASME) PVP-Vol. 381, pp. 55-62.

6. Lee, K. S., "Biomechanical Response of the Human Body inside a Military Vehicle Exposed to Mine Explosion", Master's Thesis, Naval Postgraduate School, Monterey, California, March 1999.

PVP-Vol. 415, Recent Advances in Solids and Structures — 2000
ASME 2000

COMPARATIVE STUDY OF S2 GLASS PLAIN AND TWILL WOVEN COMPOSITES UNDER FATIGUE LOADING

Ajit D. Kelkar* and Sunil S. Shenoy**
Center for Advanced Materials and Smart Structures
Department of Mechanical Engineering
North Carolina A&T State University, Greensboro, NC 27411 USA
Tel. (336) 334 7620 (Ext. 324)
(Professor*, Graduate Student**)

ABSTRACT

Woven composites have good properties in mutually orthogonal directions, more balanced properties than unidirectional laminates and have better impact resistance. The use of these composites for primary structural applications in place of conventional laminated composites has been increased considerably in the recent years. They are being manufactured by using new processes such as Resin Infusion (VARIM) and Resin Transfer Molding (RTM). These new processes are low cost, affordable and suitable for high volume manufacturing environment. One of the popular plain woven composites is fabricated using S2-Glass and SC-15 resin system components by using Resin Infusion (VARIM) process. These woven composites are being evaluated for Integral Armor applications. These components are expected to be under fatigue loading. To assess the feasibility of this material manufactured through Resin Infusion (VARIM), it is very important to understand the fatigue behavior of these composite materials. The present study provides comparison of the performance evaluation of plain and twill woven composite material for Integral Armor applications. Tension-Compression (R = -1) fatigue experiments were performed. All the fatigue tests are performed at 1 Hz frequency. S-N diagram and stiffness degradation over the fatigue life of the specimens were obtained.

INTRODUCTION

The study of any material involves finding the elastic properties of the material. These elastic properties include the Young's Modulus, the Poisson's ratios and the Ultimate Tensile strength. These properties are used in comparing various materials. In addition to these properties it is an added benefit to have the fatigue behavior of the material. The present paper focuses on the fabrication process of the S2 Glass - SC-15 Epoxy Resin plain-woven composites. VARIM (Vacuum Assisted Resin Infusion Process) was used for fabricating the panels. The fabricated panels were then cut into coupons and tested in tension. In addition, fatigue tests were performed to get the fatigue behavior of the composite material.

OBJECTIVES

- To obtain basic properties such as modulus, ultimate tensile strength, Poisson's ratio of VARIM plain woven composite.
- To generate two important curves to characterize the fatigue behavior of this material system with R ratio equal to -1:
  1. The stress-fatigue life (S/S_u-N_f) diagram
  2. Stiffness degradation over the entire fatigue life of the specimens.
- To compare fatigue behavior of S2 Glass plain weave-SC-15 Epoxy Resin composites with those of S2 Glass twill weave-C50 resin composites.

MATERIAL SYSTEM

Material system used was S2-Glass plain woven fabric and SC-15 Epoxy resin system. In all these panels there were 28 layers of S2-glass plain weave fabric. The above composite is compared with S2 glass-411 C50 resin that had 48 layers. The panel parameters are given in Table 1.The burn test was performed to get the fiber volume fraction of the fabricated composite. The fiber volume fraction was found to be about 50%.

FABRICATION PROCEDURE

The fabrication procedure used in the composite was Vacuum Assisted Resin Infusion process. This fabrication process uses the vacuum to maintain a uniform pressure over the mold. The arrangement is as shown in Fig.1. The various parts of the mold are as described below.

- Metal base : The metal base is the flat part on which the whole of the fabrication unit is set up. The base supports the fabric layers and the vacuum bagging.
- Peel Ply : The Peel Ply is the nonstick medium which helps the easy removal of the finished fabricated unit. The peel ply is laid on the top as well as on the bottom of the fabric and it helps to maintain a gap between the final fabricated unit and the metal base.
- Fabric : The fabric used in the present fabrication is S2 Glass Plain weave fabric and Twill weave fabric.
- Distribution System : The distribution system used is the wire mesh which provides the least resistance to the flow of resin. Hence it allows the easy flow of resin on the top of the fabric layup and then the resin percolates through the thickness.
- Nylon Cloth : This cloth helps to maintain the vacuum in the mold and also puts sufficient pressure on the mold.
- Breather : The breather is a cloth which helps maintain a constant vacuum across the mold.

FABRICATION PROCEDURE

The lay-up is done as shown in the Fig.1. The vacuum pump is let run for 24 hours before the resin is infused into the mold. The resin is then placed in the vacuum chamber for 20 minutes before it is let into the vacuum mold. This helps to free the resin from any of the trapped air pockets. The vacuum pressure is measured at constant intervals with a vacuum gage. The resin is infused into the mold in intervals of 1 minute with the flow at full rate for 1 minute. This helps give the resin the necessary impulse which is required to penetrate into the thickness of the panel. Then the panel is cured at room temperature for 24 hours. The cure cycle used for this plain weave composite is as shown in Fig.2.

TESTING PROCEDURE

Basic static and fatigue properties are usually obtained using tensile testing machines. For such tests ASTM has recommended specimen that is uniform in cross section. (ASTM, 1987) This specimen geometry is suitable for thin composite specimens. In the present Integral Armor applications, the woven composite panels are thick. The thickness of the panels varies from 0.7 to 0.85". In this case it was observed that the shear failure occurs near and under grip area before the axial failure in the gage section. Thus the axial ultimate tensile strength could not be obtained with ASTM standard specimen. To reduce shear stresses in the resin layer under grips, it was concluded to have higher width of the specimen in and near the grips than in the gage length (Kelkar, et al, 1998). This solution accomplished the objective of reducing the shear stress in the resin layer under the grips and the axial failure occurred in the gage length. The overall sample size was 2" wide and 10" long. Then additional machining was performed to reduce width to 1" in the gage length, making a dog-bone type specimen. The geometry of the test specimen is shown in Fig. 3.

All the tests were performed on MTS testing machine with Instron Controller 8500 and with Instron Hydraulic Grips (300 KN). Static tests were performed using the fabricated coupons to evaluate the basic properties such as tensile modulus, ultimate tensile strength, and Poisson's ratio. The loading rate used for the tensile tests was 6 kips/min.

In the fatigue tests, the applied σ_{max} was varied from 45% to 20% of the ultimate tensile strength. The R ratio ($\sigma_{min}/\sigma_{max}$) was kept -1 so that it was a complete stress reversal with $\sigma_{min} = -\sigma_{max}$. Now fatigue tests were conducted for all various σ_{max} and corresponding fatigue life cycles (N_f) were obtained. The fatigue tests were carried out on plain specimens in the tests where the fatigue load was below 18 kips. Above this load it was found that the specimens were slipping in the grip section. Hence it was imperative to have dogbone specimens. This increased the gripping surface area.

Strain was measured by clip gage extensometer and strain gages. The values measured with these two methods were compared and found to be within good accuracy of 5%. Peak values of loads and strains were recorded over the entire fatigue life. Stiffness degradations of all the specimens were monitored from those values during the testing. All the fatigue tests were performed at 1 Hz frequency.

RESULTS AND DISCUSSION

Static tests were performed for five test samples followed by fatigue tests. For these test samples, the applied σ_{max} was varied from 45% to 20% of the ultimate tensile strength. Table 2 shows the tensile modulus and ultimate tensile strength of the two composites being studied. It is seen that the moduli of both the composites are almost the same but the ultimate tensile strength of S2 Glass with SC-15 Resin is much higher than the former combined with C50 resin. Figure 4 shows the stress strain curve for S2 Glass Plain weave / SC-15 Epoxy resin composite.

Fatigue tests results are presented in Table 3. The endurance limit for unnotched VARIM material system is 20% of the ultimate tensile strength for both the composite materials. The comparison of S-N Curves for the two types is as given in the Fig.5.

Normalized stiffness degradation for S2 Glass-SC-15 resin is presented in Fig. 6. The fatigue tests clearly show this material system has three stages of fatigue life (Fujii and Amijima, 1993). The stiffness decreases rapidly in the first stage, up to about 20% of fatigue life. The stiffness decreases slowly up to about 80% of fatigue life followed by rapid decrease in stiffness for remaining life.

Comparison of fatigue test data is presented in Table 3. The fatigue strength of the S2 Glass/SC-15 resin is found to be better than that of S2 Glass/C50 resin. This can be concluded from the S-N Curves.

CONCLUSIONS

Important conclusions from this study are as follows:

- The endurance limit for both the material systems is 20% of the ultimate tensile strength. (Table 3)
- The fatigue life of S2 Glass Plain weave/SC-15 resin is better than S2 Glass Twill weave/C50 resin.
- The fatigue tests clearly show that both S2-Glass twill woven fabric and 411 C-50 resin system and S2 Glass Plain weave and SC-15 resin have three stages of fatigue life as seen in Figures 4. (Fujii and Amijima, 1993)
- The stiffness decreases rapidly in the first stage, up to about 20% of fatigue life . The stiffness decreases slowly up to about 80% of fatigue life followed by rapid decrease in stiffness for remaining life. (Fujii and Amijima, 1993)

ACKNOWLEDGEMENT

This work was supported by "Intelligent Resin Transfer Molding for Integral Armor Applications" project funded by Army Research Laboratory (ARL) under grant number DAAH04-95-1-0369.

REFERENCES

ASTM. 1987, "D 3039 (Reapproved 1982) Standard Test Method for Tensile properties of Fiber-Resin Composites", *ASTM Standards and Literature References for Composite Materials*, Philadelphia, PA.

Fujii, T. and Amijima, S, 1993, "Microscopic Fatigue Processes in a Plain-Weave Glass-Fiber Composite", *Composites Science and technology*, 49:327-333, 1993.

Kelkar, A. D.; Sankar,J. ; Chaphalkar,P; Grace,C; Yarmolenko, S. N.; Mall,S.;Vaidya,U. 1997, "Fatigue Behavior Of Resin Infusion Processed S2-Glass Woven Composites", International Mechanical Engineering Congress and Exposition, ASME International, Dallas, TX, Nov 1997.

Kelkar, A; Chaphalkar C.; and Sankar, J. 1998, "Development Of Tensile Coupons For Thick Composites Using Finite Element Method", *SECTAM, Nineteenth Southeastern Conference on Theoretical and Applied Mechanics, Deerfield Beach, FL*, May 1998

Table 1. VARIM Panels Parameters

Panel	**Average Thickness**	V_f
Plain Weave S2 Glass SC-15 epoxy resin	0.73" (18.54 mm)	0.51
Twill Weave S2 Glass-C50 Resin	0.83" (21.1 mm.)	0.50

Table 2. Static Test Results for VARIM Panels

Properties	Unit	S2 Glass-C50 Resin	S2 Glass-SC-15 Epoxy Resin
Young's Modulus E	Gpa	27.80	29.69
Ultimate tensile Strength	Mpa	429.6	522.22

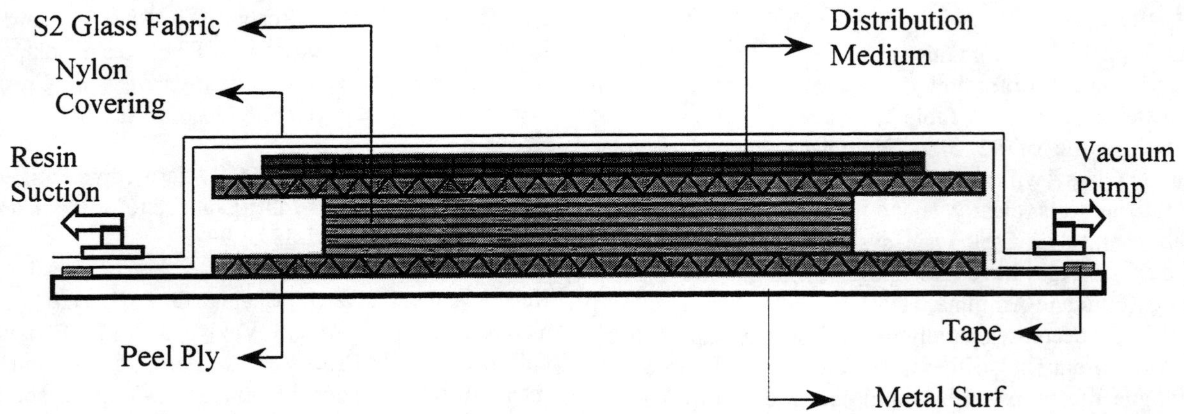

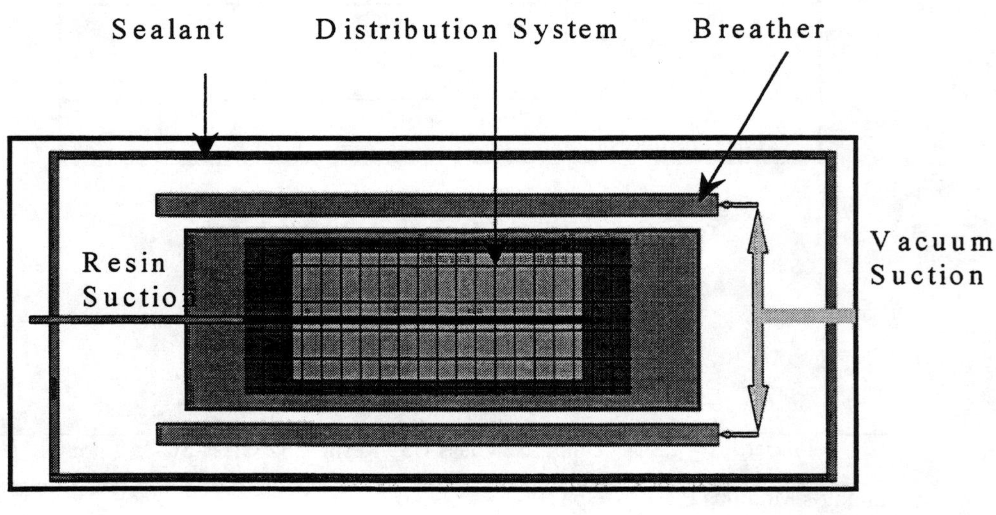

Figure 1. Arrangement of the Vacuum Assisted Resin Infusion Process for fabrication of S2 Glass Plain weave-SC-15 Epoxy resin

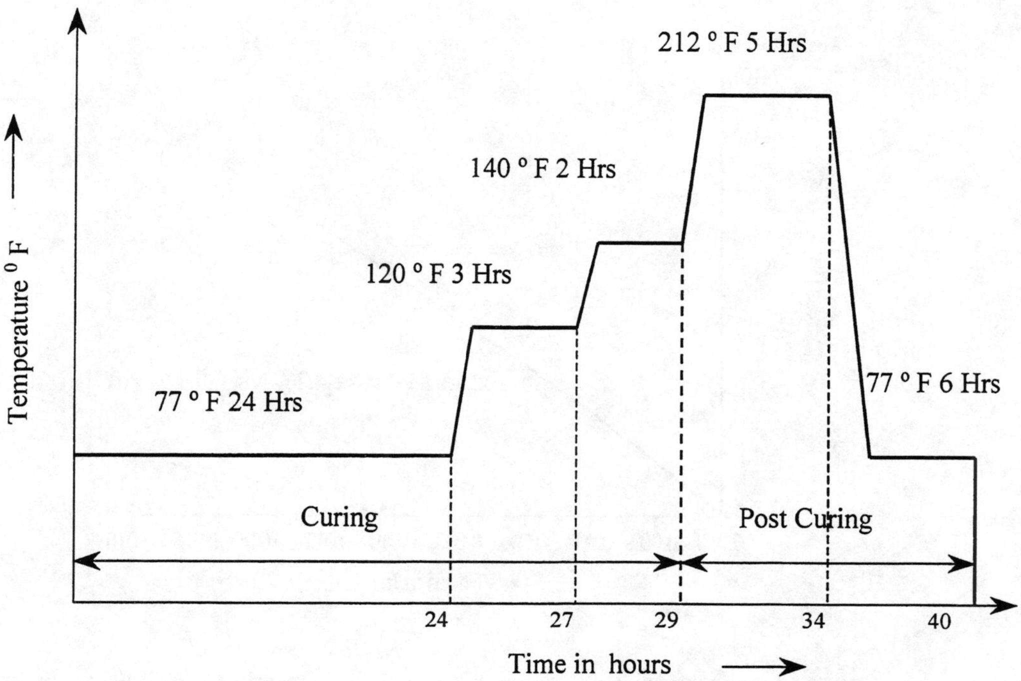

Figure 2. Cure Cycle used for S2 Glass-SC-15 Epoxy Resin system

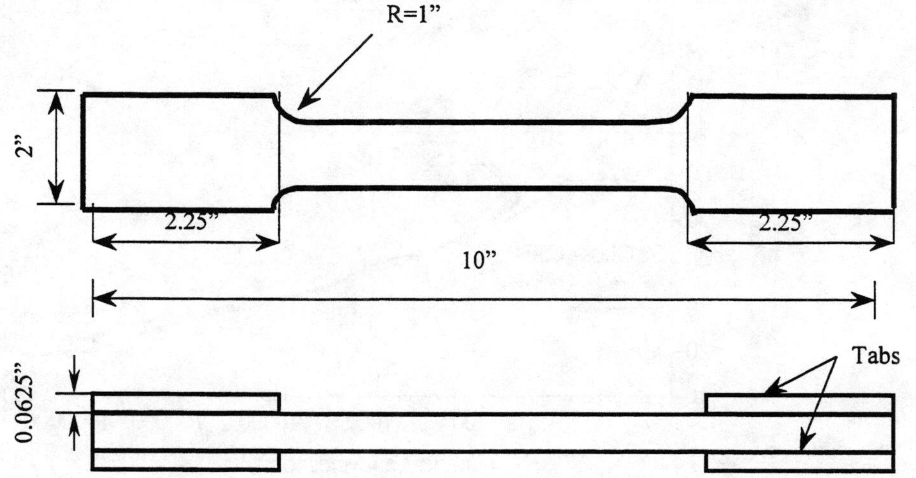

Figure 3. Test Specimen Geometry of the Thick Composite

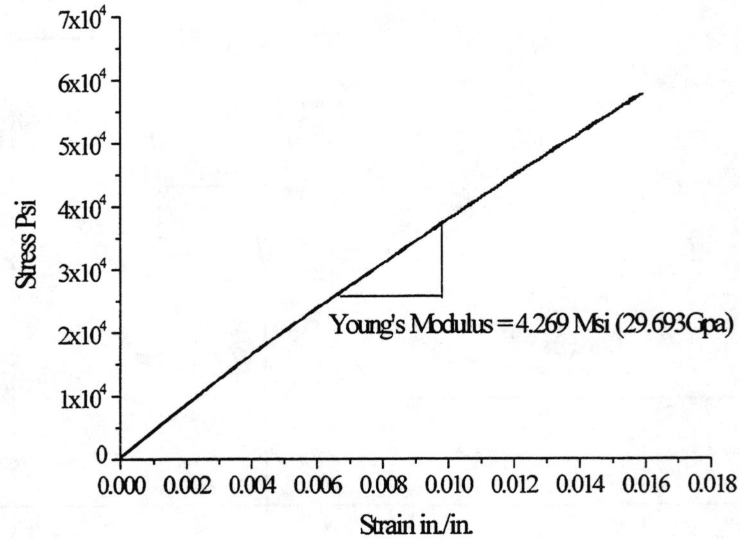

Figure 4. Stress Strain curve for S2 Glass plain weave / SC-15 Epoxy Resin

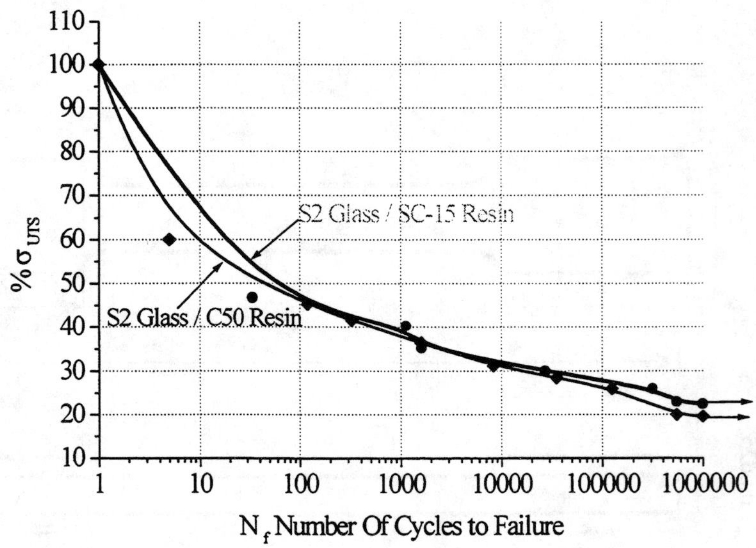

Figure 5. Comparison of S-N Curves for S2 Glass Plain Weave / SC-15 Resin and S2 Glass Twill Weave / C50 Resin

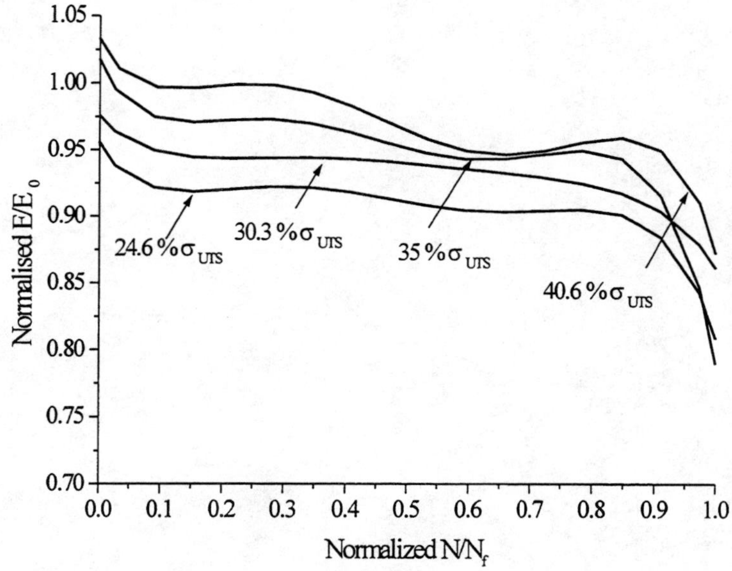

Figure 6. Stiffness Degradation of S2 Glass Plain weave / SC-15 Epoxy Resin Composite

Table 3. Fatigue Comparison of S2-Glass Plain and Twill Weave VARIM Panels

% σ_{ult}	**S2 Glass / SC-15 Resin Number of Cycles to Failure**	**S2 Glass / C50 Resin Number of Cycles to Failure**
100	1	1
60.0	N/A	8
46.8	34	N/A
45.3	N/A	150
41.6	N/A	327
40.3	1133	N/A
36.4	N/A	1,610
35.3	1625	N/A
31.2	N/A	8,272
30	27400	N/A
28.5	N/A	35,740
26	315,000	124,491
24	N/A	333,020
20	550,000*	550,000*

* The sample did not fail

MIXED MODE I/II FRACTURE AND FATIGUE CRACK GROWTH ALONG 63SN-37PB SOLDER/BRASS INTERFACE

H. Nayeb-Hashemi and Pinghu Yang,
Department of Mechanical, Industrial and Manufacturing Engineering
Northeastern University, Boston, MA 02115, US

Abstract:

Solder joints are extensively used in electronic packaging. They provide critical electrical and mechanical connections. Single edge notched sandwich specimens, which were made of two blocks of brass joined with a 63Sn-37Pb solder layer, were prepared for fatigue and fracture study of the joint under mixed mode loading. Mode I and mixed mode I/II fracture toughness, fatigue crack thresholds, and fatigue crack growth rates (FCGR) were measured at room temperature using a four point bending test setup.

It was found that the fracture toughness of the joint increased and FCGR decreased upon the introduction of mode II component. The interface fracture toughness was higher than that of reported for pure solder. The data of FCGR correlated well with the power law relation of $da/dN = C*(\Delta G)^m$. It was also observed that both fracture toughness and FCGR were a function of thickness of solder layer. When the solder layer thickness increased from 0.1mm to 1.0mm, the fracture toughness decreased substantially and FCGR increased slightly.

For mode I loading, fatigue crack propagated inside the solder layer. However, for mixed mode loading, once a crack initiated, it changed its direction toward the interface and then propagated along the interface. These observations were related to local mode I and mode II stress fields. Fracture surface showed sign of rubbing under mixed mode loading with elongated cavities at the crack tip. However, under mode I loading, fracture surface was covered with equi-ax voids.

1. Introduction

Recently there has been interest in using diffusion bonded sandwich-layered structures. Diffusion bonded materials are used in wear resistance die, cutting and measuring tools, metal-ceramic pressure seal, elastic elements of transducers, radial-turbine wheels, turbine blades, valves, pistons and cylinder liners for internal combustion engines etc. In electronic industry, diffusion bonding has been used in design of micro switches, bonding ceramics, etc.

To insure that layered material withstands service life of the components, there have been a number of research reports toward understanding microstructure of interfacial reaction zone, its thickness, and their effects on bond strength. However, there has been relatively little research toward understanding fatigue crack growth behavior at the interface of those materials and factors affecting fatigue crack growth rate under mixed mode loading.

Willams (1959) [1] first addressed a problem of interfacial cracks between dissimilar isotropic materials along a straight interface. He found an oscillating stress singularity in a region near the crack tip. This problem was further studied by Erdogan (1963) [2-3], England (1965) [4], Rice and Shi(1965) [5], Dundurs (1969) [6], and Comninou (1977) [7]. Works done by Park and Earmme (1986) [8], Hutchinson *et al.* (1987-1991) [9-12], Rice(1988) [13], Gautesen and Dundurs(1988) [14], Sih and Arson (1988) [15], Cao and Evans (1989) [16], Jesene (1990) [17], R.M. Cannon *et al* (1991) [18, 19], Thruston and Zehnder (1993) [20], J.M. McNaney *et al* (1996)[21], J.K. Shang *et al.* (1996) [22-24], and Shaw (1997) [25].

Research reports indicate that interface fracture toughness and fatigue crack growth behavior are significantly influenced by the presence of mode II stress [16, 17, 24, 26, 27]. Due to elastic mismatch of the two bonded materials at interface, interfacial crack tip generally experiences both mode I and mode II loading even when external loading is pure mode I or pure mode II [9-12, 13, 17, 22, 27]. These tensile and shear stress field at the interfacial crack tip are inherently coupled, and their relative proportion depends on both distances ahead of the crack tip and the elastic mismatches across the interface [9-12]. It has been shown that interfacial fracture toughness increases with the increase of mode II/mode I ratio [16, 17, 24, 26, 27].

Figure 1 is a sketch of an interface crack in a sandwich specimen. A measure of elastic mismatch across interface is defined by the two Dundurs' parameters α, β [6]

$$\alpha = \frac{E_1^{'} - E_2^{'}}{E_1^{'} + E_2^{'}} \quad (1)$$

$$\beta = \frac{1}{2} \cdot \frac{\mu_1(1-2\upsilon_2) - \mu_2(1-2\upsilon_1)}{\mu_1(1-\upsilon_2) + \mu_2(1-\upsilon_1)} \quad (2)$$

Where, subscript 1 and 2 refer to material # 1 and material # 2, respectively, as shown in Fig. 1

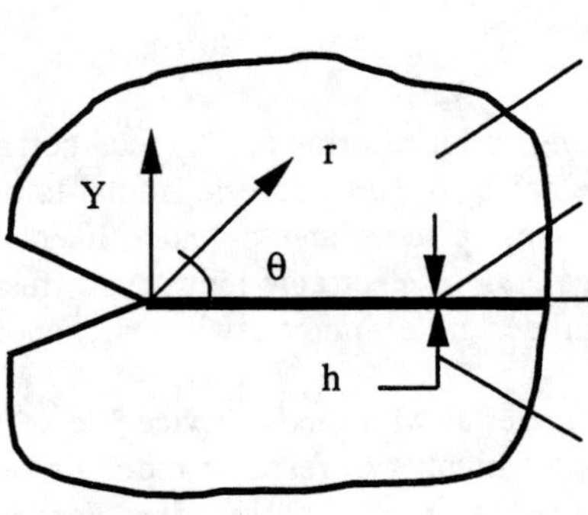

Fig.1 An interface crack in a sandwich specimen.

For plane strain and plane stress, the $E_i^{'}$ is defined as,

$$E_i^{'} = \frac{E_i}{1-\upsilon_i^2} \quad \text{(for plane strain)} \quad (3)$$

$$E_i^{'} = E_i \quad \text{(for plane stress)} \quad (4)$$

Where, E_i is modulus of elasticity and μ_i and ν_i are modulus of rigidity and Poisson's ratio, respectively. Parameters α and β are a measure of differences in elastic and shear modulus of boned joint and for a given material combination their values are tabulated in [9, 10, 12].

Oscillation index ε is another bimaterial constant that is derived from β,

$$\varepsilon = \frac{1}{2\pi} \cdot \ln\frac{1-\beta}{1+\beta} \quad (5)$$

For an interfacial crack which is parallel to the interface, the traction ahead of the crack tip ($\theta = 0$) is given by [9, 10, 12]

$$(\sigma_{yy} + i\sigma_{xy})_{\theta=0} = \frac{Kr^{i\varepsilon}}{(2\pi r)^{1/2}} \quad (6)$$

Where K is near crack tip interfacial stress intensity factor and is defined as

$$K = K_1 + iK_2 \quad (7)$$

K_1, K_2 are near crack tip mode I and mode II stress intensity factor, respectively, r is a distance to the crack tip as defined in Fig.1, and $i = \sqrt{-1}$. The corresponding phase angle is defined as [10, 12]

$$\psi = \tan^{-1}\left(\frac{\sigma_{xy}}{\sigma_{yy}}\right)_{\theta=0} = \tan^{-1}\left\{\frac{\mathrm{Im}(Kr^{i\varepsilon})}{\mathrm{Re}(Kr^{i\varepsilon})}\right\} \quad (8)$$

For mixed mode fracture and fatigue analysis, strain energy release rate is used as a crack driving force. Far field strain energy release rate under mixed mode I and II loading is determined from

$$G = \frac{K_I^{2}}{E^{'}} + \frac{K_{II}^{2}}{E^{'}} \quad (9)$$

For a sandwich specimen, Sou and Huntchinson derived a universal relation between the near tip stress intensity factor K and the far field stress intensity factor K^{∞} [10, 12].

$$Kh^{i\varepsilon} = pK^{\infty}e^{i\omega} \quad (10)$$

or

$$K_1 + iK_2 = p(K_I + iK_{II})h^{-i\varepsilon}e^{i\omega} \quad (11)$$

Where,

$$K^{\infty} = K_I + iK_{II}, \quad (12)$$

K_I and K_{II} are the far field mode I and mode II stress intensity factors, h is the thickness of sandwich layer and p is a bimaterial constant and is given as

$$p = \sqrt{\frac{1-\alpha}{1-\beta^2}}, \quad (13)$$

and ω is phase shift function and its value for different material combination are tabulated in [9, 10, 12,13]. Noting that $\left|Kh^{i\varepsilon}\right| = |K|$, and $Kh^{i\varepsilon} = |K|e^{i\xi}$, eqs. 10 and 11 can be expressed in terms of magnitude of stress intensity factor and phase angle ψ.

$$|K| = p(\alpha,\beta) * |K^{\infty}| = p(\alpha,\beta) * \sqrt{K_I^{2} + K_{II}^{2}} \quad (14)$$

$$\psi = \varphi + \omega(\alpha,\beta) + \varepsilon\ln\left(\frac{r}{h}\right) \quad (15)$$

Where φ is the far field phase angle defined as

$$\varphi = \tan^{-1}\frac{K_{II}}{K_I}. \quad (16)$$

Considering eqs. 15 and 16, it can be concluded that

1. Magnitude of interface stress intensity factor is scaled by a factor of p with respect to the magnitude of the far field stress intensity factor,

2. Near crack tip phase angle is shifted by an angle of $(\omega(\alpha,\beta)+\varepsilon\ln(r/h))$ with respect to the far field phase angle;
3. Because of phase shift function (eq. 15), interface is no longer subjected to mode I, even when the specimen is under an external mode I loading.
4. When distance to the crack tip changes from r_1 to r_2, the corresponding phase angle differential is

$$\Delta\psi=\psi_2-\psi_1=\varepsilon\ln(\frac{r_2}{r_1}) \quad (17)$$

J. R. Rice and Hutchinson *et al* [9, 10, 12, 13] postulated that from the two Dundur's parameters, α appears to be more important. Since β is very small for most material combinations, and $p(\alpha, \beta)$ and $\omega(\alpha, \beta)$ are stronger functions of α than β, they proposed that it is reasonable to take $\beta = 0$.
When $\beta = 0$, it results in $\varepsilon = 0$. Eqs.11 – 16 can then be simplified to:

$$K_1+iK_2=p(K_I+iK_{II})e^{i\omega} \quad (18)$$

$$K_1=p\sqrt{K_I^{\,2}+K_{II}^{\,2}}\cos(\varphi+\omega) \quad (19)$$

$$K_2=p\sqrt{K_I^{\,2}+K_{II}^{\,2}}\sin(\varphi+\omega) \quad (20)$$

$$\psi=\varphi+\omega \quad (21)$$

For a sandwich specimen, interface strain energy release rate is determined by [12]:

$$G=\frac{(1-\beta^2)}{E^*}|K|^2 \quad (22)$$

Where, K is the near tip interfacial stress intensity factor defined by eqs. 7, and E^* is the effective modulus determined by elastic modulus of material 1 and material 2 and is defined as,

$$\frac{1}{E^*}=\frac{1}{2}\left(\frac{1}{E_1'}+\frac{1}{E_2'}\right). \quad (23)$$

Where, E_i' is elastic modulus for and interfacial crack under plane stress or plane strain, and is given by eqs. 3 and 4.
For a sandwich specimen with small layer thickness compared to the other in plane dimensions, the influence of the sandwich layer to the remote stress field can be ignored [10, 12]. Because of the conservation of J integral and because the introduction of thin layer doesn't influence remote K field, strain energy release rate at the crack tip G_c should equal to the strain energy release rate G^∞ computed using the far field stress intensity factors [10, 12, 26].

$$G_c=G^\infty=\frac{1-\nu^2}{E_1}(K_I^{\,2}+K_{II}^{\,2}) \quad \text{(Plane strain)} \quad (24)$$

The above eqs. are used to define local crack tip stress/strain field and fatigue and fracture driving force. The purpose of the current research is to investigate fracture toughness and fatigue crack growth rates at the interface of 63Sn-37Pb solder/brass joint under mode I and mixed mode I / II loading.

2. Experiment

Both fracture toughness and fatigue crack growth rates were obtained under mode ratios R = *0, 0.25,* and *0.5*. The effect of thickness of the sandwich layer on both fracture toughness and fatigue crack growth rates was also studied. Four different groups of specimens with sandwich layer thickness of h = *0.1mm, 0.25mm, 0.5mm,* and *1.00mm* were prepared for this study. All experiments were carried out using a four-point bending test fixture. Schematic diagram of loading conditions, as well as shear and bending moment diagrams are shown in Fig.2. It is possible to achieve both mode I and mixed mode I / II loading by using the four point bending test setup.

2.1 Materials and Specimens

Figure 3 shows a schematic diagram of specimens used in this study. Two blocks of 360 free machine brass were joined together by eutectic 63Sn-37Pb solder. A horizontal and vertical soldering jigs were designed and used to manufacture specimens. In addition, different types of solder, ranging from solder wire, solder paste, and solder ribbon were used to join the brass block. It was found that horizontal soldering jig gave the best quality of specimen in terms of bond strength and specimen alignment. By using the horizontal jig, specimens can be manufactured with a good alignment and with a specific solder layer thickness of solder layer and minimum of void content. 63Sn-37Pb eutectic solder wires without flux were finally chosen as soldering material after many trails. Regular solder flux paste was applied on both matching surfaces before soldering. Processing parameters such as preparation of matching surfaces, amount of flux applied, pre-heating temperature, time to evaporate flux, time to flow and reflow molten solder and cooling rate were optimized to achieve a good bond. Manufacturing operations consisted, of preheating Oven to 180°C, putting the fixture in the oven and evaporating excessive flux for five minutes, raising temperature to 260°C and allowing solder to melt and to fill the gap and finally cooling the specimen in the oven.

Table 1 summarizes materials properties and the corresponding Dundur's parameters for 63Sn-37Pb solder/brass joint. Using the four point bending test setup as shown in Fig. 2 and loading configuration shown in Fig. 3, following equations were used to calculate stress intensity factors, phase angle, and strain energy release rate.

1. *Far Field Stress Intensity Factors* [28-29]

$$K_I=\frac{M}{BW^{3/2}}f_b(\frac{a}{W}) \quad (25)$$

$$K_{II}=\frac{Q}{BW^{1/2}}f_s(\frac{a}{W}) \quad (26)$$

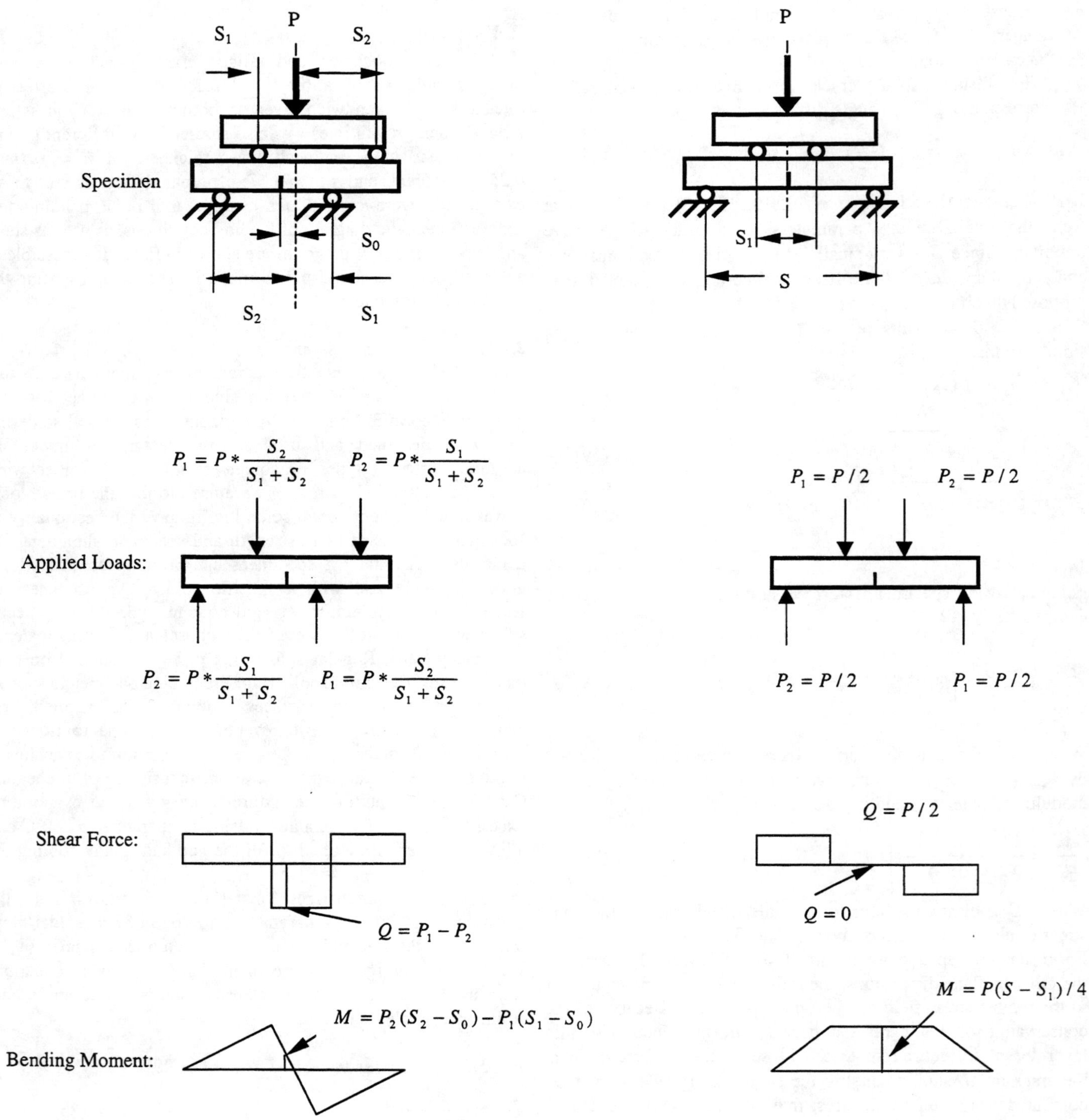

Fig. 2 Diagrams of four point bending test setup

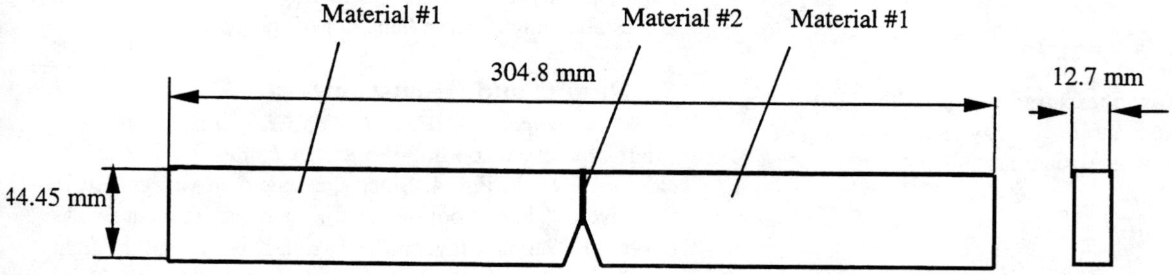

Fig. 3 Specimen geometry and composition

Table 1 Material Properties and Dundur's Parameters of 360 Free Machine Brass and 63Sn-37Pb Solder

Material Property	Brass	Solder(63Sn-37Pb)
Young's Modulus, *E(GPa)*	101	30
Shear Modulus, μ*(GPa)*	35.93	17.21
Possion's Ratio, υ	0.35	0.324
Yield Stress, σ*(Mpa)*	70	22.06
Tensile Stress, σ*(Mpa)*	241	53.1
Dundurs Parameter, α	0.3166	
Dundurs Parameter, β	0.0309	
Bimaterial Constant, *p*	0.8271	
Oscillation Index, ε	-0.0098	

2. *Stress Intensity Factors At the Crack Tip*

$$K_1 = p\sqrt{K_I^{\ 2} + K_{II}^{\ 2}}\cos(\varphi+\omega) \tag{27}$$

$$K_2 = p\sqrt{K_I^{\ 2} + K_{II}^{\ 2}}\sin(\varphi+\omega) \tag{28}$$

$$K = K_1 + iK_2 = p\sqrt{K_I^{\ 2} + K_{II}^{\ 2}}\cdot e^{i\psi} \tag{29}$$

3. *Far Field Phase Angle*

$$\varphi = \tan^{-1}\frac{K_{II}}{K_I} = \tan^{-1}\left(\frac{QWf_s}{Mf_b}\right) \tag{30}$$

4. *Near Field Phase Angle*

$$\psi = \varphi + \omega = \tan^{-1}\left(\frac{QWf_s}{Mf_b}\right) + \omega \tag{31}$$

5. *Strain Energy Release Rate (for Plane Strain)*

$$G = \frac{1-\upsilon_1^{\ 2}}{E_1}\cdot(K_I^{\ 2} + K_{II}^{\ 2}) \tag{32}$$

Where, M is the bending moment at the crack surface, Q is shear force at the crack surface, W and t are the specimen width and thickness respectively, a is the crack length, and $f_b(\frac{a}{W})$, $f_s(\frac{a}{W})$ are the geometrical coefficients and are expressed as [28]:

$$f_b(\frac{a}{W}) = 575.7576(\frac{a}{W})^4 - 823.7918(\frac{a}{W})^3 + 456.9103(\frac{a}{W})^2 - 95.4361(\frac{a}{W}) + 11.5004 \tag{33}$$

$$f_s = 14.8252(\frac{a}{W})^4 - 23.7700(\frac{a}{W})^3 + 17.6075(\frac{a}{W})^2 - 1.6054(\frac{a}{W}) + 0.2795 \quad (34)$$

For fatigue crack growth tests, ranges of the far field stress intensity factors and strain energy release rate were calculated using following equations:

$$\Delta K_I = \frac{\Delta M}{BW^{3/2}} f_b(\frac{a}{W}) \quad (35)$$

$$\Delta K_{II} = \frac{\Delta Q}{BW^{1/2}} f_s(\frac{a}{W}) \quad (36)$$

$$\Delta G = \frac{1-\upsilon_1^{\,2}}{E_1} \cdot [(\Delta K_I)^2 + (\Delta K_{II})^2] \quad (37)$$

2.2 *Experimental Procedures*

All tests were carried out at room environment *(22°C, 55%RH)* using a servo-hydraulic machine. Fatigue crack growth tests were conducted with a sinusoidal waveform at frequency of *10 Hz*. ASTM standard E647- 96 [36] was followed to obtain fatigue crack growth rates data as well as fatigue thresholds.

Fatigue crack length was continuously measured by using electrical potential drop technique. Calibration of normalized crack length *(a/W)* to normalized potential drop was obtained both experimentally as well as by finite element analyses. Fatigue crack growth rates (*da/dN*) between ~ 10^{-7} *to* 10^{-3} *mm/cycle* were obtained by load-shedding technique where the load drop at each loading step was *8 ~ 10%* of the previous load. Fatigue crack thresholds ΔK_{th} or ΔG_{th} were determined at a crack growth rate of 10^{-7}*mm/cycle*, which was consistent with the ASTM standard *E647—96* [36].

Following completion of the fatigue tests, the fracture toughness K_c and G_c were determined by loading specimens monotonically to fracture while monitoring crack growth and establishing R curve behavior of specimens.

3. Results and Discussions

3.1 Microstructure of 63Sn-37Pb Solder/Brass Joint

Microstructure of 63Sn-37Pb solder / brass joint at solder side is shown in Fig. 4. Since the specimen was cooled in the oven (slow cooling rate), lamella structure was observed. The dark layers are Pb-rich α phase and the light layers are Sn-rich β phase. An inter-metallic compound (light band) was observed at the solder / brass interface, Fig. 5. The inter-metallic compound is Cu_6Sn_5 [40]. From the picture we can see that the inter-metallic compound has a wavy profile along the interface, and it is about 10 μm wide.

3.2 Crack Growth Path and Failure Modes

An optical photomicrograph of a typical machined notch and the subsequent fatigue crack path of a specimen under pure mode I loading is shown in Fig. 6. It was observed that, for mode I loading, the crack path was parallel to the solder/brass interface and stayed in the solder layer till the specimen was broken. However, mixed mode fatigue crack path deviated from its original direction (parallel to the interface) to an angle of about 45 degrees toward the solder/brass interface and propagated about 80 μm long before changing its direction again to a direction parrallel to interface. The above process was repeated till the crack reached the region adjacent to the inter-metallic compound (interface). After reaching the inter-metallic compound, the crack path changed its direction again and took a path in solder side approximately parallel to the interface of solder/brass joint till the specimen was broken, Fig. 7.

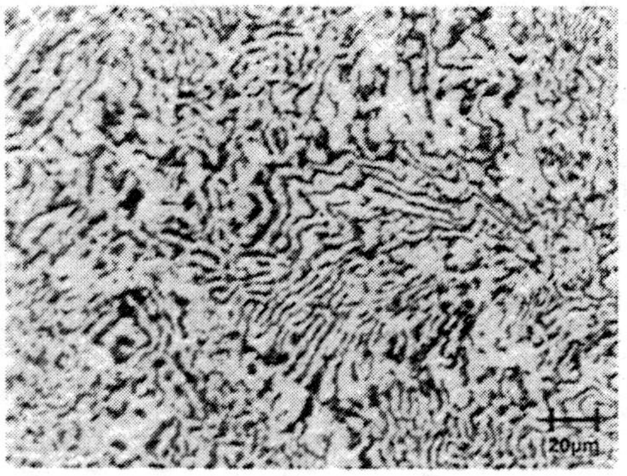

Fig. 4 Microstructure of 63Sn-37Pb solder at solder/brass joint

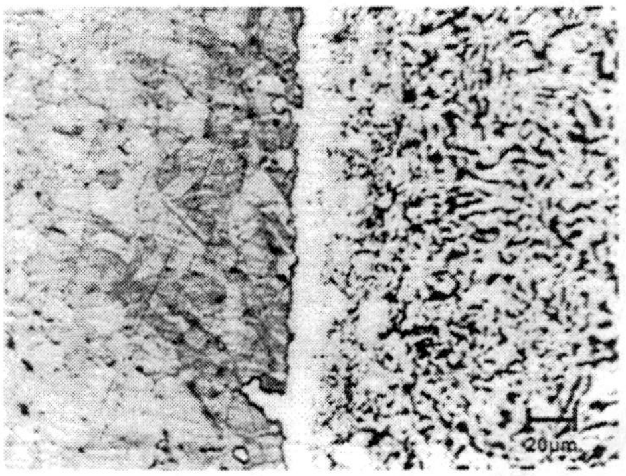

Fig. 5 Interface of 63Sn-37Pb solder/brass joint showing the formation of an inter-metallic compound

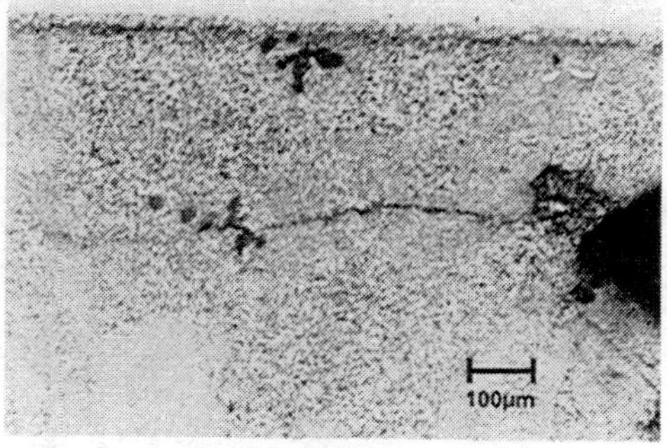

Fig. 6 Mode I fatigue crack path of solder/brass joint

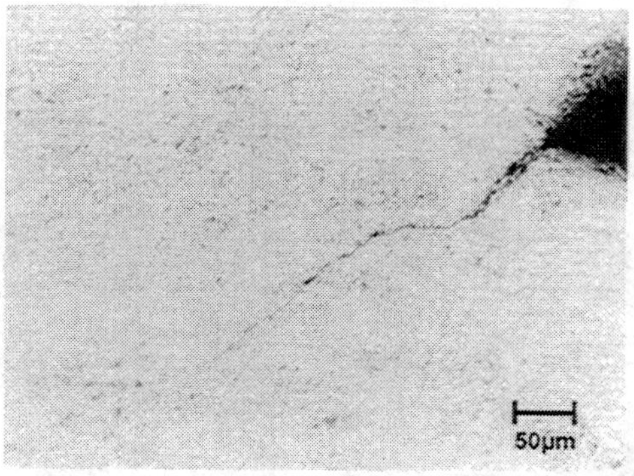

Fig. 7 Mixed mode I/II fatigue crack path of solder/brass joint (mode ratio R = 0.50)

Deviation of the crack path from its original path under mixed mode loading is analyzed by considering the local principal stress directions. The analysis of local stress field provided an excellent agreement with the experimental observation [41].

3.3 Fractographic observations

Figures 8, 9, 10, and 11 are scanning electron photomicrographs of the fracture surface of specimens under mode I and mixed mode cyclic loading. Fractographic analysis revealed that, (1) fracture surface of mode I loading (Fig. 8) was rougher than that of mixed mode loading (Fig. 9), (2) fracture surface of Mode I fatigue crack growth was covered with equi-ax voids, however, elongated voids were observed on the fatigue crack growth surfaces of specimens under mixed mode loading, Fig. 8. In addition fracture surfaces of specimens under mixed mode loading showed sign of rubbings, (4) Micromechanism of crack growth through the entire crack growth experiment was by void formation.

3.4 Fracture toughness and strain energy release rate

Mode I and mixed mode fracture toughness of the solder brass joint measured and presented in terms of K_{Ic} or energy release rate, G. Figure 12 shows mode I fracture toughness as a function of interface layer thickness. Results indicates that the fracture toughness increases with the decrease of the thickness of solder layer, and there is a big drop in fracture toughness when the thickness of the solder layer changes from 0.5mm to 1.0mm. Figure. 13 shows that there is relatively little change in strain energy release rate when the thickness of solder is between 0.1 ~ 0.5 mm. However, when the thickness of solder layer increased to 1.0 mm, the strain energy release rate was much smaller than that of specimens with thickness between 0.1 mm ~ 0.5 mm. The results also indicate that mixed mode strain energy release rates are greater than that of Mode I, Fig. 14.

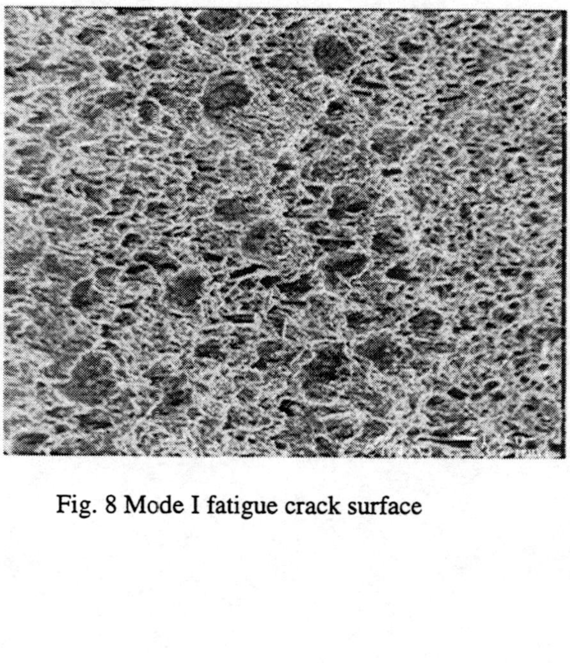

Fig. 8 Mode I fatigue crack surface

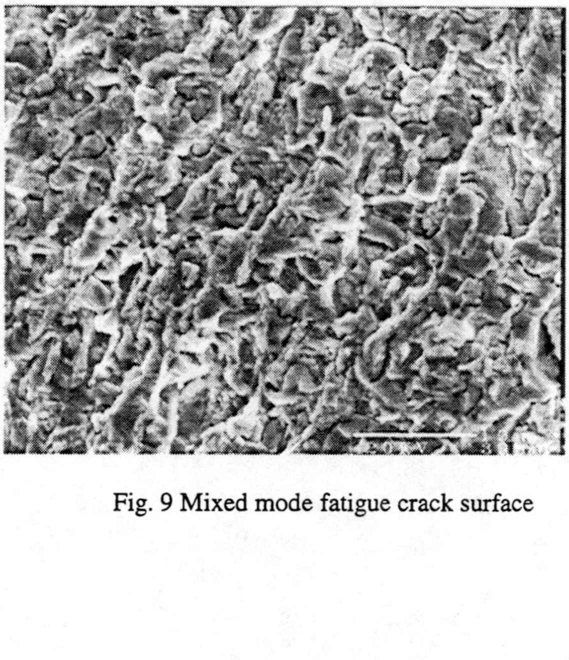

Fig. 9 Mixed mode fatigue crack surface

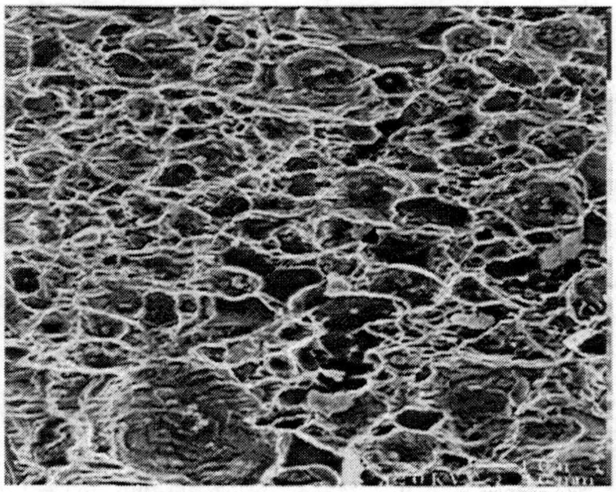

Fig. 10 Mixed mode fatigue crack surface

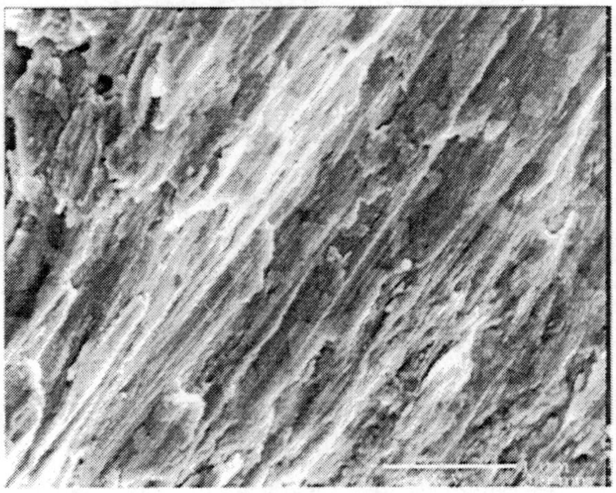

Fig. 11 Mixed mode fracture surface

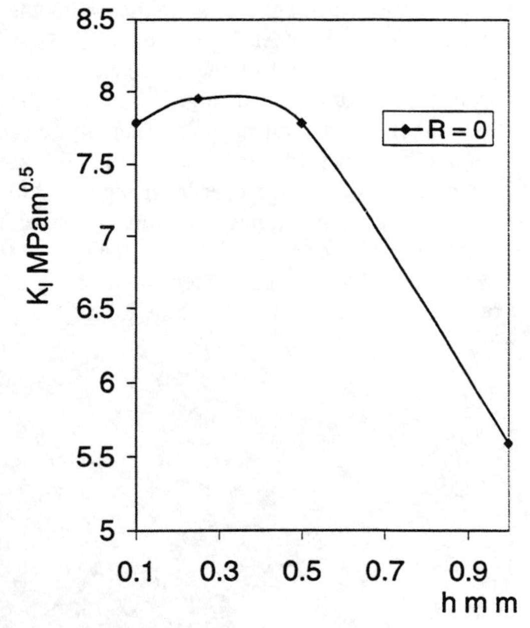

Fig. 12 Mode I fracture toughness vs. thickness of solder layer

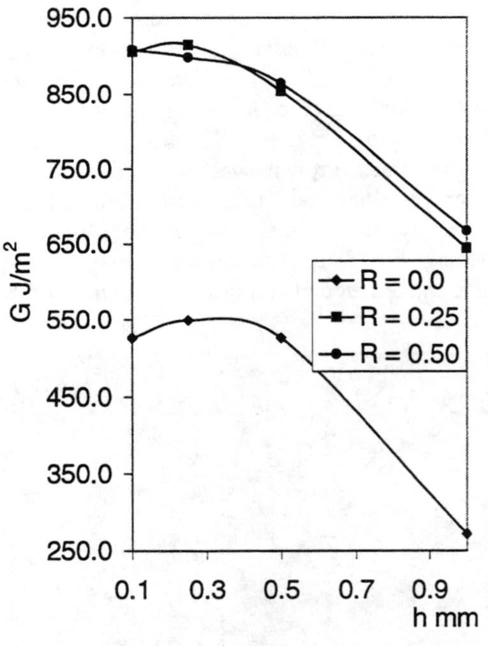

Fig. 13 Strain energy release rate vs. thickness of solder layer

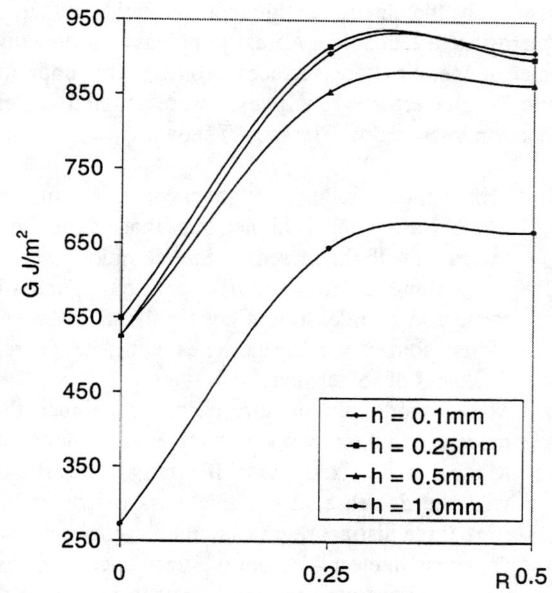

Fig. 14 Strain energy release rate vs. mode ratio

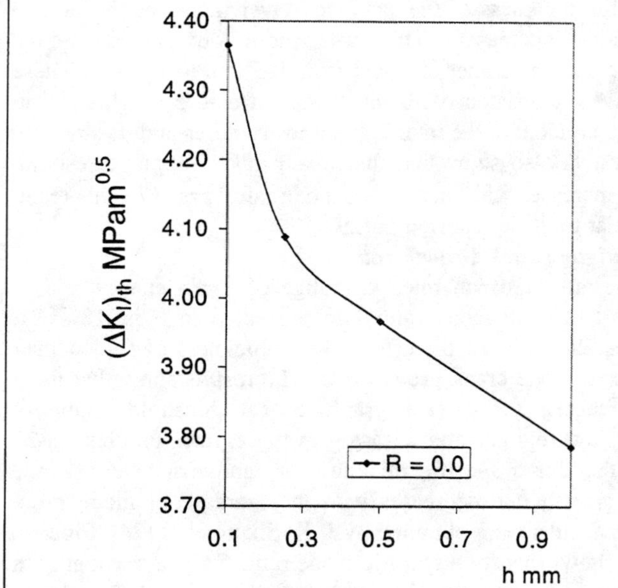

Fig. 15 Mode I fatigue threshold vs. thickness of solder layer

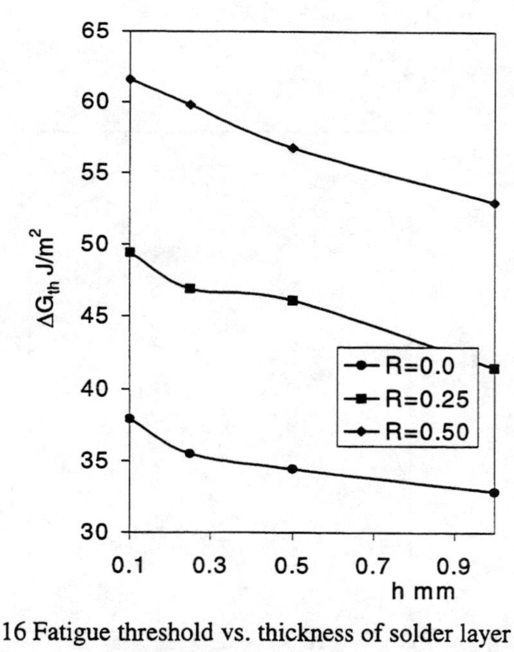

Fig. 16 Fatigue threshold vs. thickness of solder layer

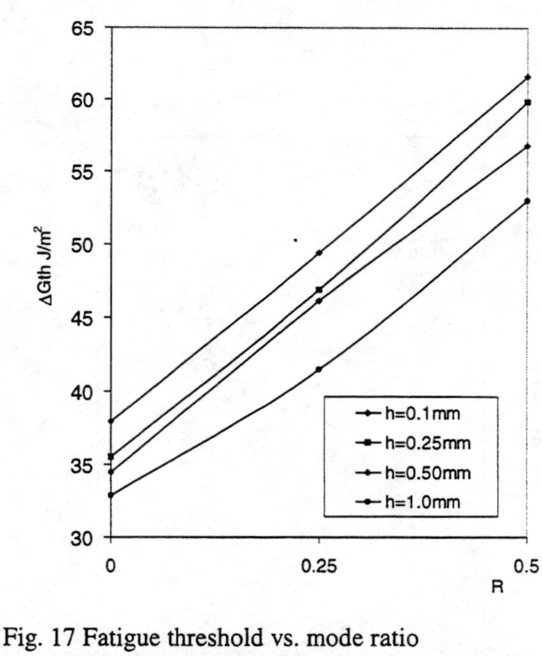

Fig. 17 Fatigue threshold vs. mode ratio

3.5 Fatigue Thresholds

Fatigue crack growth rates (*da/dN*) from ~ 10^{-7} to 10^{-3} *mm / cycle* were measured using load-shedding technique where the load drop at each loading step was *8% - 10%* of the previous load. Prior to taking data in each step, the advance of crack length was larger than the size of plastic zone determined by the previous loading (plastic zone was determined by a detail elastic/plastic finite element analyses of the sandwich layer, [41]). Fatigue crack thresholds ΔK_{th} (for mode I) or ΔG_{th} (for mixed mode) were determined at a crack growth rate of

10^{-7} *mm/cycle.* Figure 15 shows mode I fatigue threshold ΔK_{th} vs. thickness of the sandwich layer. The results indicate that when the thickness of the sandwich layer increases the fatigue thresholds decreases. This was more obvious when the thickness of the solder changed from 0.5 mm to 1.0 mm. These results are consistent with the mode I fracture toughness and could be related to the interface microstructures and its strength. The results also show that the mixed mode fatigue threshold, ΔG_{th} , increases with increase of mode ratio, Fig. 17. This result is similar to those reported in [29].

3.6 Fatigue Crack Growth rate

Fatigue crack growth rates vs. range of strain energy release rate for different mode ratios are presented in Figs. 18 – 21. Figures 22- 24 show the effect of the thickness of solder layer on mixed mode crack growth rate. The results show that there are three regimes in each graph: a near threshold regime, a power law regime, and a fast growth regime. Figures 18-21 show that for a specific thickness of sandwich layer, fatigue crack growth rate decreases with the increase of mode ratio. Similar results were obtained by J. K. Shang et al [24]. Figures. 22-24 shows that for a specific mode ratio, fatigue crack growth rate increases with the increase of thickness of the sandwich layer.

4. Conclusions

Mode I and mixed mode I / II fracture toughness, fatigue crack thresholds, and fatigue crack growth rates of 63Sn-37Pb solder / brass joint with different thickness of solder layer were measured at room temperature using four point bending sandwich specimens. Fatigue crack path, fracture path, and microstructure of solder / brass joint was observed under optical microscope. Fracture surfaces produced by both fracture test and fatigue crack growth tests were studied under scanning electron microscope. The results show,

1. Mixed mode fracture toughness in terms of strain energy release rate is much higher than that of the mode I, and it increases with the increase of mode ratio.
2. Mode I and mixed mode I/II fracture toughness in terms of strain energy release rate are significantly higher than that of the solder, and it increases with the decrease of the thickness of solder layer.
3. Mixed mode fatigue thresholds are higher than that of mode I, and it increases with increase of mode ratio.
4. Mode I and mixed mode I/II fatigue thresholds increase with the decrease of the thickness of solder layer. There were three distinct regimes in the fatigue crack growth rates vs. stress intensity factors or strain energy release rates. In the intermediate regime, fatigue crack growth rate correlated reasonably well with a power law relation of $da/dN = C*(\Delta G)^m$, and the magnitude of m was in the range of 1 ~ 3.5.

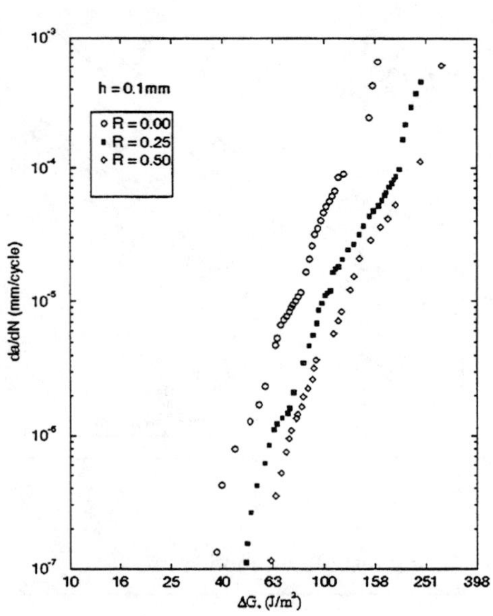

Fig. 18 Fatigue crack growth rate vs. range of strain energy release rate *(h = 0.10mm)*

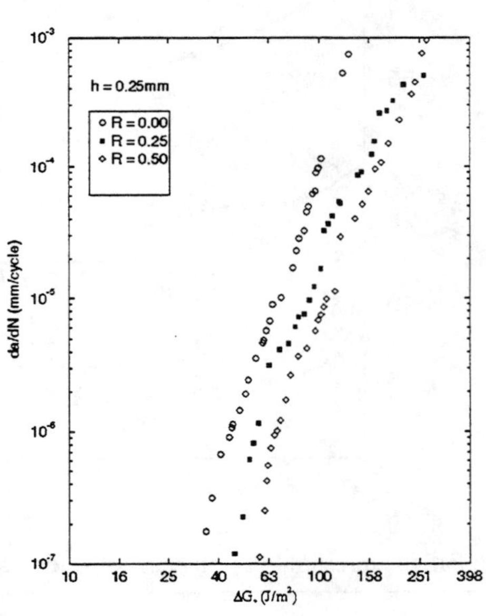

Fig. 19 Fatigue crack growth rate vs. range of strain energy release rate *(h = 0.25mm)*

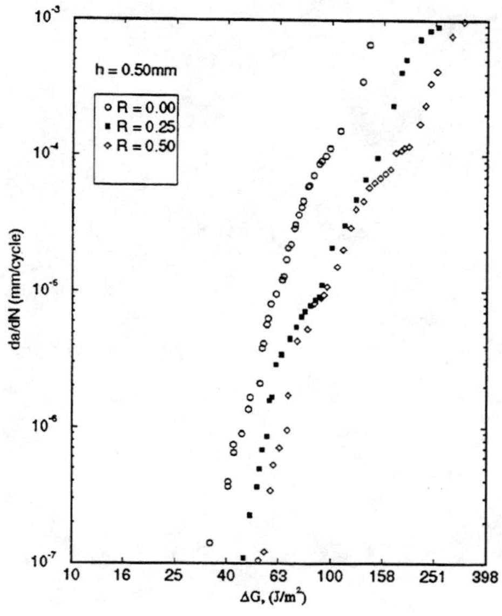

Fig. 20 Fatigue crack growth rate vs. range of strain energy release rate *(h = 0.50mm)*

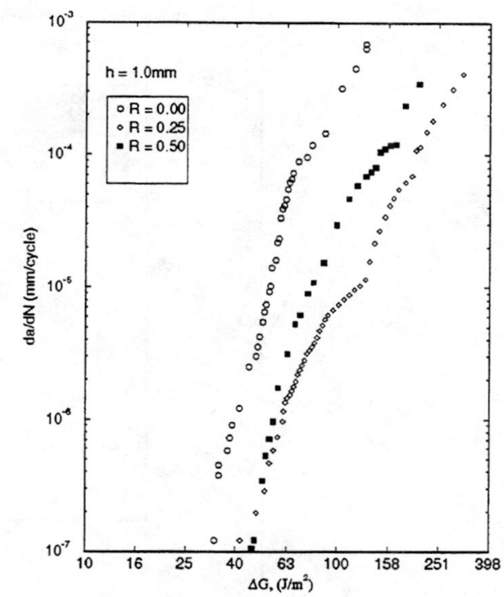

Fig. 21 Fatigue crack growth rate vs. range of strain energy release rate *(h = 1.00mm)*

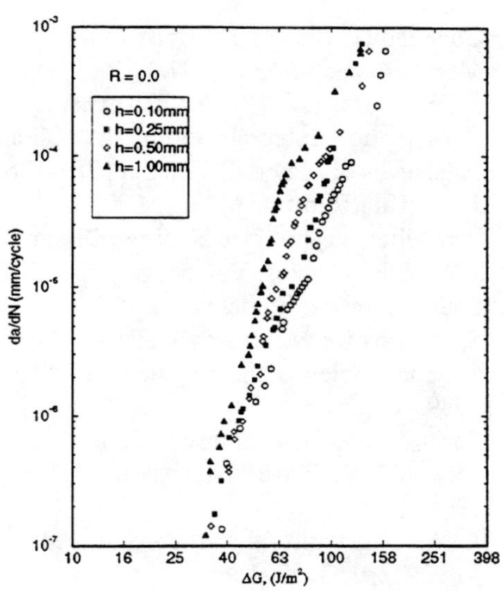

Fig. 22 Fatigue crack growth rate vs. range of strain energy release rate *(R = 0.0)*

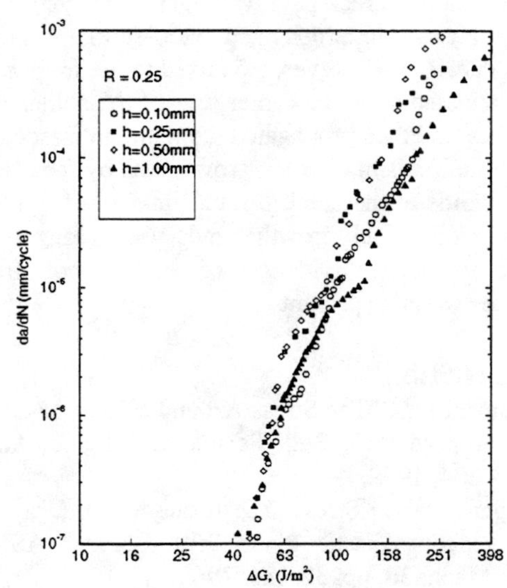

Fig. 23 Fatigue crack growth rate vs. range of strain energy release rate *(R = 0.25)*

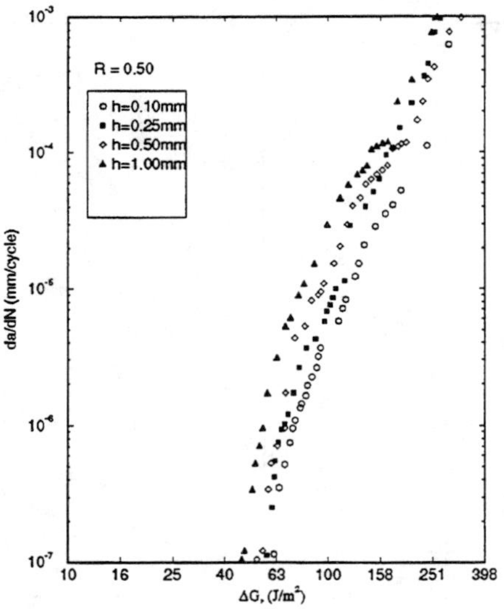

Fig. 24 Fatigue crack growth rate vs. range of strain energy release rate *(R = 0.50)*

5. Mix mode fatigue crack growth rate is smaller than that of the corresponding mode I.
6. Fatigue crack growth rates increase with the increase of sandwich layer thickness.
7. Mode I fatigue crack path was right at the center of the solder layer, and propagated parallel to the solder / brass interface. However, the mixed mode fatigue crack path deviated from the center toward the solder / brass interface and then propagated along the interface.
8. Micro-mechanism of crack growth was by formation of micro-voids at the crack tip and linkage of voids. For mixed mode crack growth, voids were elongated and linked together. Evidence of rubbing of fracture surfaces was also present.

REFERENCES

1. Willams, M. L. "The Stress Around a Fault or Crack in Dissimilar Media", Bull. Seismological Soc. Am. 49, pp199-204, 1959.
2. Erdogan, F. "Stress Distribution in a Non-homogeneous Elastic Plane With Cracks" ASME J. Appl. Mech. 30, pp232-236, 1963.
3. Erdogan, F. "Stress Distribution in Bonded Dissimilar Materials With Cracks", ASME J. Appl. Mech. 32, pp403-410, 1965.
4. England, A.H., "A Crack Between Dissimilar Media", ASME J. Appl. Mech. 32, pp400-402,1965.
5. Rice, J. R. and Shi, G.C., "Plane Problems of Cracks in Dissimilar Media", ASME J. Appl. Mech. 32, pp418-423, 1965.
6. J. J. Dundurs, J. Appl. Mech., Vol. 36, pp650-658, 1969.
7. Comninou, M. "The Interface Crack", ASME J. Appl. Mech. 44, pp631-636, 1977.
8. J. W. Park and Y. Y. Earmme, "Application of Conservation Integrals to interface Crack Problem", Mechanics of Material, Vol. 5, pp261-176, 1986.
9. J. W. Huntchinson, M. E. Mear, J. R. Rice, " Crack Paralleling an Interface Between Dissimilar Materials " J. Appl. Mechanics, Vol. 54, pp828-832, 1987.
10. Sou, Z. and Huntchinson, J.W. "Sandwich Test Specimens for Measuring Interface Crack Toughness." Materials Science and Engineering, A107, pp135-143, 1989.
11. Sou, Z. and Huntchinson, J.W. " Interface Crack between Two Elastic Layers." Int. J. Fract. 43, pp1-18, 1991.
12. J. W. Huntchinson and Z. Sou, Adv. Appl. Mech. Vol.29, pp163, 1991.
13. J. R. Rice," Elastic Fracture Mechanics Concepts for Interfacial Cracks ", J. Appl. Mech. Vol.55, pp98-103, 1988.

14. Gautense A. K. and Dundurs J., "The interfacial crack in a tension field", ASEM J. Applied Mechanics, Vol. 54, pp93-98, 1987.
15. C. F. Shih and R. J. Asaro, "Elastic Plastic Analysis of Cracks on Bimaterial Interfaces: Part I- Small Scale Yielding", ASME J. Applied Mechanics, Vol.55, pp299-316, 1988.
16. Cao H. C. and Evans A. G.," An Experimental Study of the Fracture Resistance of Bimaterial Interfaces", Mech Vol.7, pp295-304, 1989.
17. H. M. Jesene, "Mixed Mode Interface Fracture Criteria", Acta. Metall. Mater.Vol.38, No.12, pp2637-2644,1990.
18. R. M. Cannon et al. "Cyclic Fatigue-Crack Propagation Along Ceramic/Metal Interfaces ", Acta. Metall. Vol.39, No.9, pp2145-2156, 1991.
19. R.M.Cannon et al., J. Am. Ceramic Soc. Vol.79, 1996.
20. M. E. Thurston and A. T. Zehnder, " Experimental Determination of Sillca/Copper Interfacial Toughness" Acta. Metall.Mater.,Vol.41, No.12, PP2985-2992, 1993.
21. J.M.McNaney et al., " Fracture and Fatigue-Crack Growth Along Aluminium-Alumina Interfaces", Acta. Metall. Vol.44, No.12, pp4713-4728, 1996.
22. Yao. D., Zhang Z. AND J. K. Shang, "Experimental Technique for Studying Mixed-Mode Crack Growth in Solder Joints", Proc. of 1995 ASME Winter Meeting, Nov.12-17, 1995, San Francisco, CA. Paper No.95-WA/EEP-12.
23. Zhehua Zhang and J.K. Shang, "Subcritical Crack Growth at Bimaterial Interfaces:Part 1.Flexural Peel Technique", Metall. and Maters Transactions A, Vol.27A, pp205-211, Jan. 1996.
24. Yao. D., Zhang Z. AND J. K. Shang, "Effect of Load Mix on Fatigue Crack Growth in 63Sn-37Pb Solder Joints", Transaction of the ASME, Vol. 119, June, 1997
25. L. L. Shaw "Measurement of the fracture energy of interfaces in composites through sandwich-type-chevron-notched specimen", J. Material Testing and Evaluation, Vol. 25, No. 4, July 1997.
26. Wang J.S., and Sou, Z "Experimental Determination of Interfacial Toughness Curves Using Brazil-Nut-Sandwiches," Acta Metallurgic *et* Materialia, Vol.38, 1990, pp.1279-1290.
27. J. M. Howe, "Bonding, structure, and properties of metal/cermaic interfaces", Inter. Mater. Reviews, Vol.38, No.5, pp233-271, 1993.
28. Hua, G., M. W. Brown and K. J. Miller, "Mixed Mode Fatigue Thresholds", Fatigue of Engineering Materials and Structures, Vol.5, No 1, pp1-17, 1982.
29. Naybe-Hashemi, H, "Effects of Mode I and Mode II Overloads On Subsequent Mode I Crack Growth In AISI 4340 Steels," Biaxial and Multi-axial Fatigue, EGF3(ed. by M. W. Brown and K. J. Miller), Mechanical Engineering Publications, London, pp265-283, 1989.
30. Liaw P. K., Logsdon W. A. and Burke M. A., "Fatigue Crack Propagation Behavior of 63Sn-37Pb Solder", Scripta Metall. Vol.23, 1989, pp747-752.
31. Logsdon W. A., Liaw P. K.and Burke M. A., "Fracture Behavior of 63Sn-37Pb Solder", Eng. Fract. Mech., Vol36 No2, pp183-218, 1990.
32. Guo Z. and Conrad H., "Fatigue Crack Growth Rate in 63Sn-37Pb Solder Joints", ASME Journal of Electronic Packaging, Vol.115, 1993, pp159-164.
33. Guo Z. and Conrad H., "Effect of Microstructure size on Deformation Kinetics and Thermo-Mechanical Fatigue of 63Sn-37Pb Solder Joints", ASME Journal of Electronic Packaging, Vol.118, 1996, pp49-54.
34. H. U. Akay, N. H. Paydar and A. Bilgic, "Fatigue Life Prediction for Thermally Loaded Solder Joints Using a Volume-Weighted Averaging Technique", Transaction of the ASME, Vol.119, 1997, pp228-235.
35. Y. Murakami et al. "Stress Intensity Factors Handbook", Vol.2, pp909-911, 1988, The Soc. of Mater. Science, Japan.
36. ASTM Standard E647-96, Vol.3.01, pp647, 1997.
37. ASTM Standard E8-96a, Vol.3.01, pp56-76, 1997.
38. ASTM Standard E561-94, Vol.3.01, pp489-501, 1997.
39. ASTM Standard E399-96, Vol.3.01.
40. American Society for Metals, "Materials Handbook", 9th Edition, Vol.9, pp449-456, 1985.
41. P. Yang," Mixed Mode I/II Fracture and Fatigue Crack Growth along 63Sn-37Pb Solder/Brass Interface," Ms. Thesis, Department of Mechanical, Industrial and Manufacturing Engineering, Northeastern University, Boston, MA02115, USA, May 2000.

PVP-Vol. 415, Recent Advances in Solids and Structures — 2000
ASME 2000

Dynamic Response of a Heat Damaged Fiber-Resin Beam Subjected To Harmonic Forcing at the Tip

H. Nayeb-Hashemi and A. Harrison
Dept. of Mechanical, Industrial and Manufacturing Eng.
Northeastern University
Boston, MA 02115

Abstract

A cantilever beam built of fiber-resin composite material and damaged by heat is evaluated for its dynamic response using numerical methods. The goal of research is to produce a diagnostic process in which dynamic response can be used to estimate the severity of damage to the beam. Research proceeds from formulation of the continuous media equations for vibration in a multiply segmented beam, to the development of finite element models for the beam. It is discovered that the results of these two methods are qualitatively different from the predictions of a lumped system model, in that the lumped system predicts that frictional damping should reduce the dominant frequency of vibration while the more elaborated models indicate that damping may increase the dominant frequency. It is further discovered that the size and location of damage (the geometry) are equally as important as the local stiffness and damping of the damaged region (the material properties). The results indicate that the dominant frequency of dynamic response is not a sufficient symptom for complete diagnosis of damage in the beam.

Introduction

Glass-fiber / epoxy resin composite panels and beams are commonly used in aerospace structures. Damage to fiber-resin composite materials can be detected in-service by a variety of methods depending on the nature of the damage. Visual inspection is the "first-line" method of choice but is not effective for all modes of damage. The most common non-visual inspection techniques are ultrasonic testing and acoustic emission monitoring. This research is aimed toward a third non-visual technique, macroscopic vibration analysis. The three techniques, along with visual inspection, will be examined with reference to their relative advantages for detecting various types of damage.

Impact damage is sometimes apparent to visual inspection as an area of "fuzziness" where impact causes fibers to brake and degrade the resin matrix. However, impact can also cause delamination or spalling, neither of which is visible by external visual inspection. Heat damage generally causes surface blistering, discoloration of the resin, and degradation of the fiber-resin bond; excessive heat can also cause breakage of fibers deeper in the matrix due to thermal strain. Blistering and discoloration are readily apparent to visual inspection, while bond degradation and fiber breakage are generally not visible.

Acousto-ultrasonic testing measures transmission and reflection of a stress wave within the structure. Both measurements focus on locating discontinuities in the material by variations in the ultrasonic wave pattern. In transmission testing, flaws in the material cause wave front scattering, reducing the energy transmitted to the receiver. In reflection testing, the receiver picks up energy back scattered from inclusions, particulate, or voids in the material. In general, ultrasonic techniques provide a good idea of whether there is damage in a component, but do not indicate the effects of damage on component properties. The velocity and amplitude of transmitted pulses can be used with either transmission or reflection testing to obtain a rough estimate of the elastic modulus in a region; from this information, the strength of material in the region may be estimated based on a volume-fraction correlation of stiffness and strength. [1] However, this relationship may not hold in the damaged material as elastic modulus for ultrasound transmission velocity is determined by the volume fraction of all fibers, while structural strength is determined by the volume fraction of intact fibers.

Acoustic emission monitoring is an ongoing inspection technique throughout the life of the component, and relies on detection of characteristic stress waves released by fiber breakage, laminate separation, or fiber-resin slippage. By comparison to a baseline established in laboratory conditions, an acoustic emission record can provide a time trace of damage mechanisms in the component. In its most basic form the technique is not precise as far as the location of acoustic events; with multiple pick-ups and automated comparison of signal arrivals, it should be possible to pinpoint the time and location of events leading to component failure. However, like ultrasonic inspection, acoustic monitoring only detects the damage; it does not directly indicate the strength or other bulk properties of the component. [2]

Macroscopic vibration analysis offers a third possibility for non-destructive evaluation of components during routine maintenance. By applying harmonic forcing over a range of frequencies and measuring the displacement amplitude of points surrounding the load application, it should be possible to determine the dominant frequency and thereby estimate structural rigidity and density between the point of loading and the points where displacement is measured. Structural rigidity should correlate with tensile strength in fiber-resin materials, because both properties are determined by the volume fraction of unbroken fibers. Vibration analysis could therefore provide an improved indication of the strength remaining in the damaged piece.

Successful implementation of macroscopic vibration analysis requires both analytical and experimental investigation of the ways in which damage affects vibration in terms of dominant frequencies and modes. In particular, an unambiguous correlation must be drawn between dynamic response and ultimate tensile strength. The research presented here lays the analytical groundwork for such investigation.

Theoretical Investigations

The beam was modeled as a cantilever in three segments, with the middle (damaged) segment of variable length. Overall length of the beam L_{tot} was fixed at 0.3 m, with thickness t=0.00064 m and width w=0.0125 m. Initially the length L_1 of the "root" segment and the length L_2 of the damaged segment were variable. When it was found that the location of the damaged segment along the beam significantly affected its response, the root and tip segments were given equal lengths. Properties of the damaged segment were described by stiffness ratio k_2/k_1 (stiffness of the damaged material divided by stiffness of the undamaged material) and by loss factor η (fraction of vibration energy lost in each cycle).

Dynamic response of the damaged beam was obtained for L_2=0.01 m, 0.05 m, 0.1 m; k_2/k_1=0.5; η=1, 5. Response of the undamaged beam was also obtained. For all cases L_{tot}= 0.3 m. The dominant frequency of dynamic response was determined by generating a Bode diagram for each combination of damaged segment properties by the frequency scan method. Frequency of the exciting force (0.1249 N) was varied from 0 to 100 rad/s in increments of 1 rad/s for the initial scan. Once dominant frequencies were identified, a fine tuning scan was made in steps of 0.01 rad/s around 38 – 42 rad/s, the range of the lowest dominant frequency.

Beam response was examined by modeling the beam as continuous media, as a lumped system, and as finite elements.

Continuous Media Model

A continuous media model [3] was developed for the vibration under harmonic forcing of a fiber-resin beam with and without a damaged segment at the midpoint of the beam. The continuous media equations may be solved to obtain beam tip displacement as a function of beam geometry and forcing frequency. The derivation is presented first for an undamaged beam segment, and is then further developed for a damaged beam with a region of reduced stiffness and hysterisis damping. For the undamaged beam segment shown in Figure 1, modeling the segment as an Euler-Bernoulli beam of constant EI, the equation of motion can be written as

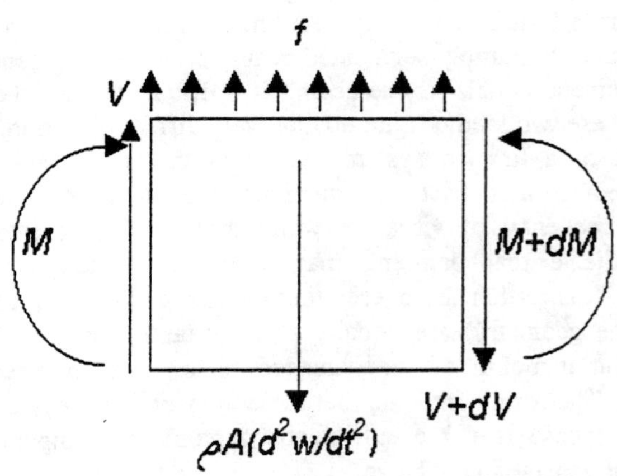

Figure 1: Free-body diagram of beam segment.

$$EI\frac{\partial^4 w}{\partial x^4}+\rho A\frac{\partial^2 w}{\partial t^2}=f \tag{1}$$

If this beam is subjected to an end load $F_0e^{i\omega t}$, the response of the beam will be in form of

$$w(x,t) = \sum_{i=1}^{4} c_i e^{s_i x} e^{i\omega t} \tag{2}$$

Where

$$s_1 = \sqrt{\omega}\sqrt[4]{\rho A / EI}$$
$$s_2 = -s_1$$
$$s_3 = is_1$$
$$s_4 = -s_3 \tag{3}$$

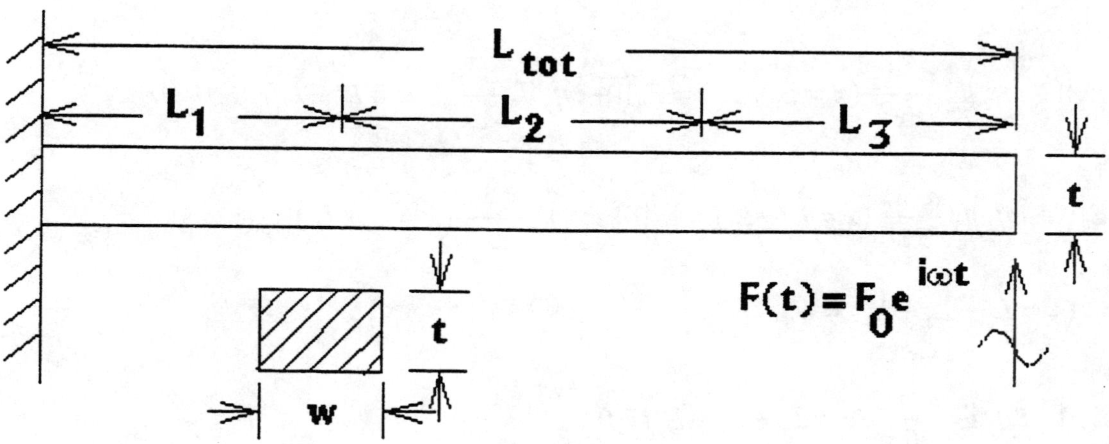

Figure 2: Cantilever beam with critical dimensions.

Assuming the distributed load f to be negligible, Eq. 1 can be duplicated to describe a cantilever beam of three segments j, each of different EI_j and L_j, with a harmonic loading $F_0 e^{i\omega t}$ at the tip as shown in Figure 2. A damaged beam can be modeled by reducing the stiffness E of the damaged segment and introducing a loss factor η. The equation of state for the damaged segment is assumed to be in a form of

$$\sigma = E\left(1 + \hat{i}\eta\right)\varepsilon \tag{4}$$

Using the above constitutive equation, the equation of motion of the beam can be presented as

$$EI(1+\eta)\frac{\partial^4 w}{\partial x^4} + \rho A \frac{\partial^2 w}{\partial t^2} = 0 \tag{5}$$

The steady state vibration of the damaged beam subjected to the harmonic tip displacements of

$$W(x = L_1 + L_2 + L_3) = W_0 e^{\hat{i}\omega t}, \tag{6}$$

where W_0 is the amplitude of the harmonic tip displacement, can be stated to be in a form of

$$w(x,t) = \sum_{i=1}^{4} c_i e^{s_i x} e^{i\omega t} \tag{7}$$

where

$$\rho A \omega^2 = s^4 EI\left(1 + \hat{i}\eta\right), \tag{8}$$

for which the solutions are

$$s_1 = \sqrt{\omega}\sqrt[4]{\frac{\rho A}{EI(1+\eta^2)}}\sqrt[8]{1+\eta^2}\, e^{\left(\frac{1}{4}\right) i \arctan(-\eta)}$$
$$s_2 = -s_1$$
$$s_3 = is_1$$
$$s_4 = -is_3 \tag{9}$$

Taking the cantilever shown in Figure 2, the boundary conditions become:

$$w_1(x=0) = 0 \tag{10}$$
$$\frac{dw_1}{dx}(x=0) = 0 \tag{11}$$
$$w_1(x=L_1) = w_2(x=L_1) \tag{12}$$
$$\frac{dw_1}{dx}(x=L_1) = \frac{dw_2}{dx}(x=L_1) \tag{13}$$

$$E_1\left(1+\hat{i}\eta_1\right)I_1\frac{d^2w_1}{dx^2}(x=L_1)=E_2\left(1+\hat{i}\eta_2\right)I_2\frac{d^2w_2}{dx^2}(x=L_1) \quad (14)$$

$$E_1\left(1+\hat{i}\eta_1\right)I_1\frac{d^3w_1}{dx^3}(x=L_1)=E_2\left(1+\hat{i}\eta_2\right)I_2\frac{d^3w_2}{dx^3}(x=L_1) \quad (15)$$

$$w_2(x=L_1+L_2)=w_3(x=L_1+L_2) \quad (16)$$

$$\frac{dw_2}{dx}(x=L_1+L_2)=\frac{dw_3}{dx}(x=L_1+L_2) \quad (17)$$

$$E_2\left(1+\hat{i}\eta_2\right)I_2\frac{d^2w_2}{dx^2}(x=L_1+L_2)=E_3\left(1+\hat{i}\eta_3\right)I_3\frac{d^2w_3}{dx^2}(x=L_1+L_2) \quad (18)$$

$$E_2\left(1+\hat{i}\eta_2\right)I_2\frac{d^3w_2}{dx^3}(x=L_1+L_2)=E_3\left(1+\hat{i}\eta_3\right)I_3\frac{d^3w_3}{dx^3}(x=L_1+L_2) \quad (19)$$

$$w_3(x=L_1+L_2+L_3)=W_0e^{\hat{i}\omega t} \quad (20)$$

$$E_3\left(1+\hat{i}\eta_3\right)I_3\frac{d^2w_3}{dx^2}(x=L_1+L_2+L_3)=0 \quad (21)$$

Transforming Eq. 8, then it is apparent that the stress is not in phase with the strain:

$$\sigma=\varepsilon E\sqrt{1+\eta^2}\,e^{\hat{i}\arctan(\eta)} \quad (22)$$

Equation 22 can also be presented as

$$\sigma=\sigma_0e^{sx}e^{\hat{i}\omega(t+\phi)} \quad (23)$$

where $\omega\phi=\arctan(\eta)$.

For consistency with the finite element approach, the above equations were also solved for a damaged beam subjected to a harmonic tip load, where,

$$E_3(1+\hat{i}\eta_3)I_3\frac{d^3w_3}{dx^3}(x=L_1+L_2+L_3)=F_0e^{i\omega t} \quad (24)$$

Harmonic loading, unlike harmonic displacement, permits vibration modes in which the point of loading is a node. For the lowest dominant frequency, in which the mode shape should be similar to the static deflection, this issue was not significant; for study of higher modes, it is very important. The differential equation involves twelve boundary and continuity conditions, and can be solved numerically for twelve constants c_{ij} to produce the function $w(x,t)=$

$$W(x,t)=\begin{cases} w_1=W_0\sum_j c_{1j}e^{s_{1j}x}e^{\hat{i}\omega t} & ,x\le L_1 \\ w_2=W_0\sum_j c_{2j}e^{s_{2j}x}e^{\hat{i}\omega t} & ,L_1\le x\le L_1+L_2 \\ w_3=W_0\sum_j c_{3j}e^{s_{3j}x}e^{\hat{i}\omega t} & ,L_1+L_2\le x\le L_1+L_2+L_3 \end{cases} \quad (25)$$

The frequency scanning method was used to evaluate the steady state response for a range of frequency ω. To implement frequency scanning, a MatLab code was developed to assemble the matrix of boundary conditions and to solve for the tip displacement as a function of time.

Lumped System Model

The dominant frequency of the undamaged beam was estimated using the standard lumped system approximation of a linear spring and a concentrated mass.

$$\omega_n=\sqrt{\frac{3EI}{0.23\rho AL^4}}\,. \quad (26)$$

As a more general check, a torsional spring model was also developed. In this model, the root segment of the beam (length L_1) was assumed clamped; the tip segment (L_3) was assumed to be a rigid body. Throughout this analysis, the x-variable is measured from the left end of the damaged segment as this is considered to be the point of lowest bending rigidity which may act as a torsional pin and its stiffness determines the vibration behavior of the beam. Using energy methods, an equivalent k_t and I_0 were developed for the beam; these values were found to be functions of L_3 and L_2.

The equivalent torsional spring constant was obtained by solving the Euler-Bernoulli equations from mechanics of materials to obtain the transverse displacement of the damaged segment as a function of tip loading. The torsional spring stiffness and equivalent I_0 were obtained

$$k_t = \frac{2L_3(EI)_2}{L_2(L_2+2L_3)} \tag{27}$$

$$I_0\dot{\theta}^2 = \int_0^{L_2} \rho A(\dot{y})^2\,dx + \int_{L_2}^{L_{total}} \rho A(\dot{y})^2\,dx \tag{28}$$

For the simplifying assumptions stated above,

$$y = \begin{cases} \dfrac{2x^2(x-3(L_2+L_3))}{L_2(2L_3+L_2)}\theta_2 & x \le L_2 \\ \dfrac{2L_2^{\,2}(L_2-3(L_2+L_3))}{L_2(2L_3+L_2)}\theta_2 + (x-L_2)\theta_2 & L_2 < x \le L_3 \end{cases} \tag{29}$$

Taking the cross -sectional density to be uniform throughout the beam, the result for I_0 was

$$I_0 = \rho A \frac{140L_3^{\,5} - 1739L_2^{\,5} - 2933L_2^{\,4}L_3 + 1540L_2L_3^{\,4} + 3451L_2^{\,3}L_3^{\,2} + 4685(\rho A)_3L_2^{\,2}L_3^{\,3}}{105(2L_3+L_2)^2} \tag{30}$$

The natural frequency for the damaged and the undamaged beam may then be estimated by the standard formula for a torsional spring and inertial body

$$\omega_n = \sqrt{k_t/I_0}\,. \tag{31}$$

The equivalent damping coefficient can be obtained from

$$c = \frac{h^*}{\omega}\ ;\ h^* = \eta k\,, \tag{32}$$

Substituting for k from Eq. 27 results in

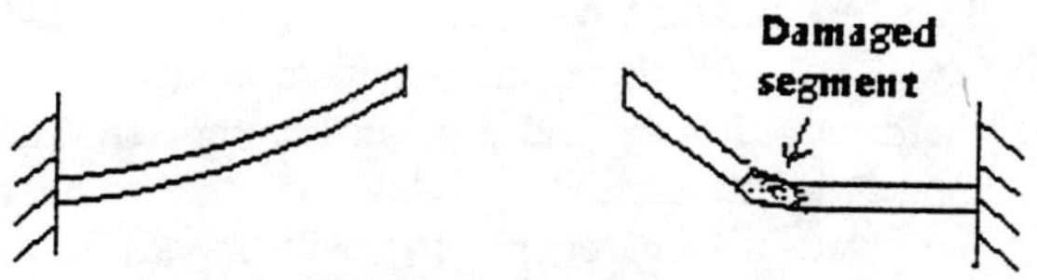

Figure 3: Continuous model (left) vs. torsional-spring

$$c_{eq} = \frac{\eta}{\omega}\frac{2L_3(EI)_2}{L_2(L_2+2L_3)}\,. \tag{33}$$

Finite Element Model

A Hermitian finite element model was developed for the damaged beam and used in a computer simulation. The model is substantially based on the standard approach presented in Dimarogonas [4], with modifications to account for the hysterisis damping of the damaged segments.
The Hermite interpolations $\mathbf{h}=[h_1; h_2; h_3; h_4]$ are defined as:

$$
\begin{aligned}
h_1(x) &= 1-3s^2+2s^3 \\
h_2(x) &= l(s-2s^2+s^3) \\
h_1(x) &= 1-3s^2+2s^3 \\
h_2(x) &= l(s-2s^2+s^3)
\end{aligned}
\qquad (34)
$$

$s = x/l$

The interpolation functions h_I can be used to describe the transverse and angular displacement of points along the element, given the nodal displacement vector $\mathbf{w}=[w(x_1)\,;\,\theta(x_1)\,;\,w(x_2)\,;\,\theta(x_2)]$:

$$
w(x) = h_1(x)w(x_1) + h_2(x)\theta(x_1) + h_3(x)w(x_2) + h_4(x)\theta(x_2)
$$

$$
\theta = \frac{dw}{dx} \qquad (35)
$$

The standard elemental stiffness matrix may be derived from the equation for elastic energy in the deformed element, using the Hermite-interpolated displacements to describe the deformation:

$$
\mathbf{K}^e = EI\int_0^l \mathbf{h}''^T\mathbf{h}''dx \qquad (36)
$$

For the undamaged segment, the element stiffness matrix becomes

$$
\mathbf{K} = EI_{undamaged}\begin{bmatrix} 12/l^3 & & & \\ 6/l^2 & 4/l & & \\ -12/l^3 & -6/l^2 & 12/l^3 & \\ 6/l^2 & 2/l & -6/l^2 & 4/l \end{bmatrix}. \qquad (37)
$$

Considering the equation of state for the damaged element (Eq. 8), then the elemental stiffness matrix for the damaged segment becomes

$$
\mathbf{K} = EI_{damaged}(1+i\eta)\begin{bmatrix} 12/l^3 & & & \\ 6/l^2 & 4/l & & \\ -12/l^3 & -6/l^2 & 12/l^3 & \\ 6/l^2 & 2/l & -6/l^2 & 4/l \end{bmatrix} \qquad (38)
$$

An energy-consistent mass matrix for both undamaged and damaged portions of the beam may be obtained by considering the kinetic energy of a transversely displaced element:

$$
T = \int \dot{\mathbf{w}}^T\mathbf{M}^e\dot{\mathbf{w}} = \int_0^l \rho A\left(\dot{\mathbf{w}}^T\mathbf{h}^T\mathbf{h}\dot{\mathbf{w}}\right)dx. \qquad (39)
$$

The equivalent elemental mass matrix $\mathbf{M}$ is then

$$
\mathbf{M} = \frac{\rho A}{420}\begin{bmatrix} 156l & & & \\ 22l^2 & 4l^3 & & \\ 54l & 13l^2 & 156l & \\ -13l^2 & -3l^3 & -22l^2 & 4l^3 \end{bmatrix}. \qquad (40)
$$

Elemental matrices are assembled to obtain the equation of motion for the beam

$$
\mathbf{M}\ddot{\mathbf{w}} + \mathbf{K}\mathbf{w} = \mathbf{F}e^{i\omega t}. \qquad (41)
$$

Assuming that $w(x,t)$ is a periodic exponential function of t, Eq. 41 can be transformed to obtain the displacement vector $\mathbf{w}$ in terms of ω and the mass and stiffness matrices:

$$
\mathbf{w} = \left[-\mathbf{M}\omega^2 + \mathbf{K}\right]^{-1}\mathbf{F}e^{i\omega t}. \qquad (42)
$$

A MatLab code was developed to form and assemble the elemental matrices based on user input. The program scans a range of frequencies to generate a Bode diagram for the damaged and undamaged beam.

Results and Discussion

It was assumed throughout the analysis that heat

damage decreases the stiffness of the fiber-resin by damaging fibers, and introduces or increases hysterisis damping by degrading the fiber-resin interfaces. Several combinations of diminished stiffness ratio and increased loss factor were investigated to determine which combinations produced detectable changes in natural frequency. It was found that significant reductions in stiffness (on the order of 50%) produced reductions on the order of 1% to 10% in the dominant frequency depending on size and location of the damaged segment.

(Fig. 4). For the damaged beam, results of the continuous and finite element models were similar for dominant frequency; finite element modeling diverged from the continuous media by a maximum of 6% for η_2 = 1.0, k_2/k_1=0.5 (Fig. 5). The lumped system model was used to confirm natural frequency of the undamaged specimen.

In the damaged beam, the dominant frequency decreases with stiffness ratio k_2/k_1 (Fig. 6). For the same value

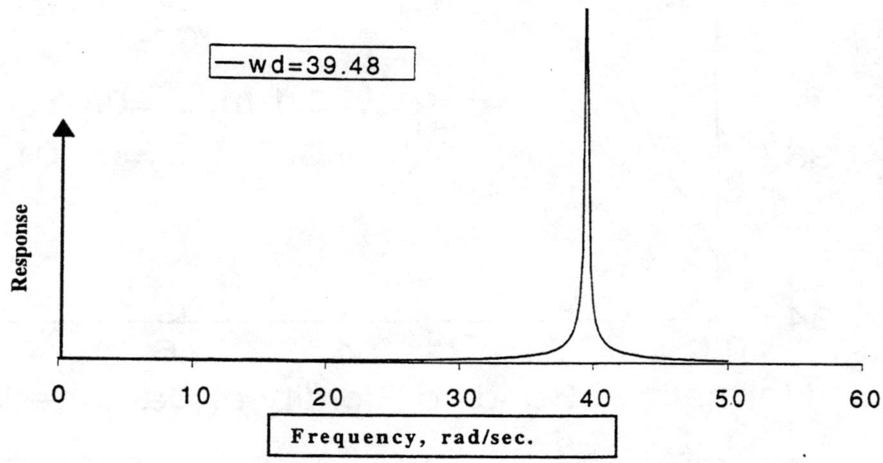

Figure 4: Dynamic response of undamaged beam.

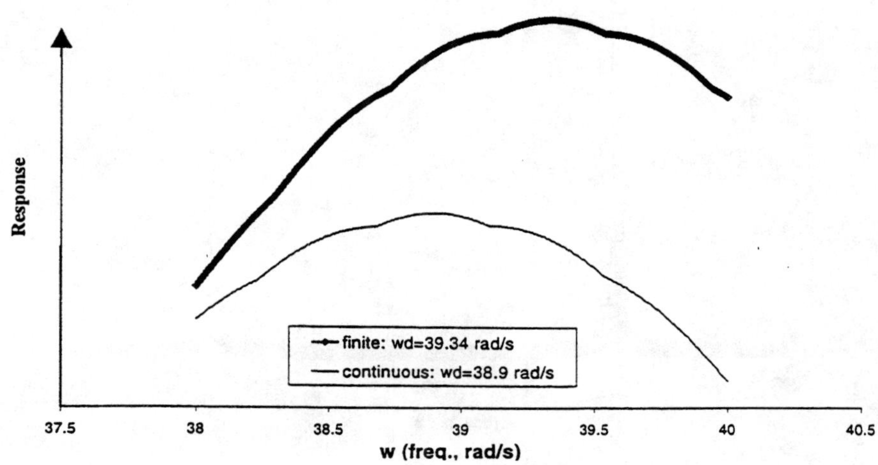

Figure 5: Finite element model vs. continuous media model for damaged segment: damping loss factor η_2 = 1.0, stiffness ratio k_2/k_1=0.5, length L_2=0.1 m.

For the undamaged beam, both continuous media and finite element methods predict natural frequency of 39.48 rad/s

of k_2/k_1, a greater length of the damaged segment will reduce the dominant frequency (Fig. 7). The size of the damaged segment amplifies the influence of the stiffness ratio; for equal stiffness ratios, a longer damaged segment will respond at lower dominant frequency (Fig. 6, Fig. 8). It is possible that two beams with different L_2 and different k_2/k_1 could respond at the same dominant frequency.

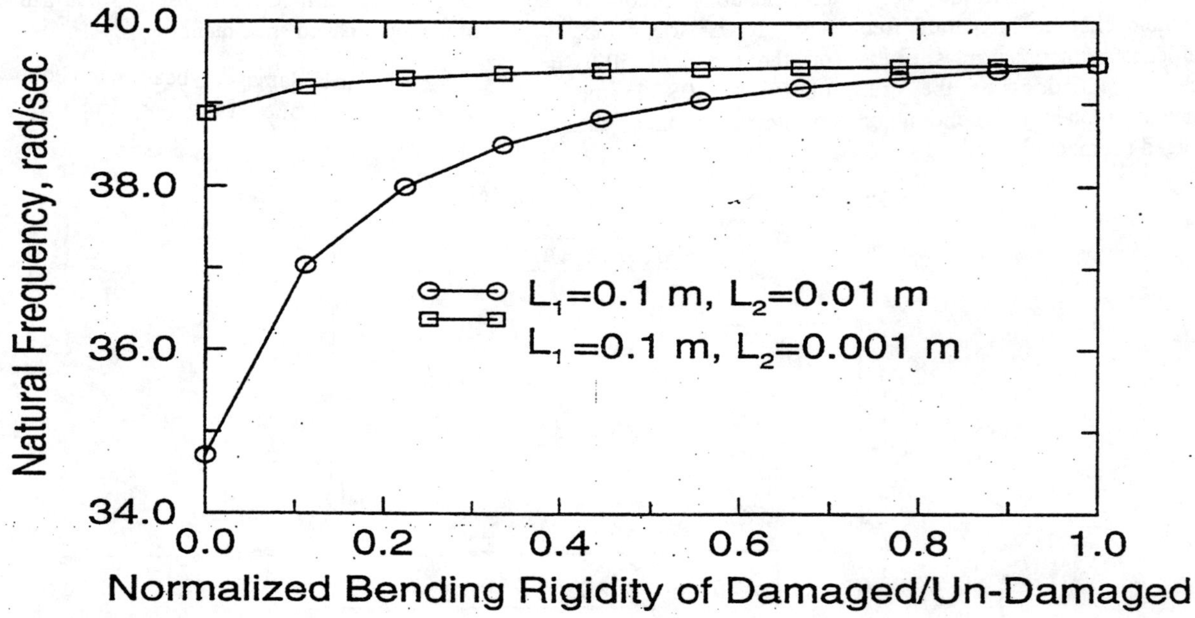

Figure 6: Dominant frequency vs. stiffness of damaged segment, damping loss factor □$_2$=0.

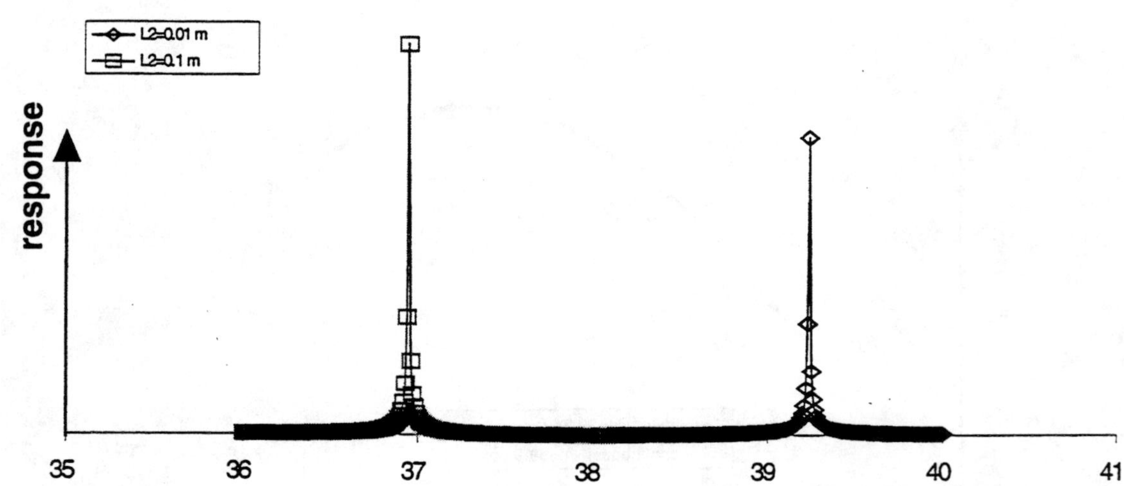

Figure 7: Dynamic response vs. frequency for L_2=0.01 m, 0.1 m; k_2/k_1=0.5; □$_2$=0.

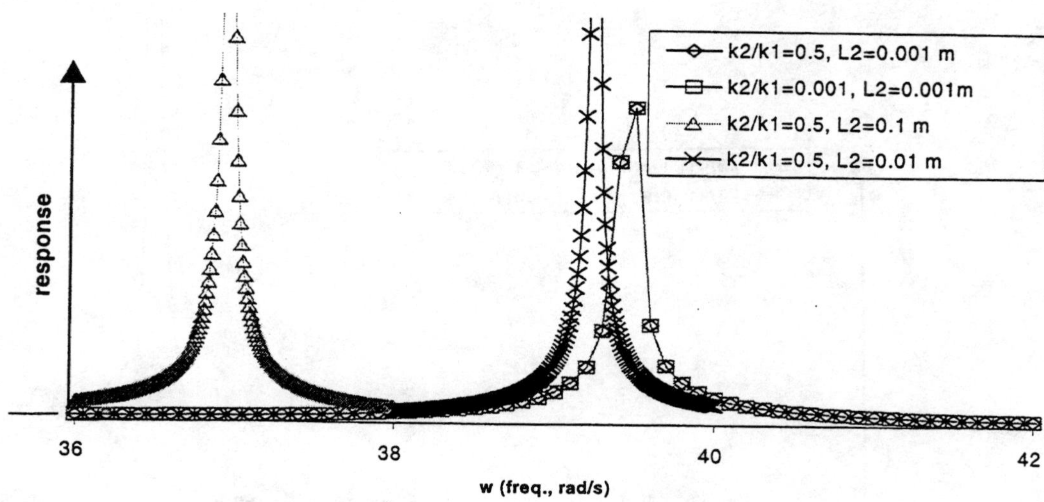

Figure 8: Dynamic response vs. frequency for several variations of L_2, k_2/k_1; □$_2$=0.

Damage produces broken fibers and degraded fiber-resin interfaces; both mechanisms reduce structural stiffness while the degraded interfaces also promote frictional damping. Frictional damping leads to an increase in the loss factor at the damaged segment, which tends to increase the dominant frequency of the beam (Fig. 9).

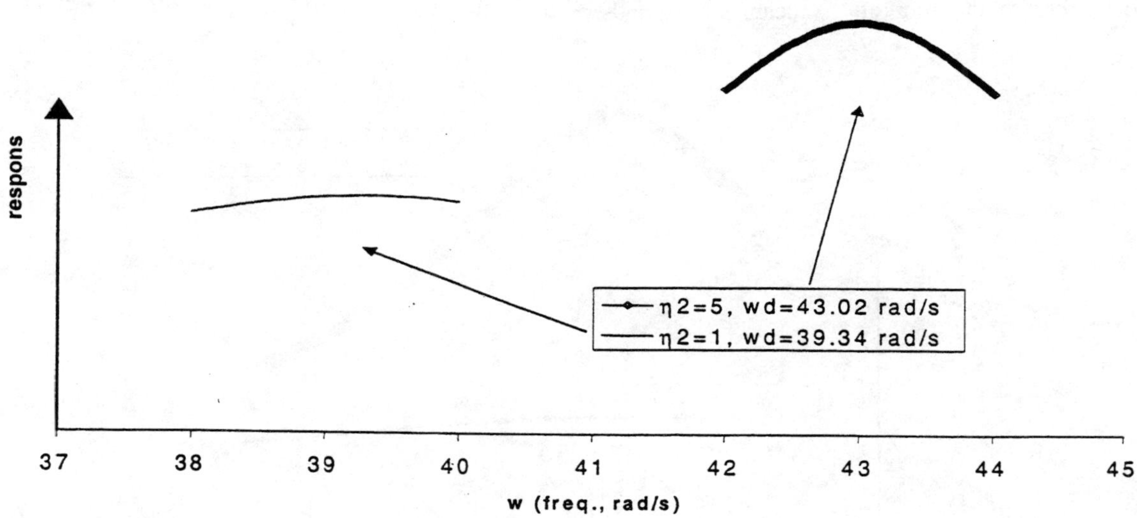

Figure 9: Dynamic response vs. frequency for□□$_2$=1, 5; k_2/k_1=0.5; L_2=0.1 m.

The increase in dominant frequency was unexpected but was confirmed by both continuous media and finite element methods. Lumped system analysis predicted a decrease in dominant frequency with increased damping; however, the lumped system analysis was divergent from the other approaches in all cases except analysis of the undamaged beam. In summary, the increase in dominant frequency due to frictional damping may offset the effects of reduced structural rigidity, and may even make the damaged beam vibrate at higher dominant frequency than the undamaged beam (Fig. 10).

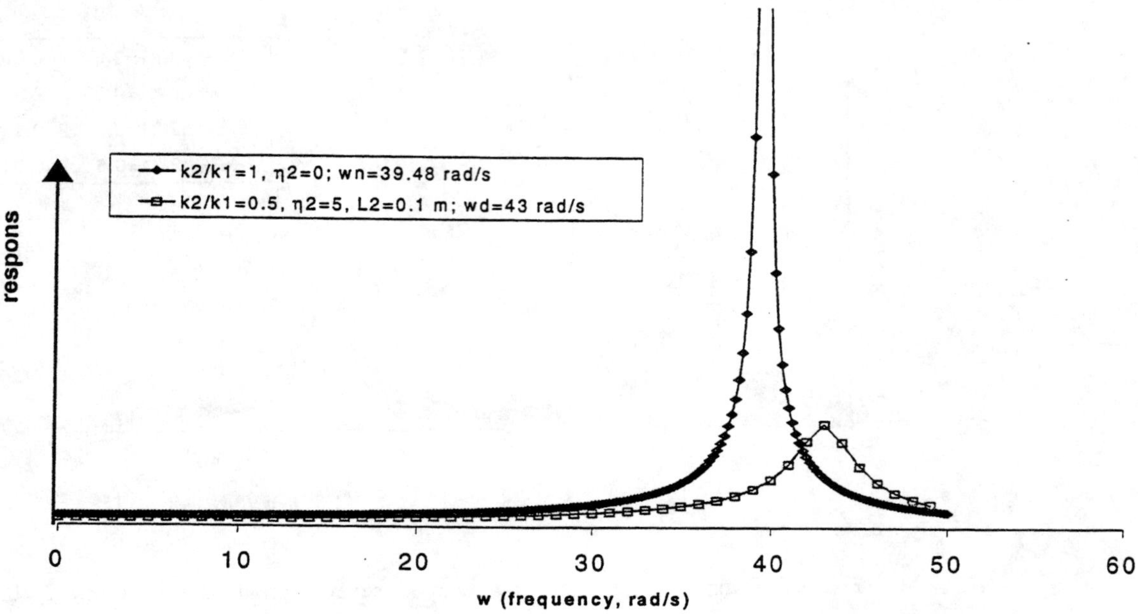

Figure 10: Dynamic response vs. frequency for damaged ($k_2/k_1=0.5$; $L_2=0.1$ m; $\eta_2=5$) and undamaged beam. Note that damaged beam responds at higher frequency.

The size and location of the damaged segment significantly modify the effects of increased damping. Location closer to the root of the beam corresponds to a reduction in the dominant frequency when all other variables are held constant (Fig. 11).

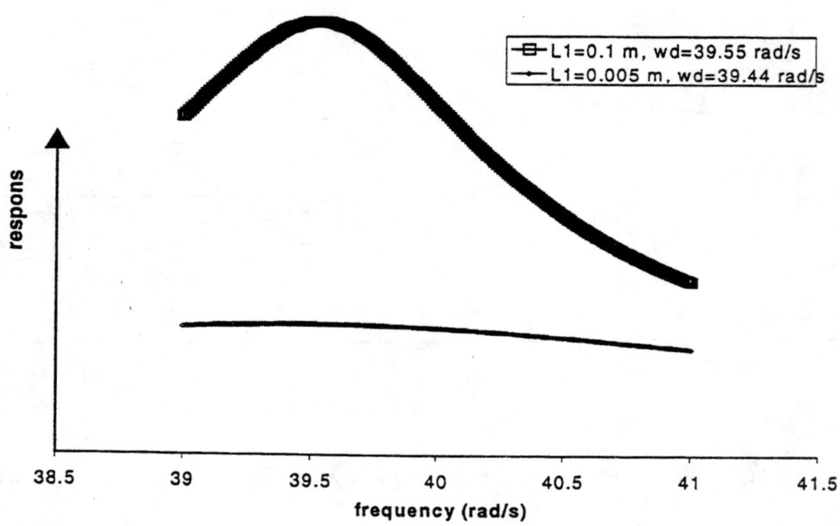

Figure 11: Dynamic response vs. frequency for root segment length $L_1=0.1$ m, 0.005 m; $k_2/k_1=0.5$; $L_2=0.01$ m; $\square_2=1$.

Conclusions

Dynamic response of a composite beam with localized heat damaged is analyzed using lumped system, continuous media, and finite element models. The damaged specimen exhibits changes in several key properties: structural rigidity, frictional damping, and tensile strength. The eventual goal of the research effort is to estimate tensile strength of a fiber-resin composite by measuring changes in the dominant frequency of dynamic response. This goal is based on a chain of reasoning in which stiffness (bulk modulus of elasticity) is closely related to tensile strength and can be easily obtained from the dominant frequency of dynamic response.

It was found that material stiffness is not easily obtained from the dominant frequency of dynamic response. A 50% reduction in structural rigidity produced only a 10% change in dominant frequency for the range of geometry considered. Further, the change in dominant frequency is more sensitive to the size and location of a damaged area than to the severity of damage in that area. Depending on the location of the damaged area, the dominant frequency of vibration may be above or below the natural frequency of the undamaged beam.

References

Birks, A.S., Green R.E., McIntire, P. *NDT Handbook, 2 ed.* v 7: "Ultrasonic testing". American Society for Non-destructive Testing. 1991.

Miller, R.K. and McIntire, P. *NDT Handbook.* v 5: "Acoustic emission testing". American Society for Non-destructive Testing. 1987.

Rao, S. *Mechanical Vibrations.* Addison-Wesley. 1995.

Dimarogonas, A. *Vibrations for Engineers.* Prentice Hall. 1992.

THE EFFECT OF AGING ON THE MICROSTRUCTURE AND PRECIPITATION RESPONSE OF AN ALUMINUM-LITHIUM ALLOY

James M. Fragomeni
Ohio University
Department of Mechanical Engineering
251 Stocker Engineering Center
Athens, Ohio 45701-2979, USA

ABSTRACT

The consequence of thermal treatment on the precipitation response of intermetallic precipitates in the microstructure microstructure and subsequent mechanical properties of an Al-2.6wt.%Li-0.09wt.%Zr alloy was studied. The alloy was solution heat treated and artificially aged for a series of aging times and temperatures at various precipitation hardening conditions. The underaged, peak-aged and overaged microstructures of the alloy were analyzed. Quantitative particle size measurements were performed to determine the size, distribution, morphology and coarsening rate for both δ'(Al_3Li) and δ'(Al_3Li)/Al_3Zr precipitates. The average particle size, distribution, spacing and volume fraction of the intermetallic precipitates were related to the heat treating process. For all of the aging times studied, the δ'(Al_3Li)/Al_3Zr particles were much larger in size than the δ'(Al_3Li) and Al_3Zr-free particles. The particle coarsening rate, determined from the Lifshitz, Slyozov and Wagner coarsening rate theory, was more accelerated for the δ'(Al_3Li)/Al_3Zr particles than for the δ'(Al_3Li) precipitates. The presence of the Al_3Zr phase was found to accelerate the aging kinetics of the alloy. Therefore, a small amount of zirconium in the alloy resulted in a faster particle coarsening rate of the overall combined particle size distribution and thus led to more rapid precipitation aging response.

Keywords: Aluminum-Lithium, precipitation hardening, intermetallic particles, microstructure

INTRODUCTION

Binary aluminum-lithium alloys are precipitation hardened by aging that causes the nucleation and growth of intermetallic strengthening precipitates randomly distributed in the grain crystalline structure. The intermetallic strengthening precipitates are know as δ'(Al_3Li) particles which nucleate homogeneously within the crystalline structure. The amount or volume fraction of the δ'(Al_3Li) in the microstructure is driven by the composition i.e., the weight percent lithium in solution, and the aging practice (aging time & temperature). The greater the composition weight percent of the lithium the greater will be the precipitation response. Increasing the aging time and or temperature will result in a larger size of the δ'(Al_3Li) particles. The δ'(Al_3Li) particle size, volume fraction, interparticle spacing and size distribution are a direct consequence of the aging practice and composition and thus affect the extent of plastic deformation. Quantitative microscopy of the intermetallic precipitates can be performed to relate the measured deformation behavior to the microstructural features. The quantitative transmission electron microscopy (TEM) study can reveal the distribution of both δ'(Al_3Li) and Al_3Zr-Al_3Li precipitates in the microstructure of the alloy as well as the dislocation subgrain structure of the alloy. The results can then correlate to the amount of mechanical deformation, the heat treatment processing parameters and composition. The plastic deformation behavior of aluminum-lithium (Al-Li) alloys is controlled by the coherent deformable δ'(Al_3Li) particles, which are randomly distributed throughout the microstructure and impede the dislocation glide motion during deformation. The δ'(Al_3Li) particles are often sheared by dislocations in the underaged condition and are usually bypassed by the dislocations in the overaged condition. In the peak-aged condition, a combination of dislocation particle shearing and particle bypassing can occur simultaneously, due to the distribution of both large and small particle sizes (Huang, 1986, Furukawa et al., 1985).

EXPERIMENTAL BACKGROUND

An aluminum-lithium-zirconium alloy having a composition of 2.6wt.% Li and 0.09wt.% Zr (Table 1) was cast by the Aluminum Company of America (ALCOA) at the ALCOA Research Center in

Pittsburgh, Pennsylvania and was used as the experimental alloy for this investigation. The complete composition analysis was also performed by ALCOA using optical emission spectrometric analysis. Due to the difficulty in reproducibility of casting several small ingots, one large ingot (2250 kg) was cast and rolled into a slab having dimensions of 30.5 cm x 96 cm x 30.5 cm. The slab was later heat treated in a gas-fired furnace for a total time of 20 hours. The first 8 hours were in a furnace temperature range of 482-500 °C and the last 12 hours in a furnace temperature range of 527-538 °C. Several billets were then machined from the heat-treated slab, having the dimensions of 15.25 cm in diameter and 25.4 cm or 50.8 cm in length, to be used for the extrusion processing of the alloy.

The aluminum-lithium-zirconium billets were extruded by the ALCOA Lafayette Extrusion and Tube Division after being reheated to two different temperatures of 466 °C and 290 °C. The billets were extruded using an instrumented 2500 ton press in the direct mode to produce 1.91 cm diameter round rods in the longitudinal grain direction. The extrusion ratio was 73:1 and the corresponding aspect ratio was 1:1 for the round rod geometry. An extrusion temperature of approximately 339 °C was used.

The Al-Li alloy was machined into samples from the extruded product. The samples were first solution heat treated for one hour at 550 °C in a molten sodium nitrate ($NaNO_3$) salt solution followed by a cold water quench to room temperature. Following the solution heat treatment, the samples were artificially aged for various lengths of time in a molten sodium nitrate salt bath. Different aging treatments were utilized by varying both the time and the temperature. The samples were aged at temperatures of 185 °C and 193 °C and at aging times of 15 minutes to 225 hours. The molten salt solution was continuously stirred throughout the solution heat treatment and aging process to insure a uniform temperature distribution throughout the bath. The samples were immediately quenched in water at room temperature, after the artificial aging treatment. The samples were then prepared and studied microstructurally using optical and transmission electron methods.

The subgrain structure, particle size distribution, and particle morphology were examined and photographed using transmission electron microscopy (TEM) from thin foil specimens obtained from samples aged at 185 °C for different aging times ranging from 24 to 225 hours. The thin foil specimens were sliced with a diamond blade saw cutter and then polished to foils of about 0.05 mm in thickness. Disks approximately 3 mm in diameter were then punched from the thin foils. The thin foil disks were then electropolished using a twin jet polisher, with the disks submerged in a 3:1 methanol-nitric acid solution cooled by liquid nitrogen to between -20 to -35 °C. The thin foil disks were observed and photographed using a JEOL-2OOCX microscope operating at 200 KV for various specimen inclinations.

Particle size measurements of the δ'(Al_3Li) and Al_3Li-Al_3Zr precipitates were made directly from TEM negatives. Centered dark field images were used since they gave good contrast between images of the δ'(Al_3Li) particles and the matrix phase. A semi-automatic EyeCom II image analyzing system was used to measure the particle sizes and their morphology. In order to perform a statistical analysis, particle size distributions were determined for each aging time by measuring the diameters of over 500 particles for each aging time. Particle size distributions were constructed for both the δ'(Al_3Li) and Al_3Li-Al_3Zr precipitates as well as the overall distributions containing both types of particles. Two diameters were measured for each particle in order to determine the aspect ratio of each particle. The aspect ratios were used to quantitatively describe the spherical morphology of the particle size distributions and any deviation from that morphology. For all of the experimentally measured particle size distributions, the coefficients of skewness, kurtosis and variation were determined in order to describe the particle size distributions. The first through the fourth moments about the mean particle size of the overall size distribution were used as the basis for calculating the statistical coefficients of skewness, kurtosis and variation of the distributions.

Specimens were flattened and mechanically polished. The sequence for mechanical polishing was as follows: (1) polishing with 240, 320, 400 and 600 grit SiC paper; (2) then polishing with 600 grit alumina, (3) followed by polishing with 0.5 μm alumina slurry and finally (4) polishing with 0.05 μm MgO slurry. The specimens were anodized and then observed under polarized light to reveal the grain structure. Anodization was performed in a solution containing 948 ml deionized H_2O, 55 ml HBF_4 and 7 grams H_3BO_3 (Boric Acid) for one minute at 18 volts and -32 °C. Optical micrographs with polarized light were taken of the partially recrystallized grain structure in the peak-aged condition after 48 hours aging time at 185 ^{0}C aging temperature.

PRECIPITATION RESPONSE FROM AGING

To develop a model for predicting the precipitation hardening response, the particle size distribution and the average particle size for a given aging time along the age hardening curve must be determined. For aluminum-lithium alloys, the precipitates coarsen according to the Lifshitz, Slyozov and Wagner (LSW) coarsening theory even at very small sizes (Glazer and Morris, 1988, Kulwicki and Sanders, 1986, Baumann and Williams, 1984). The average particle size can therefore be determined from the Lifshitz, Slyozov and Wagner (Lifshitz and Sloyozov, 1961, Wagner, 1961) cubic coarsening theory, which can be expressed by the following relation.

$$r^{-3} - r_o^3 = K_c t \tag{1}$$

where r_o is the initial particle size (at t = 0) and can be taken as approximately equal to zero for the Al-Li alloy used in this investigation, r is the average particle size radius of the size distribution of particle radius, and K_c is the particle coarsening rate constant. At small particle volume fractions, the particle coarsening growth rate constant is given by K_{co} which can be expressed by the following expression (Lifshitz and Sloyozov, 1961, Wagner, 1961).

$$K_{co} = \frac{8\gamma V_m^2 C_{eq} D}{9RT} \tag{2}$$

where T is the aging temperature, R is the universal gas constant, K_{co} is the rate constant, D is the diffusion coefficient, γ is the interfacial energy, C_{eq} is the equilibrium concentration of the solute in the matrix and V_m is the molar volume of the precipitate. It is known that this expression for K_{co} is valid for very small volume fractions, while supposedly more realistic rate constants are dependent on $f_{\delta'}$ i.e., the volume fraction of δ'(Al_3Li) (Huang, 1986, Ardell, 1972). There have been modifications (Ardell, 1972, Brailsford and Wynblatt, 1979, Shercliff and Ashby, 1989) to the LSW model given in equation 1 to take into account the volume fraction dependence of the coarsening rate. For example, Ardell (Ardell, 1972) modified the LSW theory to account for the high volume fraction effect on K_{co} in some alloys. The modified version of the LSW theory is referred to as the Modified LSW (MLSW) theory and can be expressed in terms of the rate constant as (Ardell, 1972):

$$K_{fv} = \frac{6\gamma V_m^2 C_{eq} D \rho_{fv}^{-3}}{RT \nu_{fv}} \quad (3)$$

$$\frac{\rho_{fv}^{-3}}{\nu_{fv}} = \frac{4K_{fv}}{27K_{co}} \quad (4)$$

where K_{fv} is the growth rate constant at a given volume fraction of f_v, K_{co} is the growth rate at very small volume fractions, $\rho = \bar{r} / r_c$ is the relative radius, ν_{fv} is a function related to volume fraction and a constant times dt / dr_c^3, where r_c is the critical particle radius of the polydisperse system (Ardell, 1972). When $f_v = 0$ then $\nu_{fv} = 27/4$ and equation 3 reduces to the LSW theory (equations 1 and 2) (Ardell, 1972, Gu 1985). The growth rate can also be expressed in terms of the activation energy for diffusion given by the following relation (Shercliff and Ashby, 1989):

$$K_c = \frac{C_k}{T} \exp\left[\frac{-Q_A}{RT}\right] \quad (5)$$

where Q_A is the activation energy, T is the aging temperature, C_k is the kinetic constant and R is the universal gas constant. The cubic coarsening expression can therefore be expressed as follows (Lifshits and Slyozov, 1961, Wagner 1961, Shercliff and Ashby, 1989):

$$r^{-3} - r_o^{-3} = \frac{C_k t}{T} \exp\left[\frac{-Q_A}{RT}\right] \quad (6)$$

Alternatively, the growth rate can be written in terms of the composition and aging practice for the Al-Li alloy used this investigation. Based on the microstructural model (Jha et al., 1987, Jha, 1987, Gu, 1985, Gu et al., 1985a, Gu et al., 1985b, Gu et al., 1986a, Gu et al., 1986b, Gu et al., 1986c, Jha et al., 1986) along with some data generated by some other investigators (Jensrud and Ryum, 1984, Williams and Edington, 1975), a simple empirical expression relating lithium content, aging temperature, and aging time may be determined as (Mahalingham et al., 1987):

$$\ln\{K_c T\} = \frac{b}{T} + c \quad (7)$$

where b and c can be expressed as functions of lithium content given by (Mahalingham et al., 1987)

$$b = -2545.73\ (\text{wt.\% Li}) - 4749.06 \quad (8)$$

$$c = 5.88(\text{wt.\% Li}) - 36.87 \quad (9)$$

These simple expressions permit the calculation of the particle size for the demonstration alloy based on the aging practice and composition parameters (Mahalingham et al., 1987).

For Al-Li alloys, the particle size distribution can be modelled with the Weibull distribution(Jha et al., 1987, Jha, 1987, Gu, 1985, Gu et al., 1985a, Gu et al., 1985b, Gu et al., 1986a, Gu et al., 1986b, Gu et al., 1986c, Jha et al., 1986, Mahalingham et al., 1987, Mahalingham 1989), which has the form given by

$$P(x) = ghX^{h-1} \exp(-gX^h) \quad (10)$$
$$\text{for } X > 0,\ g > 0 \text{ and } h > 0$$

and

$$P(x) = 0 \text{ elsewhere} \quad (11)$$

$$\int_0^\infty P(X)dX = 1 \quad (12)$$

where $P(X)$ is the Weibull probability density function, g and h are the Weibull parameters and X, which is the normalized particle size diameter, is the ratio of the ith particle diameter d_i to the average particle size diameter $\bar{d}$ for each determ(6)ed particle size distribution along the precipitation hardening curve. X is also equivalent to the ratio of the ith particle radius r_i to the average particle size radius $\bar{r}$ of each particle size distribution along the precipitation-hardening curve. For the Al-Li alloy used in this research, the values for the Weibull parameters can be determined from the empirical relations given by (Jha et al., 1987, Jha, 1987, Gu, 1985, Gu et al., 1985a, Gu et al.,

1985b, Gu et al., 1986a, Gu et al., 1986b, Gu et al., 1986c, Jha et al., 1986, Mahalingham et al., 1987, Mahalingham 1989):

$$g = 0.94 - 0.03(\text{wt.\% Li}) \tag{13}$$

$$h = 5.80 - 0.52(\text{wt.\% Li}) \tag{14}$$

where g and h are the Weibull distribution parameters given in terms of the weight percent lithium. These equations can also be expressed in terms of the volume fraction of precipitates given as:

$$g = 0.88 - 0.12 f_v \tag{15}$$

$$h = 4.69 - 2.17 f_v \tag{16}$$

where g and h are expressed in terms of the volume fraction (f_v) of the δ'(Al_3Li) precipitate phase in the matrix. These Weibull parameters (g and h) are most applicable for the heat treating of binary Al-Li alloys such as the alloy described in this research. Figure 1 shows the experimentally measured particle size distribution for the aluminum-lithium alloy aged at 96 hours. For aluminum-lithium alloys, which have ordered, coherent, spherical particles of uniform diameter, which are distributed in a triangular array on the slip plane, the interparticle spacing λ is given by the following expression (Dinsdale et al., 1981):

$$\lambda = 1.075/(N_A)^{0.5} - \pi d/4 \tag{17}$$

where N_A is the number of particles intersected per unit area of slip plane. In terms of the average particle size diameter (d) and volume fraction of the precipitates (f_v), interparticle spacing λ is given by (Dinsdale et al., 1981):

$$\lambda = 0.538d(2\pi/3f_v)^{0.5} - \pi d/4 \tag{18}$$

As the precipitates coarsen, their spacing increases concurrently with their size. Eventually they will reach a critical particle size where the dislocation looping of particles becomes easier than the dislocation shearing of particles. Table 3 shows the calculated values for the interparticle spacing based on equation (18) for the Al-Li-Zr alloy at all the various different aging times. The minimum radius for Orowan looping is achieved when the particle size reaches the point of equality in the force balance between the stress for precipitate shearing and Orowan looping. At the limit between these two processes, the strengthening for some alloy systems goes through a maximum. However, for some systems, the maximum in strength occurs prior to the particle shearing-looping transition. Before there is Orowan looping for all of the particles of the size distribution, there is a transition in which both Orowan looping and particle shearing simultaneously occur for a given distribution. The precipitate size distributions are not monodispersed because of the statistical distribution of particle sizes where the large precipitates are bypassed and looped by the dislocations, and where the small precipitates are sheared by the dislocations. The Orowan looping radius can be measured from the size of the smallest Orowan loops observed by TEM on plastically deformed alloys or approximated from the mean precipitate size of the δ'(Al_3Li) particle size distribution at maximum strength.

ANALYSIS OF THE PRECIPITATION RESPONSE

The Al-2.6wt.% Li-0.09wt.% Zr alloy utilized in this investigation was strengthened primarily by the δ'(Al_3Li) precipitates and to a small extent by some Al_3Zr-δ'(Al_3Li) precipitates, where the δ'(Al_3Li) phase coats the Al_3Zr precipitates. The δ'(Al_3Li) is known (Gayle and Van der Sande, 1984) to nucleate directly on the existing Al_3Zr particles. Both the δ'(Al_3Li) and a few Al_3Zr-δ'(Al_3Li) precipitates can be seen in the microstructure of the ternary Al-2.6wt.%Li-0.09wt.%Zr alloy shown by the TEM micrograph of Figure 2. This figure shows the micrograph of a representative TEM dark field image corresponding to the peak-aged condition of a 48 hour at 185 °C artificial aging treatment. Only a relatively small number of composite Al_3Zr-δ'(Al_3Li) precipitates exist in the microstructure of the Al-2.6wt.%Li-0.09wt.% Zr alloy compared to the large number of δ'(Al_3Li) precipitates in the microstructure. This is attributed to the small amount of zirconium (0.09wt.%) available in the alloy to form the Al_3Zr precipitates. Huang (Huang, 1986) and Huang and Ardell (Huang and Ardell, 1988a, Huang and Ardell, 1988b) had found this to be the case for two different Al-Li-Cu-Zr alloys that contained 0.12 wt.% Zr. Based on the kinetics of the δ'(Al_3Li) particle growth, the ductility and strength of the alloy can be related to the particle size distribution in terms of the dislocation particle shearing and the dislocation particle bypassing/looping interaction mechanisms.

As illustrated by Figure 3, the TEM micrograph revealed that the subgrain structure was generally not uniform. The subgrain structure consists of subgrains that lie inside the grains and are separated by sub-boundaries i.e., low angle boundaries. It was also determined from the TEM study that subgrain size had a negligible effect on the tensile properties. The elimination of the sub-boundaries is an important part of the recrystallization process.

The particle size distribution and particle morphology were obtained from samples aged at 185 °C for various aging times ranging from 24 to 225 hours. Centered dark field images the δ'(Al_3Li) particles and the matrix phase are shown in Figures 2 and 4. Figure 1 shows an example of the experimentally measured particle size distribution for a 96 hours aging practice, corresponding to the TEM micrograph of Figure 4. Table 2 summarizes the results of the particle size measurements. Table 3 summarizes the morphology of the δ'(Al_3Li) precipitates. In order to estimate any deviation from the spherical morphology, an aspect ratio of greater than 0.8 was arbitrarily chosen as indicative of spherical morphology. As shown in Table 3, an increased aging time resulted in a smaller percentage of δ'(Al_3Li) and Al_3Zr-free precipitates that deviated from an approximately spherical morphology. However, this trend was not as evident with the δ'(Al_3Li)/Al_3Zr particles. In addition, all of the aging times that were studied produced a very high percentage, greater than

80%, of spherical precipitates. Figure 5 shows the comparison between the experimentally measured average particle size radius and the experimentally measured interparticle spacing for the Al-2.6wt.%Li-0.09wt.%Zr alloy aged at 185°C. It can be seen from this figure that as the average δ'(Al_3Li) particle size increases so does the average δ'(Al_3Li) interparticle spacing also increases. Figure 6 shows the experimentally measured δ' particle interparticle spacing distribution for all δ' precipitates of the Al-2.6wt.%Li-0.09wt.%Zr alloy aged at 185 °C for 48 hours.

For all of the aging times studied the δ'(Al_3Li)/Al_3Zr particles were much larger in size than the δ'(Al_3Li) and Al_3Zr-free particles. The δ'(Al_3Li)/Al_3Zr particles were found to coexist (Figures 1 and 3) directly with the Al_3Zr-free and δ'(Al_3Li) particles in the matrix. As seen from Table 3, there were a very small number of the δ'(Al_3Li)/Al_3Zr precipitates in comparison to the δ'(Al_3Li) and Al_3Zr-free precipitates. The particle size distributions of the δ'(Al_3Li)/Al_3Zr precipitates were, however, more skewed toward larger particle sizes. These particles shifted the overall particle size distribution to the right of the central maximum and the overall particle size distribution was determined to be bimodal. However, the particle size distributions for the δ'(Al_3Li)/Al_3Zr precipitates were generally more narrow than those of the δ'(Al_3Li) particles and the calculated coefficients of variation for the former particles were smaller. In addition, the particle-size distributions for δ'(Al_3Li) particles were found to resemble the Weibull distribution.

The presence of the Al_3Zr phase was found to accelerate the aging kinetics of the alloy as seen from the coarsening rate constants given in Table 4, and act as preferential sites for the growth of the δ'(Al_3Li) phase as can be seen by Figure 2. Therefore, the small amount of zirconium (0.09wt.%) in the alloy resulted in a faster particle coarsening rate of the overall combined particle size distribution and thus causes the alloy to precipitation harden more rapidly. The particle coarsening rate, determined from the Lifshitz, Slyozov and Wagner (LSW) coarsening rate theory, for the composite δ'(Al_3Li)/Al_3Zr particles was more accelerated than that of the δ'(Al_3Li) precipitates. A linear regression analysis was performed on the experimentally measured average particle size versus aging time data to estimate the critical particle radius at the onset of coarsening, R_{co} and the growth rate constant K_c. These results are summarized in Table 4 and were determined by the kinetic law defined by equation (1) determined from the LSW theory given by equations (1) and (2) for the three different particle distribution types. The particle size distribution types were composite δ'(Al_3Li)/Al_3Zr particles, Al_3Li (free δ') precipitates only, and size distributions consisting of both composite Al_3Li/Al_3Zr particles and Al_3Li precipitates.

Polarized light optical micrographs of the Al-2.6Li-0.09Zr alloy in the solution heat treated and quenched condition, shown in Figures 7 and 8, were obtained from the as-extruded longitudinal and transverse grain structures. The longitudinal section, shown in Figure 7 shows that the grains are elongated with well-defined boundaries. The transverse section, shown in Figure 8, had grain boundaries that were much more irregular. The anisotropic grain structure was a result of the direct extrusion processing of the alloy from a billet and the mechanical properties were seen to vary with direction.

SUMMARY OF CONCLUSIONS

Quantitative particle size measurements were performed to determine the size, distribution, morphology and coarsening rate for both δ'(Al_3Li) and δ'(Al_3Li)/Al_3Zr precipitates. The average particle size, distribution, spacing and volume fraction of the intermetallic precipitates were related to the heat treating process which included the aging times and temperature.

The transmission electron microscopy revealed the presence of Al_3Zr-free, δ'(Al_3Li) and δ'(Al_3Li)/Al_3Zr particles that coexisted together in the α matrix phase. The Al_3Zr-free and δ'(Al_3Li) particles were found to have a much larger density than the δ'(Al_3Li)/Al_3Zr particles (see Figure 1). However, the δ'(Al_3Li)/Al_3Zr particles were much larger in size and had a much greater value for the growth rate constant (see Figure 1). The particle coarsening rate, determined from the Lifshitz, Slyozov and Wagner (LSW) coarsening rate theory for the δ'(Al_3Li)/Al_3Zr particles, was more accelerated than that of the δ'(Al_3Li) precipitates. Also, increasing the aging time resulted in a smaller percentage of δ'(Al_3Li) precipitates that deviated from an approximately spherical morphology, as shown by Figure 9. The presence of the Al_3Zr phase was found to accelerate the aging kinetics of the alloy. Therefore, a small amount of zirconium in the alloy resulted in a faster particle coarsening rate of the overall combined particle size distribution and thus led to more rapid precipitation aging more rapidly.

ACKNOWLEDGMENTS

This work was supported by the National Science Foundation under Grant CDR 8803017. The aluminum-lithium alloy used for this study was provided by the Aluminum Company of America (ALCOA) in Pittsburgh Pennsylvania, USA. The extrusion and composition analyses were performed at the ALCOA Lafayette Extrusion and Tube Division in Lafayette, Indiana, USA.

REFERENCES

A. J. Ardell, *Acta Metallurgica*, vol. 20, 1972, pp. 61-71.

S. F. Baumann and D. B. Williams, *Scripta Metallurgica*, vol. 18, 1984, pp. 611-616.

A. D. Brailsford and P. Wynblatt, *Acta Metallurgica*, vol. 27, 1979, pp. 489-497.

C. K. L. Davies, P. Nash and R. N. Stevens, *Acta Metallurgica*, vol. 28, 1980, pp. 179-189.

K. Dinsdale, S. J. Harris and B. Noble, Aluminum-Lithium Alloys, The Metallurgical Society of A.I.M.E., Philadelphia, PA, USA, 1981, p. 101.

M. Furukawa, Y. Miura, and M. Nemoto, Transactions of the Japan Institute of Metals, vol. 26, no. 4, 1985, pp. 225-229.

F. W. Gayle and J. B. Vander Sande, *Scripta Metallurgica*, vol. 18, 1984, pp. 473-478.

J. Glazer and J. W. Morris, *Acta Metallurgica*, vol. 36, no. 4, 1988, pp. 907-915.

B. P. Gu, "Coarsening Kinetics of δ'(Al_3Li) Precipitates in Al-Li Alloys," Ph.D. Dissertation, Purdue University, 1985.

B. P. Gu, S.C. Jha, G. L. Liedl, K. Mahalingam and T. H. Sanders, Jr., "Microstructure and the Behavior of Aluminum-Lithium Alloys," *Light Metals*, 1985a, pp. 35-43.

B. P. Gu, G. L. Liedl, T. H. Sanders, Jr. and K. Welpmann, *Material Science and Engineering*, vol. 76, 1985b, pp. 147-157.

B. P. Gu, K. Mahalingam, G. L. Liedl and T. H. Sanders, Jr., "The δ'(Al_3Li)Particle Size Distribution in a Variety of Al-Li Alloys," Aluminum-Lithium Alloys III, edited by C. Baker, P.J. Gregson, S.J. harris, and C.J. Peel, Proceedings of the Third International Aluminum-Lithium Conference, The Institute of Metals, London, 1986a, pp. 360-368.

B. P. Gu, G. L. Liedl, K. Mahalingam and T. H. Sanders, Jr., *Material Science and Engineering*, vol. 78, 1986b, pp. 71-85.

B. P. Gu, G. L. Liedl, K. Mahalingam and T. H. Sanders, Jr., "Coarsening Phenomenon in Aluminum-Lithium Alloys," Unusual Techniques and New Applications of Metallography, edited by R.J. Gray and L.R. Cornwell, ASM Metals Park, Ohio, vol. 24, 1986c.

J. C. Huang, "Microstructural Evolution and Strengthening Mechanisms in two Ternary Aluminum-Lithium-Copper Alloys," Ph.D. Dissertation, University of California, Los Angeles, 1986.

J. C. Huang and A. J. Ardell, *Material Science and Engineering*, vol. A104, 1988a, pp. 149-156.

J. C. Huang and A. J. Ardell, *Acta Metallurgica*, vol. 36, 1988b, pp. 2995-3006.

O. Jensrud and N. Ryum, *Materials Science and Engineering*, vol. 64, 1984, pp. 229-236.

S. C. Jha, "Precipitation Processes in Al-Li and Al-Li-Mg Alloys," Ph.D. Dissertation, Purdue University, 1987.

J. H. Kulwicki and T. H. Sanders, Jr., "Coarsening of δ'(Al_3Li) Precipitates in an Al-2.7Li-0.3Mn Alloy," Aluminum-Lithium Alloys II, edited by T.H. Sanders, Jr., and E.A. Starke, Jr., Proceeding of the Second International Aluminum-Lithium Conference, The Metallurgical Society of Mining and Metallurgical Engineers, Warrendale, PA, 1986, pp. 31-51

I. M. Lifshitz and V. V. Slyozov, *Journal Physical Chemical Solids*, vol. 19, 1961, pp. 35-50.

K. Mahalingam, B. P. Gu, G. L. Liedl and T. H. Sanders, Jr., *Acta Metallurgica*, vol. 35, 1987, pp. 483-498.

K. Mahalingam, "Precipitation Behavior of δ'(Al_3Li) in Binary Aluminum-Lithium Alloys," Ph.D. Dissertation, Purdue University, 1989.

H. R. Shercliff and M. F. Ashby, "Modelling of the Response of Heat-Treatable Aluminum Alloys to Thermal Processing", Report CUED/C-Mat./TRl 56, Cambridge University, Engineering Department, 1989.

C. Wagner, *Zeitschrift fur Elekrochemie*, vol. 3, no. 7, 1961, pp. 581-591.

D. B. Williams and J. W. Edington, *Metal Science Journal*, vol. 9, 1975, pp. 529-532.

Table 1.Composition analysis determined by optical emission spectrometric analysis for the Al-Li-Zr research alloy.

Al	Li	Zr	Cu	Mg	Si
bal	2.59	.09	.11	.07	.04

Fe	Ti	B	Na	Ca
.03	.01	<.001	<.001	<.001

Table 2: Average particle size diameter of different precipitate distributions.

Aging Time (hours)	Al_3Zr – free δ' particles (Al_3Li) (Angstroms)	composite particles (Al_3Li/Al_3Zr) (Angstroms)	All δ' particles (Al_3Li and Al_3Li/Al_3Zr) (Angstroms)
48	294	644	303
72	306	670	315
96	303	682	472
225	590	985	600

Table 3: Experimentally Measured Size ``and Morphology for Al_3Zr-free δ' precipitates.

Aging Time (hours)	Particles with Aspect Ratio ≤ 0.8	Percentage Of Total Measured	Average Particle Diameter Measured (Angstroms)	Average Particle Diameter Calculated From Model (Angstroms)
48	97	19.6	294	277
72	73	12.6	306	317
96	47	7.5	303	349
225	24	4.8	590	463

Table 5: Total number of precipitates analyzed for the different particle size distributions.

Aging Time (hours)	Number of TEM foils examined	Al_3Zr – free δ' particles (Al_3Li)	Composite particles (Al_3Li/Al_3Zr)	All δ' particles (Al_3Li and Al_3Li/Al_3Zr)
48	3	495	17	512
72	4	578	31	609
96	4	626	14	640
225	3	502	106	608

Table 4: Growth rate constant and critical radius at the onset of Coarsening from the LSW theory for the experimental particle Size data of the Al-2.6Li-0.09Zr alloy aged at 185°C.

TYPE OF PRECIPITATE	K $10^{-24}\left(\frac{cm^3}{sec}\right)$	R_{co} (Angstroms)
Al_3Zr – free δ' particles	32.8	-182.9
Composite particles (Al_3li-Al_3Zr)	143.6	95.6
All δ' particles (Al_3Zr and Al_3li-Al_3Zr)	35.1	-146.1

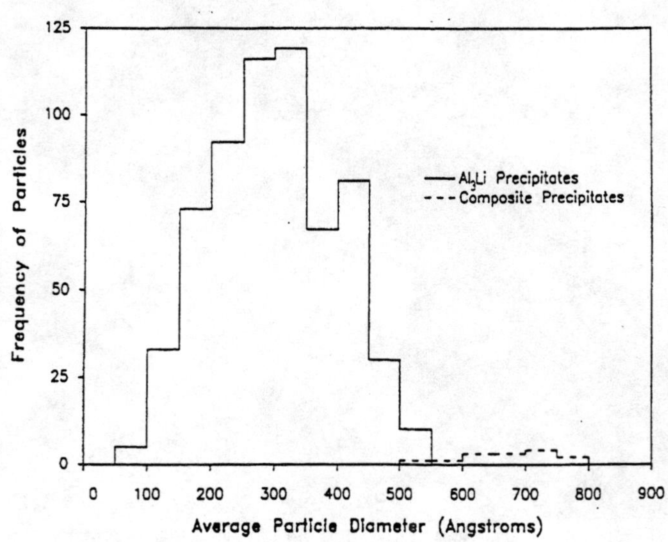

Figure 1: Histogram showing the experimentally measured Al_3Zr-free δ' particle size distribution and the Al_3Li/Al_3Zr composite particle size distribution for the Al-2.6wt.%Li-0.09wt.%Zr alloy aged at 185 °C for 96 hours.

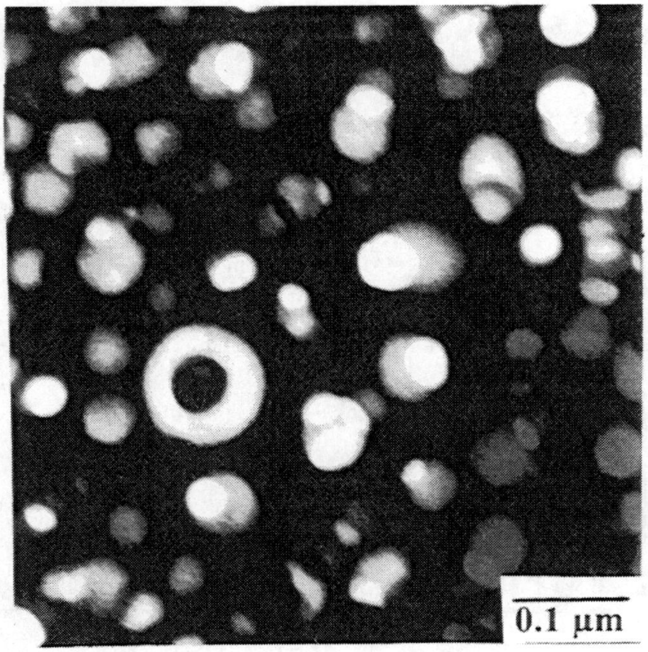

Figure 2. Dark field transmission electron micrograph showing the microstructure of the Al-2.6wt.%Li-0.09wt.%Zr alloy in the peak-aged condition, artificially aged at 185°C for 48 hours, (240000X), from TEM thin foil specimens.

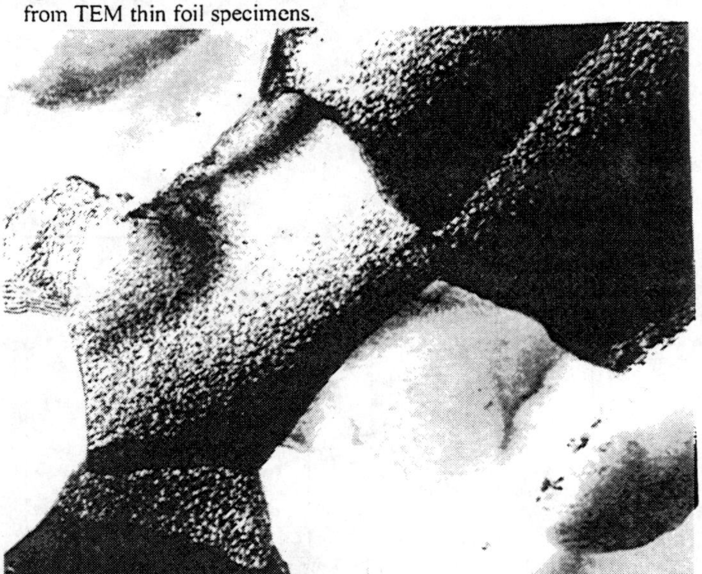

Figure 3. Dark field TEM micrograph showing the dislocation subgrain structure of the Al-2.6wt.%Li-0.09wt.%Zr Alloy in the peak-aged condition, artificially aged at 185°C for 48 hours, (12000X), from TEM thin foil specimens.

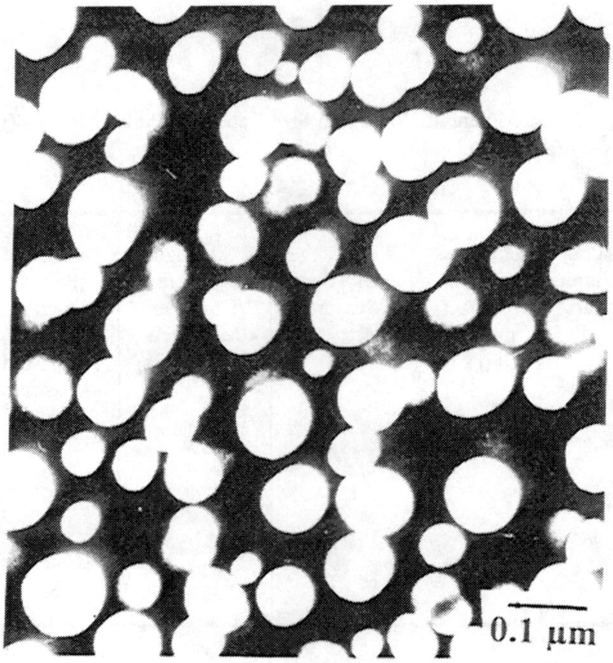

Figure 4: Transmission electron micrograph dark field image showing the microstructure of the Al-2.6wt.%Li-0.09wt.%Zr demonstration alloy in the overaged condition, aged at 185°C for 96 hours, (200,000X), from thin foil specimens.

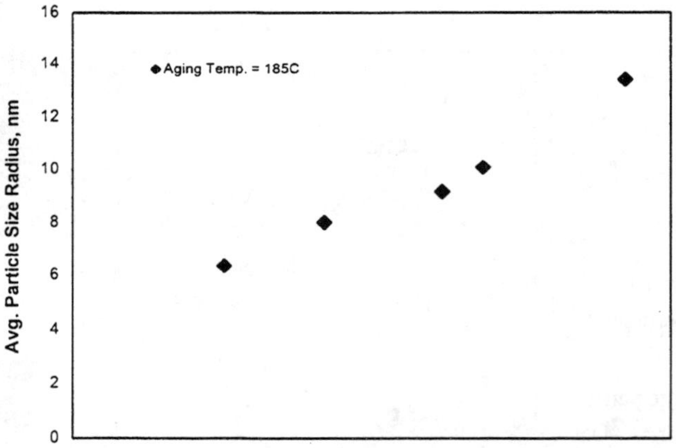

Figure 5: Comparison Between the Experimentally Measured Average δ'(Al_3Li) Particle Size Radius and the Experimentally Measured δ'(Al_3Li) Interparticle Spacing for the Al- 2.6wt.%Li-0.09wt.%Zr Alloy Aged at 185°C.

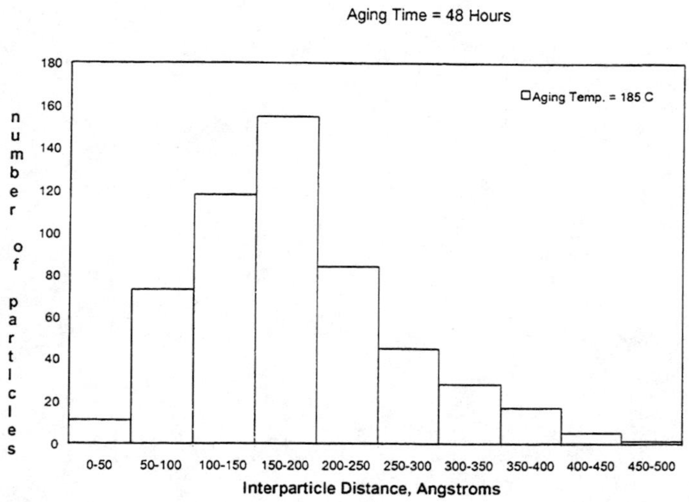

Figure 6: Histogram showing the experimentally measured δ' particle interparticle spacing distribution for all δ' precipitates of the Al-2.6wt.%Li-0.09wt.%Zr alloy aged at 185 °C for 48 hours.

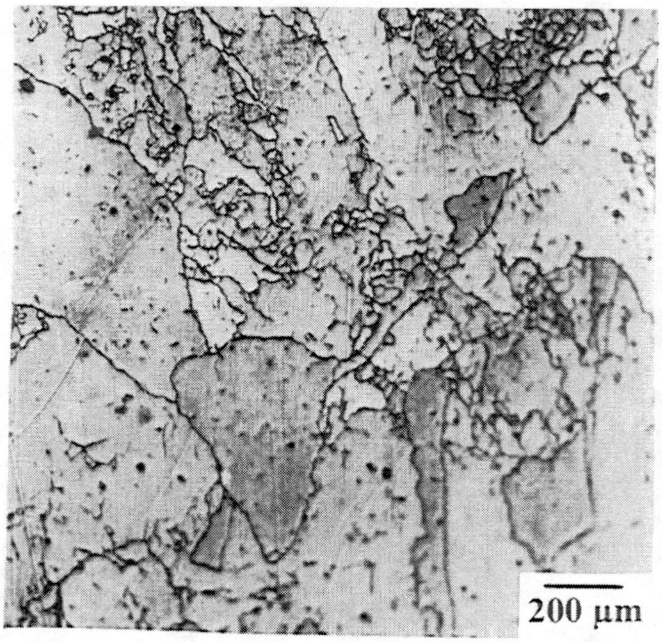

Figure 8: Light Optical Micrograph Showing the Grain Structure for the Solution Heat Treated Condition of the Al-2.6wt.%Li-0.09wt.%Zr Alloy Showing the Transverse Direction.

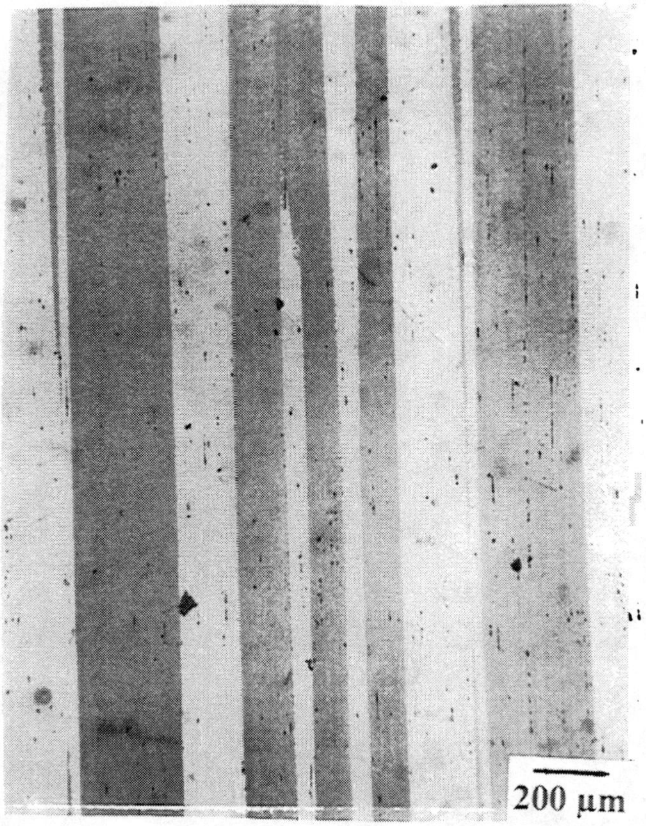

Figure 7: Light Optical Micrograph Showing the Grain Structure for the Solution Heat Treated Condition of the Al-2.6wt.%Li-0.09wt.%Zr Alloy Showing the Longitudinal Direction.

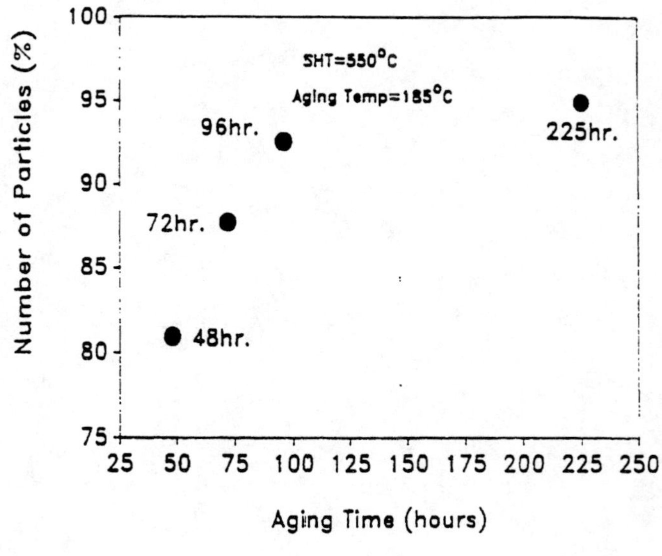

Figure 9: Percentage of Spherical Al_3Zr-free δ' Precipitates with Aging Time for the Al-2.6wt.%Li-0.09wt.%Zr Alloy.

PVP-Vol. 415, Recent Advances in Solids and Structures — 2000
ASME 2000

PROBABILISTIC FAILURE ANALYSIS FOR COMPOSITE LAMINATES

Paul D. Herrington
Paul J. Schilling
Department of Mechanical Engineering
University of New Orleans
New Orleans, LA 70148

ABSTRACT

Considering the increasing applications of composites as structural components, the reliability of these materials is an important issue. Given the inherent variability of the material properties of composites, a probabilistic approach to the design of these materials is required. In this paper, the first-order second-moment method, coupled with the Tsai-Wu interactive failure criterion, is used to determine the reliability of composite laminates under various plane stress conditions. Uncertainty in material and loading parameters are considered. The method considered requires only the first two moments of the basic variables as input. This factor, along with its ease of use, makes this technique attractive to designers. Numerical examples, along with plots of the effect of coefficient of variation on the safety index, are given for a typical glass/epoxy composite.

INTRODUCTION

Due to their lightweight, high strength, and high corrosion resistance, composite materials are increasingly being used in structural applications. However, due to the heterogeneous nature of these materials, significant scatter exists in their strength properties. This scatter in material properties and performance requires the use of probabilistic methods for the proper design of composite materials and structures. A number of studies have been undertaken to apply probabilistic and statistical techniques to polymer composite strength data. Among these studies, Hwang and Han (1987), and Tenn (1981), have characterized composite strength data, while studies concentrating on the reliability of composites include Cederbaum *et al.* (1990), Thomas and Wetherhold (1991, 1986), Liu and Mahadevan (2000), Murotsu and Miki (1994), and Gao (1993).

In this study, the reliability of composite laminates will be determined based on the first-order second-moment method. Variabilities in material properties and loading states are included. The limit state function developed is based on the Tsai-Wu interactive failure criterion. While this method is approximate in cases of nonlinear limit state functions and non-normally distributed variables, it does provide insight into the important effects of material and loading uncertainties. Numerical examples and plots of the reliability versus applied load are given for several composite laminates.

NOMENCLATURE

σ_1 = normal stress along principle axis.

σ_2 = normal stress along principle axis.

τ_{12} = shear stress along principle axis.

M = safety margin.

S_i = tensile strength parameter.

S_i' = compressive strength parameter.

F_{12} = interactive strength parameter.

$\overline{X}_i$ = mean values of the random variable X

$\hat{\sigma}_{X_i}$ = standard deviation of the random variable X

α_{X_i} = partial derivative of the random variable X

UR_i , = usage ratio

β = safety index

p_f = probability of failure

Φ = cumulative distribution function

N_x = axial load per unit length

N_y = transverse load per unit length

N_{xy} = shear load per unit length

RELIABILITY ANALYSIS

This analysis considers an orthotropic material in a state of plane stress, where 1 and 2 represent the fiber and transverse directions (principal axes). Stresses in the principal coordinate are related to the stresses in the reference axes $x-y$ by the transformation,

$$\begin{Bmatrix}\sigma_1\\ \sigma_2\\ \tau_{12}\end{Bmatrix}=\begin{bmatrix}\cos^2\theta & \sin^2\theta & 2\cos\theta\sin\theta\\ \sin^2\theta & \cos^2\theta & -2\cos\theta\sin\theta\\ -\cos\theta\sin\theta & \cos\theta\sin\theta & \cos^2\theta-\sin^2\theta\end{bmatrix}\begin{Bmatrix}\sigma_x\\ \sigma_y\\ \tau_{xy}\end{Bmatrix} \tag{1}$$

where θ is the angle between the applied stress and the principal axes, and σ_x, σ_y, and τ_{xy} are the applied stress components. The mean value of principal stress components are related to the mean applied stresses via the same transformation.

In order to calculate the reliability of a composite laminate, we must first consider a valid failure criterion. For composite materials, failure criteria are generally grouped into two categories, namely interactive and non-interactive. Given the non-homogeneous nature of composites, in general, the use of interactive criteria are preferred. A widely used interactive failure criterion is the tensor polynomial theory proposed by Tsai and Wu (1971). This criterion can be written in the form of a safety margin M as,

$$M=1-\left[\sigma_1\left(\frac{1}{S_1}-\frac{1}{S_1'}\right)+\frac{\sigma_1^2}{S_1S_1'}+\sigma_2\left(\frac{1}{S_2}-\frac{1}{S_2'}\right)+\frac{\sigma_2^2}{S_2S_2'}+\frac{\tau_{12}^2}{S_6'^2}+\frac{2F_{12}\sigma_1\sigma_2}{\sqrt{\left(S_1S_1'S_2S_2'\right)}}\right] \tag{2}$$

where S_1, S_2, and S_6 are the tensile strength components along the principal axes, and the prime denotes compressive strength components. F_{12} is the interaction parameter and is taken to be -0.5. Failure is assumed to occur when the safety margin is less than or equal to zero.

The first-order second-moment method is used to determine the reliability of the composite. This method approximates the mean and variance of M by expanding the function in a Taylor series about the means of the variables (Kapur and Lamberson, 1977). Higher order terms are truncated giving,

$$\overline{M}=f\left(\overline{X}_1,\overline{X}_2,\ldots\overline{X}_n\right) \tag{3}$$

where $\overline{X}_i$ represents the mean values of the random variables, which in this case are the five strength components and the three applied stresses. For the case of statistically independent basic variables, the first order approximation of the variance is given by,

$$\hat{\sigma}_M^2=\sum_{i=1}^{n}\left(\frac{\partial f}{\partial X_i}\right)^2\hat{\sigma}_{X_i}^2 \tag{4}$$

where $\hat{\sigma}_{X_i}$ is the standard deviation of the i-th basic variable. The partial derivatives of f are evaluated at the mean value of the parameters. Applying this technique to Equation 2 results in the mean given by,

$$\overline{M}=1-\left[\sum_{i=1}^{2}\left\{\overline{\sigma}_i\left(\frac{1}{\overline{S}_i}-\frac{1}{\overline{S}_i'}\right)+\frac{\overline{\sigma}_i^2}{\overline{S}_i\overline{S}_i'}\right\}+\frac{\overline{\tau}_{12}^2}{\overline{S}_6'^2}+\frac{2F_{12}\overline{\sigma}_1\overline{\sigma}_2}{\sqrt{\left(\overline{S}_1\overline{S}_1'\overline{S}_2\overline{S}_2'\right)}}\right] \tag{5}$$

where the overbar denotes mean values and the variance of M is given by the following equation,

$$\hat{\sigma}_M^2=\left(\alpha_{S_1}\hat{\sigma}_{S_1}\right)^2+\left(\alpha_{S_2}\hat{\sigma}_{S_2}\right)^2+\left(\alpha_{S_6}\hat{\sigma}_{S_6}\right)^2+\left(\alpha_{S_1'}\hat{\sigma}_{S_1'}\right)^2+\left(\alpha_{S_2'}\hat{\sigma}_{S_2'}\right)^2+\left(\alpha_{S_6'}\hat{\sigma}_{S_6'}\right)+\left(\alpha_{\sigma_1}\hat{\sigma}_{\sigma_1}\right)^2+\left(\alpha_{\sigma_2}\hat{\sigma}_{\sigma_2}\right)^2+\left(\alpha_{\tau_{12}}\hat{\sigma}_{\tau_{12}}\right)^2 \tag{6}$$

where α_{X_i} denotes the partial derivative of M with respect to X_i. Defining a usage ratio UR_i, which is a ratio of the applied stress to the ultimate strength as,

$$UR_i = \frac{\overline{\sigma}_i}{\overline{S}_i} \qquad i = 1,2,6$$

or,

$$UR_i' = \frac{\overline{\sigma}_i}{\overline{S}_i'} \qquad i = 1,2 \tag{7}$$

and letting,

$$\lambda = \frac{1}{\sqrt{\frac{S_1' S_2'}{S_1 S_2}}} \tag{8}$$

allows the partial derivatives α_{X_i} to be written as,

$$\alpha_{S_1} = UR_1\left(\frac{1}{S_1} + \frac{1}{S_1'}UR_1 + \lambda F_{12} S_1 UR_2\right) \tag{9}$$

$$\alpha_{S_2} = UR_2\left(\frac{1}{S_2} + \frac{1}{S_2'}UR_2 + \lambda F_{12} S_2 UR_2\right) \tag{10}$$

$$\alpha_{S_1'} = UR_1'\left(\frac{1}{S_1'} + \frac{1}{S_1}UR_1' + \frac{F_{12} S_1'}{\lambda} UR_2'\right) \tag{11}$$

$$\alpha_{S_2'} = UR_2'\left(\frac{1}{S_2'} + \frac{1}{S_2}UR_2' + \frac{F_{12} S_2'}{\lambda} UR_1'\right) \tag{12}$$

$$\alpha_{S_6} = \frac{2UR_6^2}{S_6} \tag{13}$$

$$\alpha_{\sigma_1} = \frac{1}{S_1} - \frac{1}{S_1'} + \frac{2}{S_1'}UR_1 + \frac{2F_{12}S_1}{\lambda}UR_2 \tag{14}$$

$$\alpha_{\sigma_2} = \frac{1}{S_2} - \frac{1}{S_2'} + \frac{2}{S_2'}UR_2 + \frac{2F_{12}S_2}{\lambda}UR_1 \tag{15}$$

$$\alpha_{\sigma_6} = \frac{2}{S_6}UR_6 . \tag{16}$$

Once the mean and variance of M have been obtained, a safety index β, which represents the reliability can be established as,

$$\beta = \frac{\overline{M}}{\hat{\sigma}_{\overline{M}}} \tag{17}$$

where $\overline{M}$ and $\hat{\sigma}_{\overline{M}}$ are given by Equations 5 and 6, respectively. The probability of failure p_f, is given by,

$$p_f = \text{Prob}(M < 0) = F_M(0) \tag{18}$$

where F_M is the cumulative distribution function of M. If M is assumed to be normally distributed, then the probability of failure is,

$$p_f = 1 - \Phi(\beta) \tag{19}$$

where Φ is the cumulative distribution function of the standard normal variate. The probability of failure is exact in cases where the distribution of M is normal and the limit state function is linear.

RESULTS AND DISCUSSION

The material used for the calculations is the well-known Scotchply® 1002 E-glass/epoxy. Material properties were assumed to be E_1 = 38.6 GPa (5.6×10^6 psi), E_2 = 8.27 GPa (1.2×10^6 psi), G_{12} = 4.14 GPa, (0.6×10^6 psi) and ν_{12} = 0.3. Two different laminates were evaluated, one oriented as $[0]_S$, and the other as $[0/90]_S$. The thickness of each laminate is 0.1 mm (3.94×10^{-3} in). Material strength properties and their corresponding standard deviation are given in Table 1.

Table 1. Material strength parameters (Gibson, 1987).

Strength Parameter	μ (MPa)	$\hat{\sigma}$ (MPa)
S_1	1103	110
S_1'	621	62
S_2	27.6	2.7
S_2'	138	13.8
S_6	82.7	8.27

Figure 1 shows the safety index versus the nondimensional ratio N_x/FPF, for the unidirectional $[0]_S$ laminate. Here, N_x is the laminate applied load per unit length, and FPF is the first ply failure load, as calculated by a deterministic failure analysis using the Tsai-Wu criterion. The three curves shown represent variables with coefficients of variation of 5%, 10%, and 15%, where the coefficient of variation is defined as the standard deviation divided by the mean. Each curve represents a global coefficient of variation for all of the present variables. For the case of Figure 1, the variables include the relevant strength and loading parameters. From Figure 1, it can be seen that as the variability between parameters is reduced, an increase in the safety factor, or in effect, the reliability is found. The normalizing factor in Figure 1 is a deterministic analysis of FPF. Here first ply failure would occur at 1103 MPa, and any applied load N_x less than this value should be considered safe. Figure 2 is plotted in the same form and shows the laminate under a transverse normal load N_y. In this case, the weaker matrix material dominates the laminate performance, hence to obtain the same reliability, the applied load must be decreased by approximately one order of magnitude. Figure 3 show the results for the [0]s laminate under the application of an in-plane shear load N_{xy}.

Figures 4, 5, and 6 show the safety factors obtained versus various applied loads for the cross ply laminate $[0/90]_S$. Again three curves are shown having coefficients of variation of 5%, 10%, and 15%. Figure 4 illustrates the results for the case of an applied load N_x. In this case, a deterministic analysis would yield a first ply failure of N_x = 138 MPa. Here again, the safety index curves are shown relative to first ply failure, and in this case would imply failure of the 90-degree ply. The remaining 0-degree plies can continue to support additional loading. The reliability of the cross ply laminate under an applied load N_y is given in Figure 5, and the results for an applied shear load N_{xy} is shown in Figure 6.

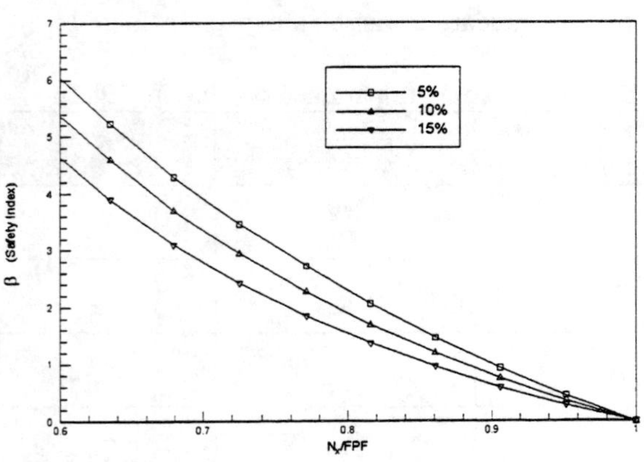

Figure 1. Safety index for the $[0]_S$ laminate under an applied load N_x.

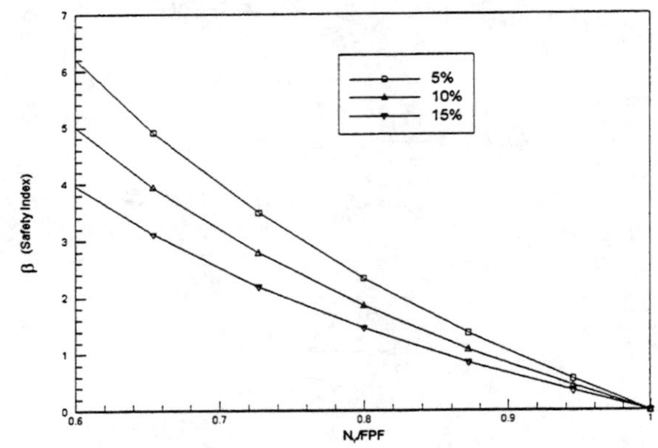

Figure 2. Safety index for the $[0]_S$ laminate under an applied load N_y.

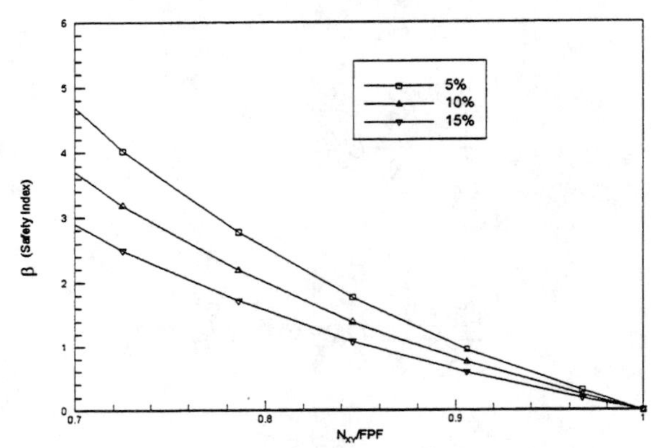

Figure 3. Safety index for the $[0]_S$ laminate under an applied load N_{xy}.

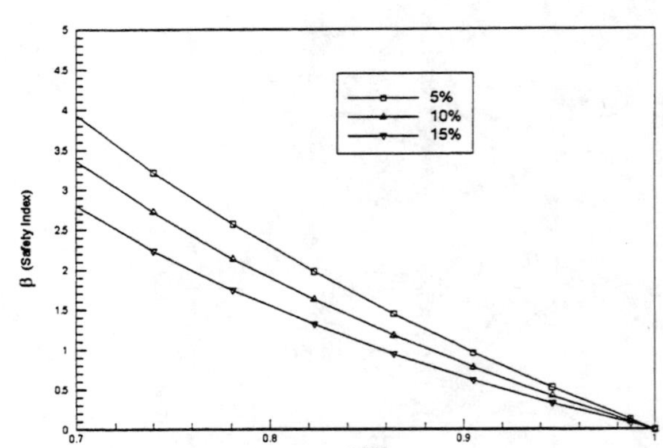

Figure 4. Safety index for the $[90]_S$ laminate under an applied load N_x.

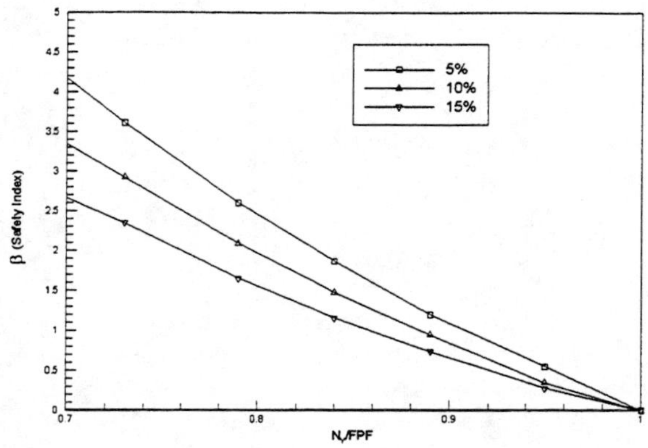

Figure 5. Safety index for the $[90]_S$ laminate under an applied load N_y.

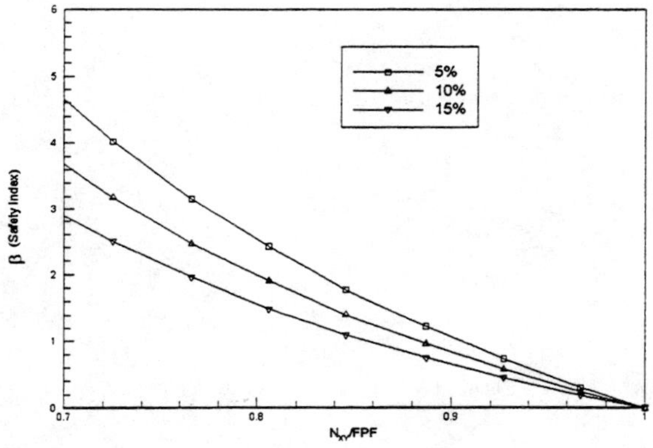

Figure 6. Safety index for the $[90]_S$ laminate under an applied load N_{xy}.

CONCLUSIONS

The first-order second-moment method has been outlined and applied to the interactive Tsai-Wu failure criterion. Examples using two glass/epoxy laminates are presented. Equations are developed that allows determination of the safety index for any type of composite laminate under plane stress conditions. While serving as an approximation in the case of nonlinear limit state functions and non-normally distributed variables, the method provides important insight into the effects of material and loading uncertainties. The method is especially attractive to designers since the input required is limited to only the first two moments of the basic variables. Seldom is the case where more information concerning additional moments and the distributional form of the variables is available.

REFERENCES

Cederbaum, G., Elishakoff, I., and Librescu, L., 1990, "Reliability of Laminated Plates via the First-Order Second-Moment Method," *Composite Structures*, Vol. 15, pp. 161-167.

Gao, Z., 1993, "Reliability of Composite Materials under General Plane Loadings," *Journal of Reinforced Plastics and Composites*, Vol. 12, pp. 430-456.

Gibson, R. F., 1994, "*Principles of Composite Material Mechanics*," McGraw-Hill, Inc.

Hwang, W., and Han, K. S., 1987, "Statistical Study of Strength and Fatigue Life of Composite Materials," *Composites*, Vol. 18, No. 1, pp. 47-53.

Kapur, K. C., and L. R. Lamberson, 1977, "*Reliability in Engineering Design*," J. Wiley & Sons, Inc.

Liu, X., and Mahadevan, S., 2000, "Ultimate Strength Failure Probability Estimation of Composite Structures," *Journal of Reinforced Plastics and Composites*, Vol. 19, No. 5, pp. 403-426.

Murotsu, Y. and Miki, M., 1994, "Reliability Design of Fiber Reinforced Composites," *Structural Safety*, Vol. 15, Nos. 1-2, pp. 35-49.

Tenn, L. F., 1981, "Statistical Analysis of Fibrous Composite Strength Data," *Test Methods and Design Allowables for Fibrous Composites*, ASTM STP 734, C. C. Chamis, Ed., pp. 229-244.

Thomas, D. J., and R. C. Wetherhold, 1991, "Reliability Analysis of Continuous Fiber Composite Laminates," *Composite Structures*, 17, pp. 277-293.

Tsai, S. W., and Wu, E. M., 1971, "A General Theory of Strength for Anisotropic Materials," *Journal of Composite Materials*, Vol. 5, pp. 55-80.

Wetherhold, R. C., 1986, "Statistical Distribution of Strength of Fiber-Reinforced Composite Materials," *Polymer Composites*, Vol. 7, No. 2, pp. 116-123.

PVP-Vol. 415, Recent Advances in Solids and Structures — 2000
ASME 2000

THE CONTOUR METHOD: SIMPLE 2-D MAPPING OF RESIDUAL STRESSES

Michael B. Prime
Engineering Sciences and Applications Division
Los Alamos National Laboratory
Los Alamos, NM 87545 USA
email: prime@lanl.gov

ABSTRACT

An entirely new method for measuring residual stress that is extremely simple to apply yet more powerful than existing techniques is presented. In this method, a part is carefully cut in two. The contour of the resulting new surface is measured, which gives the displacements caused by the release of the residual stresses. By Bueckner's superposition principle, analytically forcing the surface back to its original flat state gives the residual stresses that originally existed normal to the plane of the cut. The main advantage of this method is that the measured data can be used to solve directly for the stresses, whereas other methods require a complex inversion process.

INTRODUCTION

In relaxation methods, the residual stresses are determined from deformations measured after material removal. Analytical complexity generally limits such techniques to measurement of one-dimensional (1-D) stress variations. For example, in the hole drilling [1] and crack compliance methods [2] strains measured at incremental hole or slot depths, z, may be expressed as

$$\varepsilon(z) = \int_0^z A(z,Z)\sigma(Z)dZ\,, \qquad (1)$$

where A is a function of geometry and material properties that is determined using an extensive series of finite element calibrations [1]. The techniques used to *invert* this equation for the residual stress as a function of depth, $\sigma(z)$, are often prohibitively complex. Furthermore, the inversion quickly becomes unstable because measuring surface strains is insufficient to uniquely determine internal stresses.

It is even more difficult to measure complex 2-D or 3-D spatial variations of residual stress. Fundamentally, the analytical complexity occurs because the deformations are measured remote from the location of stress relief, e.g. on a pre-existing free surface.

The new relaxation technique presented in this work [3] can determine arbitrary, 2-D variations in residual stress *directly* from the measured deformations. The analysis to solve for the stresses from the measurements is exceedingly simple. These advances are possible because deformation is measured on the surface created by a cut, the location of stress relief, rather than on a pre-existing free surface.

The only common methods that can measure similar 2-D stress maps have significant limitations [4]. The neutron diffraction method is nondestructive but sensitive to microstructural changes [5], time consuming, and limited in maximum specimen size, about 50 mm, and minimum spatial resolution, about 1 mm. Sectioning methods [e.g., 6] are experimentally cumbersome, analytically complex, error prone, and have limited spatial resolution, about 1 cm.

THEORY

The contour method for measuring residual stresses is based on a variation of Bueckner's superposition principle [7]. Figure 1 presents an illustration in 2-D for simplicity, although the principle applies equally in 3-D. In **A**, one starts with the undisturbed part containing the residual stresses to be determined. In **B**, the part has been cut in two and has deformed because of the relieved residual stresses. In **C**, the free surface created by the cut is forced back to its original configuration. Superimposing the stress state in **B** with the change in stress from **C** gives the original residual stresses. This superposition principle assumes that the material behavior is linearly elastic and that the material removal process does not introduce significant stresses.

Proper application of this superposition principle allows one to experimentally determine the residual stresses along the plane of the cut. Experimentally, the contour of the free surface is measured after the cut. Analytically, the surface of a stress-free model is forced back to its original configuration as in step **C**. Because the stresses in **B** are unknown, one cannot obtain the original stresses throughout the body. However, the stresses normal to the free surfaces in **B** must be zero. Therefore, step **C** by itself will give the correct stresses along the plane of the cut.

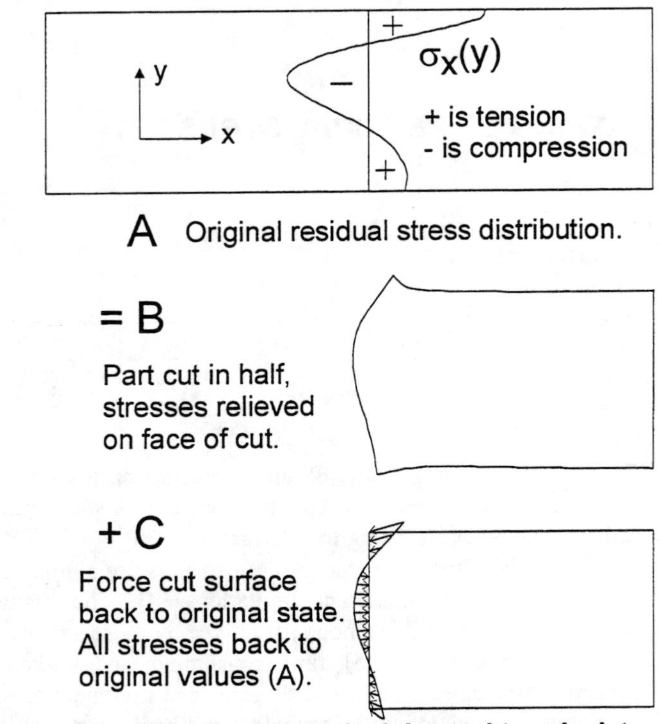

Figure 1. Superposition principle used to calculate original residual stress from measurement of surface contour after cutting a part in two.

There is generally no need to model the deformed shape of the body before analytically performing step C. A flat surface can be used as the starting point for this step because the deformations are quite small for engineering materials, and the analysis is linear. The results will be the same, and the analysis is simpler.

Experimentally, there is an arbitrary displacement in the contour measurement, i.e., the zero is arbitrary. There is also an arbitrary rotation in 2-D and two arbitrary rotations in 3-D. However, the arbitrary motions can be determined analytically by the need for the residual stresses to satisfy force and moment equilibrium. In fact, an FEM model used to solve for the stresses automatically accounts for the arbitrary motions, as will be demonstrated.

Measurement of the surface contour provides information about the displacements in the normal (x) direction only, not about those in the transverse (y) direction. Therefore, the analytical approximation of step **C** will force the surface back to its original configuration in the x-direction only, leaving the y-displacements unconstrained. If the residual shear stresses were zero along the plane of the cut, the approximation is exact: Poisson contractions will return the surface to its original y-position, and the stresses will be correct. In the general case, one assumes that the shear stresses on the cut plane are small. Similar assumptions are made by other proven measurement techniques, including hole drilling and sectioning methods [4]. Qualitatively, this assumption is reasonable because the shear stress magnitude is limited by the free surface condition, the 2-D local equilibrium condition,

$$\partial\sigma_x/\partial x + \partial\tau_{xy}/\partial y = 0, \qquad (2)$$

and the need to satisfy net cross-section equilibrium. Furthermore, if σ_x is uniform in the x-direction, which is a reasonable assumption for many situations, the shear stresses must be zero. The errors caused by non-zero shear stresses will be assessed by numerical simulation in the next section of this paper.

A few researchers have applied a similar superposition principle before (8,9,10). However, they measured 1-D *approximations* of the contour on *pre-existing* free surfaces. A 3-D FEM simulation (11) showed that large errors resulted from such simplifications.

NUMERICAL VERIFICATION

An FEM simulation demonstrated the validity of the contour method and quantified the errors caused by the possible presence of shear stresses. A 2 × 1 beam was modeled using the ABAQUS commercial finite element code and a 40 × 20 mesh of 8-noded, quadratic shape function, plane stress elements (CPS8). The material behavior was isotropic linear elastic with an elastic modulus of 1000 and Poisson's ratio of 0.3. Residual stresses were initialized using a user subroutine, and one analysis step was performed to ensure initial equilibrium. A second analysis step removed the elements on the right half of the beam to simulate separating the part.

The first simulation considered the beam having no shear stresses along the plane of the cut. The axial residual on the plane of the cut, where the beam thickness goes from $y = 0$ to 1, were given by a simple parabolic distribution that satisfied equilibrium:

$$\sigma_x(y) = 6y^2 - 6y + 1 \ . \qquad (3)$$

Figure 2 shows that applying the contour method to the beam simulation gave the correct results. To apply the superposition principle, a model of the undeformed and unstressed half of the beam was taken from the full mesh. The displacements of the cut surface from the initial FEM model were applied with opposite sign as constraints to the nodes along the cut surface. Applying only the x-displacements, as would be the case with experimental implementation of the contour method, gave the correct stresses along the cut plane. Applying both x- and y-displacements gave identical results because the y-constraints gave zero constraint forces and the Poisson contraction automatically resulted in the correct transverse displacements. Note that this simulation started with an undeformed body, rather than with the body deformed in the shape of the measured contour. As predicted in the Theory section, the correct results are still obtained.

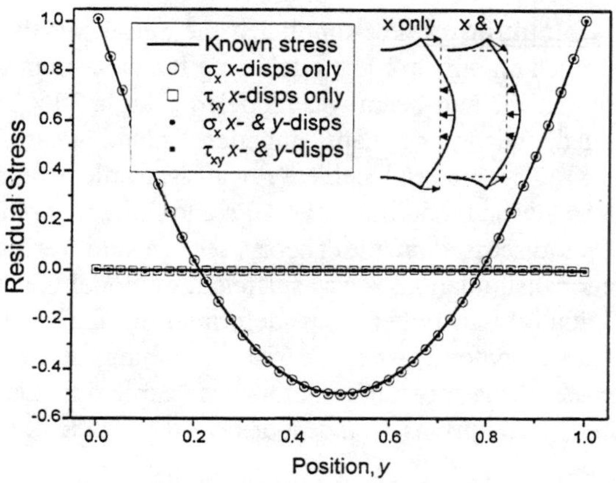

Figure 2. Simulation results, no shear stress.

The second simulation considered a beam with shear stresses along the plane of the cut. The axial stresses along the plane of the cut were still given by Eq. (3) but were varied in the x-direction to give $\partial\sigma_x(x,y)/\partial x = -\sigma_x$. Combined with the equilibrium condition, Eq. 2, and free surface conditions, this results in a shear stress distribution on the cut plane of

$$\tau_{xy}(y) = \left(2y^3 - 3y^2 + y\right). \quad (4)$$

Figure 3 shows the small errors that resulted from applying the contour method when shear stresses were present on the cut plane. Applying both the *x*- and *y*-displacements to the surface gave the correct results for both normal and shear stresses. However, experimentally we would only be able to determine the normal (*x*) displacements. Applying just the *x*-displacements gave a root-mean-square error in the σ_x distribution of 0.059, or 5.9% of the peak value. As expected, application of only *x*-displacements gave zero shear stresses.

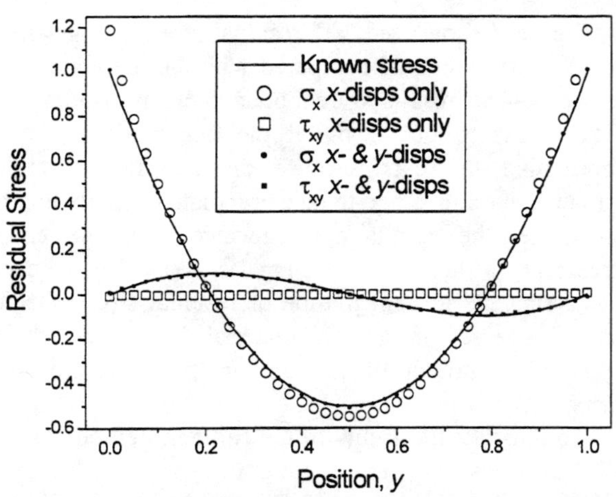

Figure 3. Simulation results, with shear stress.

The shear stress simulation is a pessimistic case, and yet the errors were quite small. Because of the free surface conditions and equilibrium condition [Eq (2)], the magnitude of $\partial\sigma_x(x,y)/\partial x$ determines the maximum value of shear stress. For our simulated beam, σ_x decays from its peak value to zero in about one beam thickness. By St. Venant's principle, one expects such changes near the ends of a body. However, the stresses can reasonably be expected to be much more constant in the central region of many bodies, resulting in lower shear stresses.

EXPERIMENTAL VALIDATION

Known Residual Stress Specimen. A plastically bent beam was carefully prepared in order to provide a specimen with a known residual stress profile. 43 mm square stock of forged 21Cr-6Ni-9Mn austenitic stainless steel was annealed at 1080°C for one hour and argon quenched. Next, the beam was machined to final shape with a 30 mm × 10 mm minimum cross section. Then the beam was thoroughly stress relieved by heating in a vacuum to 1080°C for 15 minutes and slow cooling at 100°C per hour. The beam was then plastically bent and unloaded in a carefully designed four-point bend fixture [12] to a maximum outer fiber strain of about 5670 με. Strain and load measurements during bending were used to calculate independent stress-strain curves for loading and unloading in both tension and compression. Finally, superposition of these curves gave the residual stress profile. The elastic modulus determined during these tests was 194 GPa.

Experiment. For the contour method, the ideal machining process for separating the part would make a precisely straight cut, would not remove any further material from already cut surfaces, and would not cause any plastic deformation. Wire electric discharge machining (wire EDM) is probably the choice closest to the ideal. In wire EDM, a wire is electrically charged with respect to the workpiece, and spark erosion causes material removal. The cutting is noncontact, whereas conventional machining causes localized plastic deformation from the large contact forces. The wire-control mechanisms can achieve positional precision of a fraction of a micrometer, especially for a straight cut. For this test, the beam was cut with a Mitsubishi SX-10 machine and a 100 μm diameter zinc-coated wire. "Skim cut" settings, which are normally used for better precision and a finer surface finish, were used because they also minimize any recast layer and cutting-induced stresses [13]. Including the overcut, the slot was about 140 μm wide.

The beam had to be fixed firmly enough to prevent unwanted movement during cutting, which required an unconventional clamping arrangement. Usually with wire EDM only one side of the workpiece is clamped. Figure 4 shows how the beam was clamped on both sides for this test. Before clamping, the beam and all the clamps were allowed to come to thermal equilibrium in the water tank to assure that no thermal stresses would arise during the cutting.

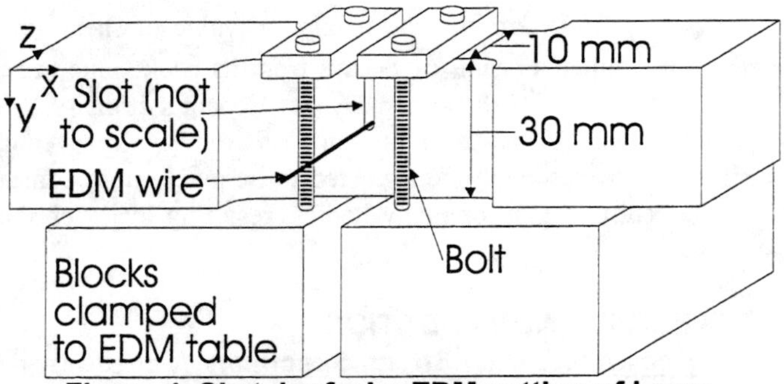

Figure 4. Sketch of wire EDM cutting of beam.

After unclamping the beam, the contour of the cut surface was measured using a Brown & Sharpe XCEL 765 coordinate measuring machine (CMM), which resides in a temperature and humidity controlled inspection laboratory. This CMM uses a touch trigger probe to register mechanical contact. An optoelectric system using glass scales gives the probe location, which is combined with machine coordinates to locate the surface. Because the CMM uses a probe tip with a finite radius, surface roughness and porosity are at least partially filtered out from the measured contour. A 4 mm diameter ruby tip was selected after trial measurements using tips with diameters from 1 mm to 8 mm revealed no significant measurement differences.

Figure 5 shows 1-D contours measured on the cut surfaces from both halves of the beam. The zeros are arbitrary, so the *x*-direction offset between the two curves is irrelevant. The measurements were taken in the center of the surface, and $y = 0$ represents the beginning of the cut. Additional traces taken off from the center showed very similar data. For both surfaces, positive x is taken in the outward normal direction; therefore, values that are more positive represent higher regions of the surface. 2-D measurements were also taken over the entire surface using the CMM and a 50 by 300 grid.

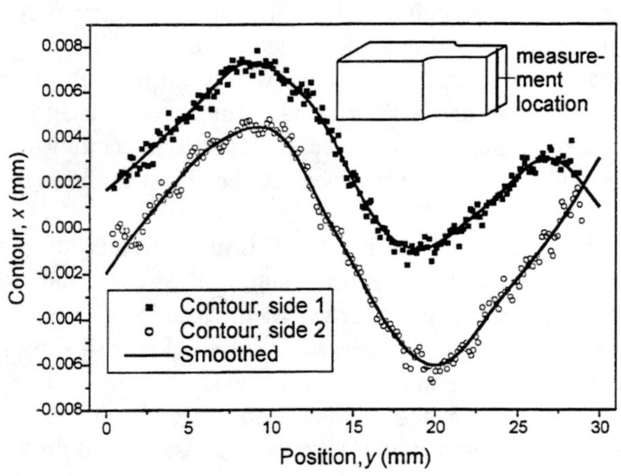

Figure 5. Surface contour measured on both halves of the beam.

Calculations. For simplicity, the initial calculations were performed on only a 2-D model using the measurements shown in Fig. 5. The half-beam was modeled using a 30 × 78 mesh of 8-noded, quadratic shape function, plane stress elements (CPS8). The material behavior was isotropic linearly elastic with an elastic modulus of 194 GPa and Poisson's ratio of 0.28. As was discussed in the Theory section and verified by the numerical simulation, it was sufficient to model the cut surface as initially flat rather than deformed in the shape of the measured contour. After careful smoothing of the data, the opposite of the measured contour was applied as displacement boundary conditions to the nodes on the surface representing the cut.

Figure 6 shows the exaggerated deformed shape of the FEM model. Only one constraint additional to the imposed contour was necessary to prevent rigid-body motions. As discussed previously in the Theory section, the FEM solution easily handled the arbitrary displacement and rotation in the measured contour. The movement of the free end of the beam illustrates the rotation necessary to satisfy moment equilibrium and the slight contraction to satisfy force equilibrium.

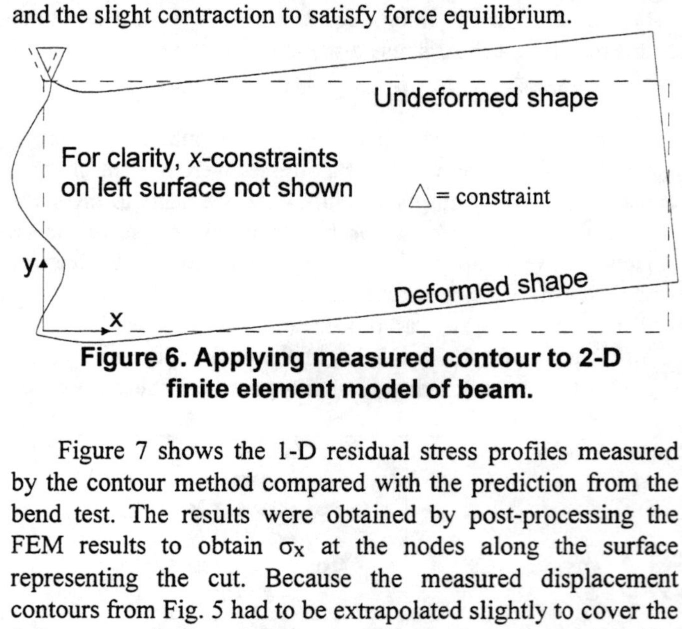

Figure 6. Applying measured contour to 2-D finite element model of beam.

Figure 7 shows the 1-D residual stress profiles measured by the contour method compared with the prediction from the bend test. The results were obtained by post-processing the FEM results to obtain σ_x at the nodes along the surface representing the cut. Because the measured displacement contours from Fig. 5 had to be extrapolated slightly to cover the full range, the results are truncated near the ends. The agreement with the prediction is very good, especially considering the low magnitude of residual stresses. Such low residual stresses in a stiff material like steel result in a decreased magnitude of the contour and, therefore, increased errors.

To illustrate the ability of the contour method to measure a full 2-D stress map, the contour measured on the entire cut surface was applied to a 3-D model. The half-beam was modeled in ABAQUS using 16,200 20-noded, quadratic shape function brick elements (C3D20). This mesh gave a 10 × 30 mesh of elements on the 10 mm × 30 mm cut surface. The measured surface contour data was fit to a bivariate Fourier series, averaged between sides one and two, and then applied as displacement boundary conditions in the FEM model. Figure 8 shows the deformed finite element model and the three additional displacement constraints used to prevent rigid body motions. As before, the movement of the free end of the beam illustrates the rotation necessary to satisfy moment equilibrium and the contraction to satisfy force equilibrium.

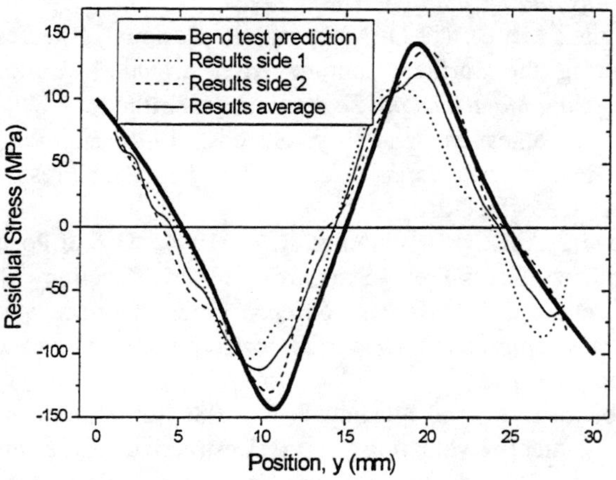

Figure 7. Results from applying 1-D contour to 2-D finite element model.

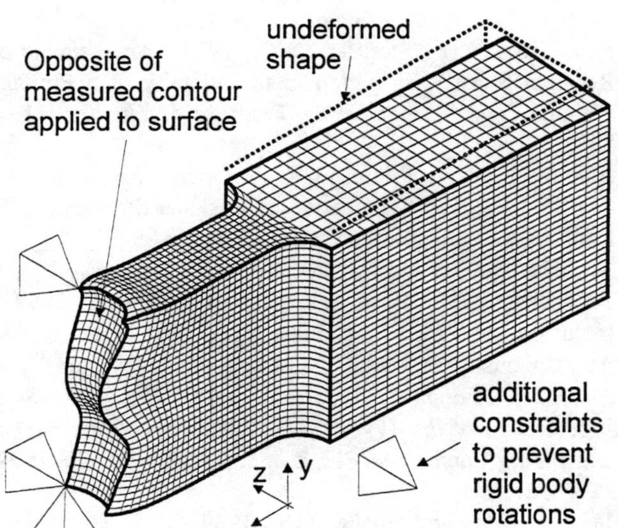

Figure 8. Applying 2-D measured surface contour to 3-D beam model.

Figure 9 shows the measured 2-D map of residual stresses compared to the bend test prediction. Because the surface contour measured on the CMM had to be extrapolated slightly to the edges of the surface, the contour lines are not plotted all the way to the edges. The agreement between measurement, Fig. 9b, and prediction, Fig. 9a, is good. However, the z-direction curvature in the measured contour, see Fig. 8, resulted in a small shift in the calculated stresses towards more compressive values along the center of the surface (z = 5). As will be discussed, that curvature results from using sub-optimal settings during the wire EDM cutting. Therefore, it was corrected for in a simple manner by fitting the curvature at y = 15 and then subtracting this curvature from the entire surface. Figure 9c shows the corrected results. The agreement with prediction is now very good, similar to the results in Fig. 7.

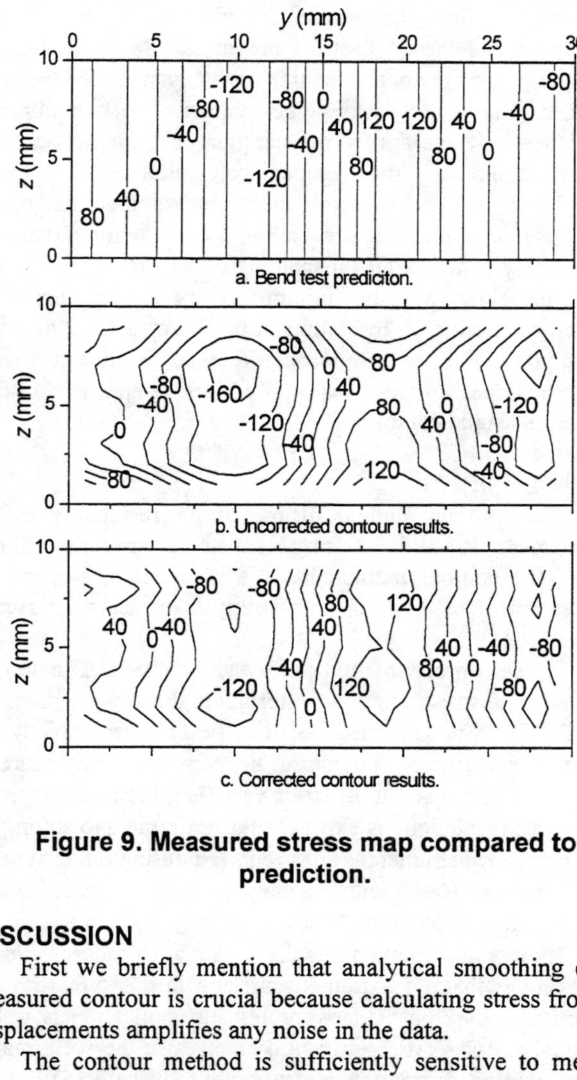

Figure 9. Measured stress map compared to prediction.

DISCUSSION

First we briefly mention that analytical smoothing of the measured contour is crucial because calculating stress from the displacements amplifies any noise in the data.

The contour method is sufficiently sensitive to measure residual stress maps of interest. The beam specimen in this study could be considered a sensitivity test because the stresses were less than 150 MPa and resulted in a contour with only about a 10 μm peak-to-peak magnitude. Residual stresses are often several times greater, the parts are also often larger, and many materials of interest have lowed elastic moduli, any of which would give larger contours. For example, some preliminary measurements on a 38-mm thick butt-welded steel plate gave a contour of about 120 μm peak-to-peak magnitude. However, larger residual stresses could also lead to yielding as the stresses are released during cutting. Future work will need to consider the effect of plasticity on the contour method.

Errors can be greatly reduced by correctly clamping the specimen during the cutting. The cut is assumed to occur along a flat plane in the original configuration. However, stress relief causes this plane to move slightly as the cut progresses. Such movement can be quantified using a simple FEM model. When the specimen is clamped on both sides of the cut, the resulting error in the stress distribution is less than 10%. In fact, this is the reason that the results in this paper, e.g., Fig. 7, underestimate the peak stress magnitude. More sophisticated clamping arrangements could probably reduce the errors further. Initial simulations suggest that errors can be minimized in general by measuring the contour on both halves of any specimen and using the average in calculations.

The z-direction curvature in the measured surface contour, see Fig. 6, appears to be caused by a common source of inaccuracy in wire EDM machining called "barreling" (13 14). The barreling can be minimized by optimizing cutting parameters, such as by cutting more slowly. In addition, the barreling will have a reduced influence on measurement of larger residual stresses because it will make up a lesser portion of the measured contour.

CONCLUSIONS

The contour method of measuring residual stress was experimentally validated using a bent beam specimen. In many ways, the contour method has the potential to surpass other measurement methods in both ability to measure stresses and ease of use:

1. The contour method can measure a full 2-D map of the residual stress component normal to the cross section.
2. The residual stress map can be obtained directly from the measured contour; no inverse procedure or assumptions about the stress variations are necessary.
3. The technique is experimentally simple. No strain gages or other instrumentation is required during the testing. The necessary equipment is widely available in machine shops and inspection laboratories.

The contour method has the potential to measure residual stress maps that are extremely difficult to measure with other techniques. One particularly promising application is welding residual stresses. Microstructural changes in the weld material make neutron diffraction measurements difficult (5) but have relatively small effects on the elastic constants that would affect the contour method. Other exciting applications include parts with geometrically complex cross sections, like railroad rails, forgings, I-beams, extrusions, and castings.

ACKNOWLEDGMENTS

This work was performed at Los Alamos National Laboratory, operated by the University of California for the U. S. Department of Energy under contact number W-7405-ENG-36.

REFERENCES

1. Schajer, G. S., 1988, "Measurement of Non-Uniform Residual Stresses Using the Hole-Drilling Method," *Journal of Engineering Materials and Technology*, **110**, 338-349.

2. Cheng, W., Finnie, I., 1986, *Journal of Engineering Materials and Technology,* **108**, 87–92.

3. Prime, M. B., "2-D Mapping of Residual Stresses by Measuring the Surface Contour After a. Cut," *Journal of Engineering Materials and Technology* (submitted).

4. Lu, J., James, M., and Roy, G., eds., 1996, *Handbook of Measurement of Residual Stresses*, The Fairmont Press, Inc., Lilburn, Georgia, USA.

5. Krawitz, A. D., and Winholtz, R. A., 1994, "Use of Position-Dependent Stress-Free Standards for Diffraction Stress Measurements," *Materials Science and Engineering A-Structural Materials Properties Microstructure and Processing*, **185**, pp. 123–130.

6. Rybicki, E. F., and Shadley, J. R., 1986, "A 3-Dimensional Finite-Element Evaluation of a Destructive Experimental Method for Determining Through-Thickness Residual Stresses in Girth Welded Pipes," ," *Journal of Engineering Materials and Technology*, **108**, 99–106.

7. Bueckner, H. F., 1958, "The Propagation of Cracks and the Energy of Elastic Deformation," *Transactions of the American Society of Mechanical Engineers*, **80**, 1225–1230.

8. Williams, J. F., Stouffer, D. C., 1979, "An Estimate of the Residual Stress Distribution in the Vicinity of a Propagating Fatigue Crack," *Engineering Fracture Mechanics*, **11**, 547–557.

9. Johnson, M. R., Robinson, R. R., Opinsky, A. J., Joerms, M. W., Stone, D. H., "Calculation of Residual Stresses in Wheels From Saw Cut Displacement Data," paper 85-WA/RT-17, The American Society of Mechanical Engineers (1985).

10. Dickson, T. L., Bass, B. R., and McAfee, W. J., 1998, "The Inclusion of Weld Residual Stress in Fracture Margin Assessments of Embrittled Nuclear Reactor Pressure Vessels," *PVP-373, Fatigue, Fracture, and Residual Stresses, Proceedings of the 1998 ASME/JSME Joint Pressure Vessels and Piping Conference,* San Diego, CA, July 26–20, 1998, pp. 387–395.

11. Lin, K.Y., and Huang, J. S., 1989, "Analysis of Residual Stresses in Railroad Car Wheels Based on Destructive Test Measurements," *Theoretical and Applied Fracture Mechanics*, **12**, 73–86.

12. Mayville, R. A., and Finnie, I., 1982, "Uniaxial Stress-Strain Curves from a Bending Test," *Experimental Mechanics*, Vol. 22, pp. 197–201.

13. Cheng, W., Finnie, I., Gremaud, M., and Prime, M. B., 1994, "Measurement of Near Surface Residual Stresses Using

Electric Discharge Wire Machining," *Journal of Engineering Materials and Technology*, Vol. 116, pp. 1–7.
14. Sommer, C. and Sommer, S., 1997, *Wire EDM Handbook*, Advance Publishing, Inc., Houston, Texas, USA.

PVP-Vol. 415, Recent Advances in Solids and Structures — 2000
ASME 2000

PREDICTION FOR POST-FORMING SPRING BACK OF CICC IN THE SUPERCONDUCTING TOKAMAK FUSION DEVICE VIA VIRTUAL MANUFACTURING

Yeong Sung Suh, Jong Sung Ahn, and Sun Woong Choi
Department of Mechanical Engineering
Hannam University
Taejon, Republic of Korea

Hyun Ki Park, Yong Jin Kim, and Kee Man Kim
Energy Laboratory
Samsung Advanced Institute of Technology
Taejon, Republic of Korea

ABSTRACT

To construct the CICC for the superconducting Tokamak fusion device, the 3-roll bending, that inherently has a difficulty to form the coil with accurate radius of curvature, is used for continuous winding. In order to obtain precise dimension, a trial-an-error operation is inevitable. To reduce the effort of tryout, a relation between travel of the bending roller and spring back displacement was obtained via virtual manufacturing. The radius of CICC after forming was expressed as a function of the bend-roll travel. Next, the variation of the CICC cross-section (reduction of the conduit cross-section) was investigated during the first turn and during conduit bending with largest curvature. With largest curvature, the cross-sectional area was not much reduced. Finally, the residual stress on the CICC before roll bending was measured in order to examine the influence of the original residual stress on the final deformation behavior. The principal stress and von Mises stress were measured at the surface of CICC using specially designed strain gauge. The measured values were considered in the virtual forming. The results indicate that the residual stresses generated during the fabrication of the CICC (before coiling) do not have much influence on the final stress state.

INTRODUCTION

The CICC(Cable-In-Conduit Conductor) is a central solenoid model coil used in the superconducting Tokamak fusion device. The KSTAR magnet system configuration is shown in Fig. 1. Several coil configuration used in device is shown in Figs. 2 and 3. To construct the CICC, 3-roll bending machine displayed in Fig. 4 is used in order to wind the coil continuously. The 3-roll bending is very simple to manipulate to form a straight pipe into a hoop. But, because of the spring back, it inherently has a difficulty to form the coil with accurate radius of curvature. In order to obtain precise dimension, a trial-an-error operation is inevitable. With a rapid progress of computer and software technology, virtual manufacturing can drastically reduce this difficulty. The virtual manufacturing involves the use of a computer to simulate not only a product but also the manufacturing processes. Usually, virtual manufacturing utilizes nonlinear finite element analysis to provide optimization of factors such as manufacturability, final shape, residual stress, and life-cycle estimations. In this work, the CICC was virtually manufactured by simulating the three-roll bending process with ABAQUS so that the final geometry of CICC can be accurately obtained with a minimum effort. Fig. 5 is a schematic of the three-roll bending process. As the drive rollers push the CICC, it proceeds in the forming direction. With the bend roller the CICC is formed into a circle with a constant radius of curvature. There are various sizes of CICC in the Tokamac device, the radius of which ranges from 431mm(PF1-4) to 3850mm(PF7). The downward travel of bending roller vs. final radius of CICC after spring back was monitored. Two different materials, Incoloy 908 and SUS 304 were employed in the calculation. In addition, the cross-section with the minimum radius of curvature that may encounter in the real manufacturing was examined in order to check the state of area reduction. If the reduction is too high, the strand that will be packed inside the conduit may be damaged. It is regulated that the area of reduction may not be greater than 5%[1]. Also, in order to predict the possibility of SAGBO in the conduit, the maximum von Mises stress was also monitored. The SAGBO, an abbreviation of Stress Accelerated Grain Boundary Oxidation, occurs when the following three conditions satisfy simultaneously [1] :

Temperature: 550°C~800°C
Tensile strength: Greater than 200 MPa
Oxygen pressure: Greater than 10^{-5} torr (at 1 atm, 0.14ppm)

The superconducting strand is jacketed with a square tube before it is formed by three-roll bending process. First, a strip is formed into circular tube with multi-rolling process and, after strands are inserted, the tube is closed by seam welding. Next, the cross-section of circular tube is reduced by 8%. Finally, the tube cross-section is changed into square shape. Because of this sequence of large plastic deformation, the final tube has inherent residual stresses that may significantly affect subsequent plastic deformation. In this work, the residual stress of the square tube was experimentally measured. With this data, the influence of residual stress on the final spring back deformation was investigated.

APPROACH

Virtual Manufacturing

As a virtual manufacturing engine, ABAQUS, a commercial non-linear FEM package was used on Compaq-Digital workstation 433AU and SGI Origin2000. To reduce cpu time, only the longitudinal half of the entire configuration was modeled as shown in Fig. 6. The shell element was preferred to the brick element since the former is less stiff, accommodating shear deformation through thickness due to bending. Therefore, S4R element was used to model the CICC. Five integration points were assigned in the thickness direction to take account of an appropriate shear deformation during bending. For rollers, rigid definition was used using *RIGID SURFACE option which defines rigid surface analytically. As for the surface condition, the friction was assumed only between driving rollers and the CICC. Fig. 6 shows the schematic that shows boundary and surface conditions. The main computing step is as follows:

STEP 1. A point load (10kN) on the center of the upper roller is applied. The lower roller was fixed at this time.

STEP 2. The bending roller starts to move downward to bend the CICC. At a given displacement of the roller, it stops.

STEP 3. The upper driving roller rotates counterclockwise and lower one rotates clockwise so that the CICC may proceed to the forwarding direction.

STEP 4. The upper master and bending rollers are removed so that spring back deformation is allowed.

After spring back, the coordinate of three-points of the CICC was monitored. These numbers were put into the Intelli-CAD (a commercial 2D CAD program) and the radius of the virtually manufactured CICC was automatically calculated. The relation between the travel of the bending roller vs. radius of CICC was obtained by fitting the calculated data. The cross-sectional area and the von Mises stress at the deformed CICC with minimum radius of curvature were monitored. The mechanical properties of two materials used are included in the Table 1.

Residual Stress Measurement

The hole-drilling strain gage method of stress relaxation was used[(3), (4), (5)]. The introduction of a small hole into a residually stressed specimen relaxes the stresses at that location. This takes place because every perpendicular to a free surface (hole surface, in this case) is necessarily a principal axis on which the shear and normal stresses are zero. The elimination of these stresses on the hole-surface changes the stress in the immediately surrounding region, causing the local strains on the surface of the specimen to change correspondingly. There are two drilling methods: blind-hole drilling and through-hole drilling. The former, which is generally used, measures stress changes as the hole depth increases and generally used. Meanwhile, the latter, although it is well established mathematically, is confined to use for the thin sheet. For both methods, however, the accurate relation between relaxed strain and stress should be established. In this work, the assumption that there are no shear stresses along the direction of hole depth was made, since the CICC wall thickness is just 2.6mm.

Fig. 7 shows the whole system for the residual stress measurement. The strain gage used was Tokyo Sokki Kenkyujo FRS-3-11. To drill the hole, 1.5mm end mill was used. In order for the pipe specimen not to be deformed during drilling of the hole, the fixture as shown in Fig 8 was devised. The CICC was cut by an appropriate size such that the length was determined as more than ten times of hole-diameter to minimize edge and end effects. Four tests have been carried out by attaching strain gage on four different locations: welding, wall, opposite-to-welding and fillet surfaces.

RESULTS AND DISCUSSIONS

Initial tuning

Most of the initial computation was performed with Incoloy 908. After obtaining the spring back data, the case with SUS304 was added. Therefore, most of this section includes the computation with Incoloy 908 unless otherwise specified. The 3-dimensional FEM modeling is presented in Fig. 9. Stepwise-deformed shape of CICC with downward bending roller travel of 250mm is presented in Fig. 10. The driving rollers were set to rotate with 4.02rad in order for the CICC precedes 450 mm. Since the driving roller is the only pushing engine, it was necessary to check if the pipe proceeds without slipping during forming. In order for the CICC move forward without slipping, the total friction force between driving rollers and surfaces of upper and lower CICC should be greater than the horizontal reaction force at the bending roller including other reaction forces at the master rollers.

Vertical displacement of the bending roller vs. Final radius of CICC after spring back

The selected distance of bending roller was 50, 75, 100, 150, and 200 mm for SUS304 and 50, 75, 100, 125, 150, 180, 200, 220, 240, and 250 mm. On Compaq-Digital workstation, it took more than 15 hours for one run.

In order to predict the radius of curvature closer to real PF coils, the monitoring points should be taken after nodes pass the bending roller where continuous spring back takes place without arbitrary pulling back of the roller. In this region, the contour of von Mises and longitudinal stresses are very consistent and the magnitude of stresses did not change much when the bending roller was pulled back, as shown in Figs. 11-12. To compute the radius, however, the region after the bending roller was too narrow to pick three monitoring points. In addition, if the points are too close, the local radius may not be consistent. Therefore, the three-monitoring points were selected across wider arc length, which includes the nodes before the bending roller. Sacrificing the accuracy of the prediction was inevitable because of inconsistent stress distribution in that region. To predict the radius more accurately, longer CICC has to be modeled. However, in order to carry out parametric study as with the present work, which requires many runs of various cases, current modeling seemed to be fairly efficient. Fig. 13 shows the predicted radius of formed CICC after spring back vs. distance of the bending roller for both Incoloy 908 and SUS304. The calculation of radius was based on the three-nodes near the bending roller. The data shows that in order to get the radius corresponding to PF1 coil (431mm), it will be very difficult to form with the present layout configuration. It is suggested that the bending roller should move towards master roller. By doing so, the smaller radius of curvature can be obtained with ease. In any way, present work has shown that the data from the virtual manufacturing can be beneficial in reducing manufacturing time and human efforts by providing useful figures for initial work. The maximum von Mises on the unloaded and curved CICC was approximately 500~600MPa and the maximum tensile stress was 500MPa, which is greater than the SAGBO criteria, see Figs. 14 and 12

Thickness variation of the CICC

Thickness variation was negligible during forming. Fig. 15 shows the thickness contour when the travel distance of the bending roller was 250mm (Material: Incoloy 908). The thickness varied from 2.40mm to 2.59mm (original thickness was 2.5mm)

Cross-sectional area of CICC with minimum radius

The cross-sectional area of CICC with minimum radius was examined since the area reduction during forming may damage the strands inside the pipe. The report indicates that the area reduction should not be greater than 5%. The CICC with minimum radius was observed to be the location where the pipe is bent with nearly right angle, see Fig. 2 (a). The minimum radius was assumed to be 250mm.

To check the area reduced from severe condition, the FEM modeling was modified so that tighter radii can be obtained, see Fig. 16. The radius of master roller was increased to 200mm to accommodate 250mm-radius of CICC after spring back. The monitored nodes were selected near the lower master roller. The travel of the bending roller was 250mm. With Incoloy 908, the radii before and after the spring back were 197.7mm and 234.3 mm. Therefore, the section on the arc radius 197.7mm (*i.e.*, before spring back) was examined. In order to calculate the sectional area that consists of 21 nodes, a program was specially written. The area here is defined as the one surrounded by the perimeter of centerline through the thickness direction. The reduced area was 375.6mm^2, that is 0.63% reduction with respect to the original area, 378mm^2. Therefore, the area reduction during forming may be neglected. Similar results were obtained with SUS304. In Fig. 17, contour of von Mises stress near the region with tight radius (234.3mm) after spring back is presented. The maximum residual von Mises stress was 546MPa and the maximum tensile stress was 487MPa.

Measurement of Residual Stresses

From the depth of 0mm(before drilling) to 2.6mm(after completion) the strain was measured at every increase of 0.2mm. Table 2 includes measured principal stresses and their direction with the schematic of strain gage arrangements and locations. Table 3 includes von Mises stress from each measured surface. The results show that the magnitude of von Mises stress is greater than the minimum bound of SAGBO criteria in most measured locations.

Overall, the results indicate that the residual stresses generated during the fabrication of the CICC cannot be overlooked. The lowest residual stresses on the surface 3 may be attributed to the active accumulation of plastic deformation on that region. As shown in Table 2, admitting some experimental errors, the maximum and the minimum values were observed to be in the longitudinal and the transverse directions, respectively.

Influence of Initial Residual Stresses

The measured stress was taken into the input deck as an initial stress. The assumed magnitudes for longitudinal and transverse direction were 300MPa and 70MPa, respectively. The results showed that there was little difference between the cases with and without consideration of these initial stresses, see Fig. 18.

CONCLUSIONS

A relation between travel of the bending roller and spring back radius was obtained via virtual manufacturing with three-roll bending process. The radius of CICC after forming was expressed as a function of the bend-roll travel. The calculation was performed for Incoloy and SUS304. The maximum von Mises stress after spring back was also monitored for the SAGBO prediction. Finally, the variation of the CICC cross-section (reduction of the conduit cross-section) was investigated during the first turn and during conduit bending with largest curvature. With largest curvature, the cross-sectional area was not much reduced. The residual stress of CICC before roll bending was actually measured and the results were taken into calculation. The results showed that there were no significant influence of initial residual stress.

For more accurate prediction, the longer CICC should be modeled. This requires much computing time. Also, it is necessary to modify forming configuration in order to manufacture CICC more efficiently. Since present work was entirely based on the virtual manufacturing, experimental calibration is also required.

ACKNOWLEDGMENTS

This work was supported by Samsung Advanced Institute of Technology

REFERENCES

(1) Sang Ho Kim, 1999, "Development of the Central Solenoid (CS) Model Coil with Nb_3Sn Cable-in conduit (CIC) Superconductor for the Superconducting Tokamak Fusion Device, *PhD Dissertation*, Seoul National University. (in Korean)

(2) L. S. Toma, M. M. Steeves, R. P. Reed, 1994, "Incoloy Alloy 908 Data Handbook," *PFC/RR-94-2*, Superconducting Magnet Development Group, Plasma Fusion Center, MIT.

(3) ASTM, 1981, "Determining Residual Stresses by the Hole-Drilling Strain-Gage Method" *Annual Book of ASTM*, E837-89,730

(4) Measurement Group, Inc., 1993, "Measurement of Residual Stresses by the Hole-Drilling Strain Gage Method," *Tech Note TN-503-5*, Measurement Group, Inc.

(5) Jin W. Joo and Chan K. Park, 1996, "Determination of Non-uniform Residual Stress by the Hole Drilling Method", *KSME*, vol. A-22, no. 2, 276 (in Korean)

Table 1 Mechanical Properties of Incoloy 908[2] and SUS 304(measured)

Young's modulus	180 GPa (Incoloy 908) 195 GPa(SUS 304)
Poisson's ratio	0.3 (same for both materials)
Yield stress [MPa]	848 MPa (Incoloy 908) 321 MPa (SUS 304)
Uniaxial Stress[MPa] – Plastic strain relation	$\sigma = 1848.3(0.00334 + \varepsilon^{p})^{0.137}$ (Incoloy 908) $\sigma = 1525.5(0.03978 + \varepsilon^{p})^{0.483}$ (SUS 304)

Table 2 Calculated principal stresses and their direction with the schematic of strain gage arrangements and locations

	σ_{max}	σ_{min}	α_{min}
surface 1	203.7	78.6	2.3
surface 2	296.2	36.5	-9.1
surface 3	175.7	111.1	-40.3
surface 4	305.7	65.3	0.5

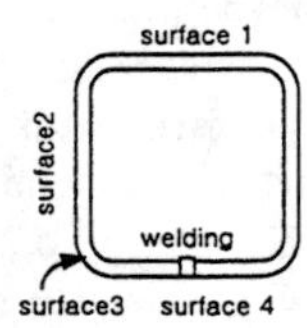

Cross-section of CICC specimen

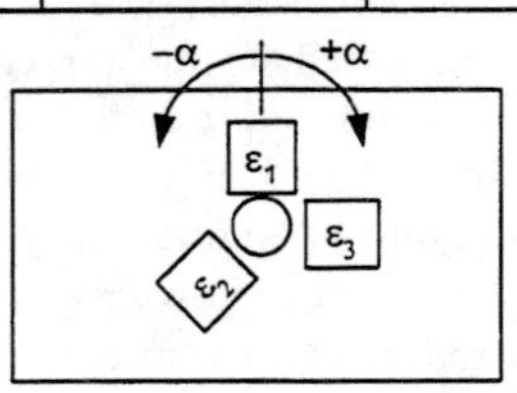

Strain gage arrangement on the surface

Table 3 von Mises stress [MPa] on the measuring surface (When hole-depth was 2.0mm)

Surface	von Mises(MPa)
1	177.9
2	279.7
3	151.7
4	278.8

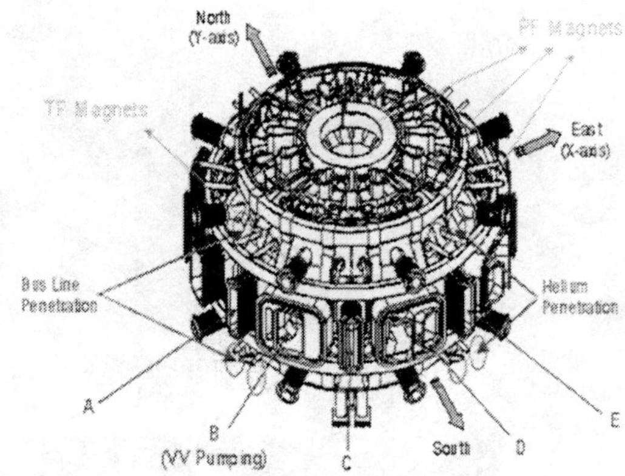

Fig. 1 KSTAR magnet system configuration

(a) (b)

Fig. 2 PF coils configuration and winding scheme used in the KSTAR superconducting magnet: (a) PF1 coil (b) PF7 coil

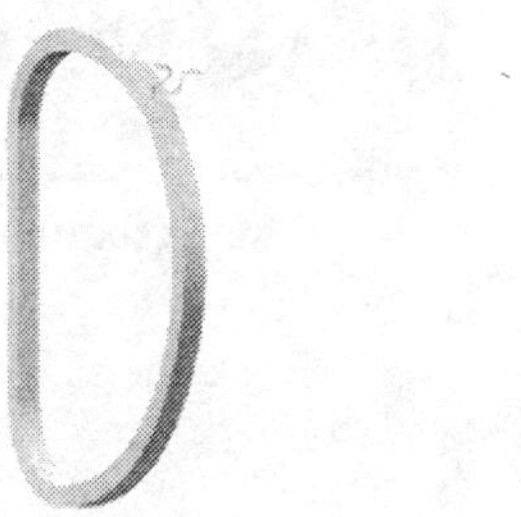

Fig. 3 TF coils configuration and winding scheme used in the KSTAR superconducting magnet

(a)

(b)

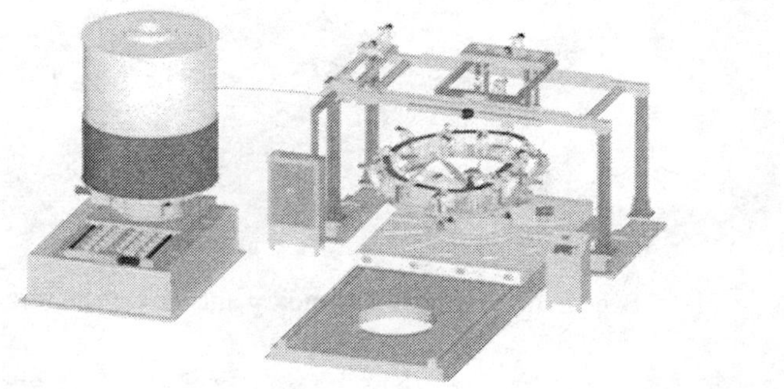

Fig. 4 KSTAR Magnet winding machine using 3-roll bending process Upper: Installed state (Energy Lab, SAIT, Taejon) Lower: Schematic layout

(a) Before forming

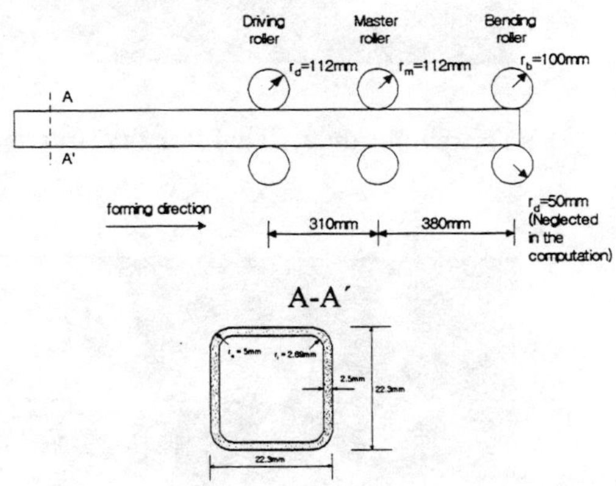

(b) During forming

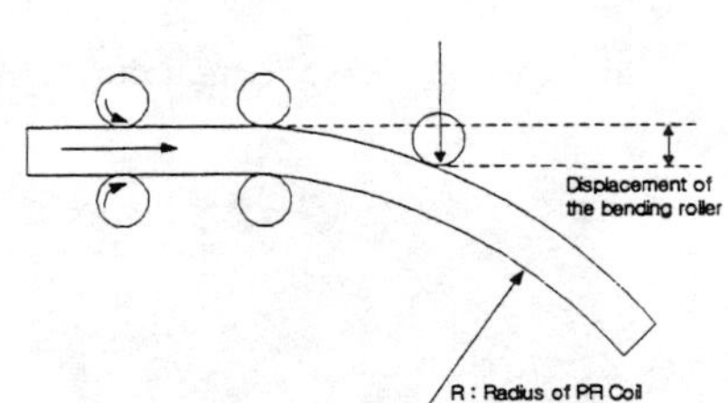

Fig. 5 Schematics of three-roll bending process to form CICC: (a) Before and (b) during forming

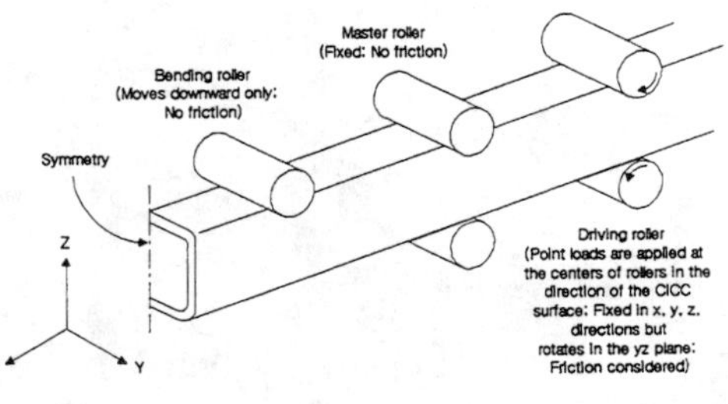

Fig. 6 Schematic that shows boundary and surface conditions

Fig. 7 Whole system for the residual stress measurement

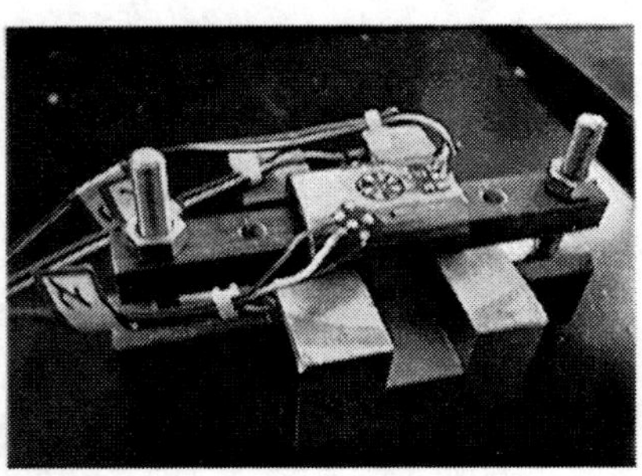

Fig. 8 Fixture to fix the CICC specimen during drilling

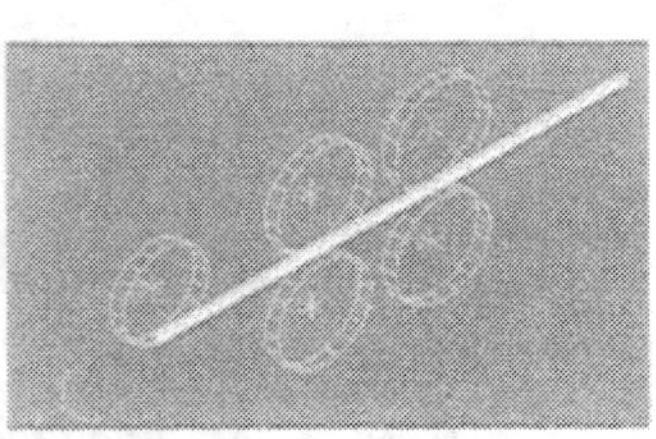

Fig. 9 Three-dimensional FEM model of the 3-roll bending process

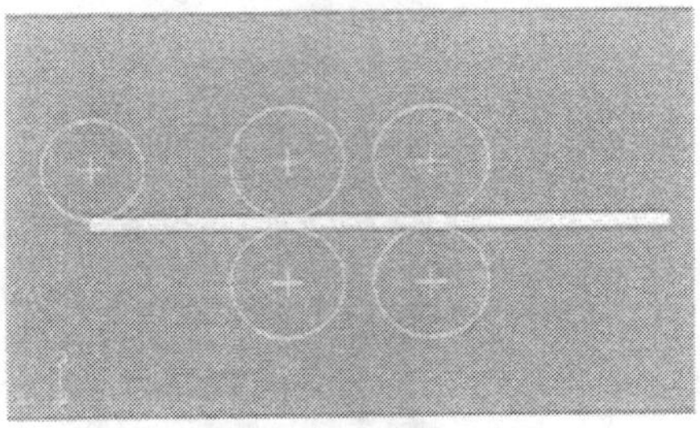

STEP 1 (clamping)

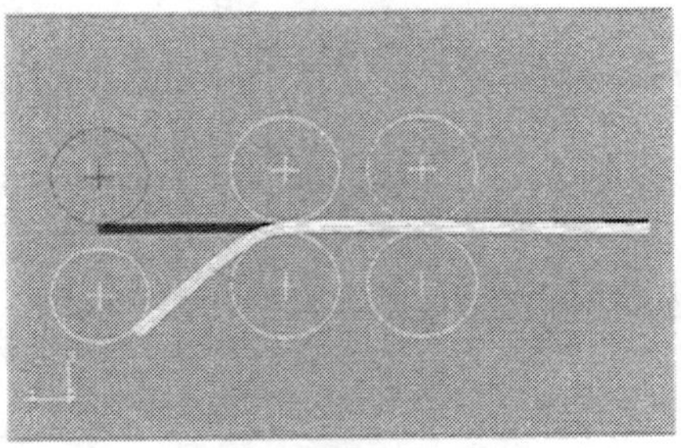

STEP 2 (bending)

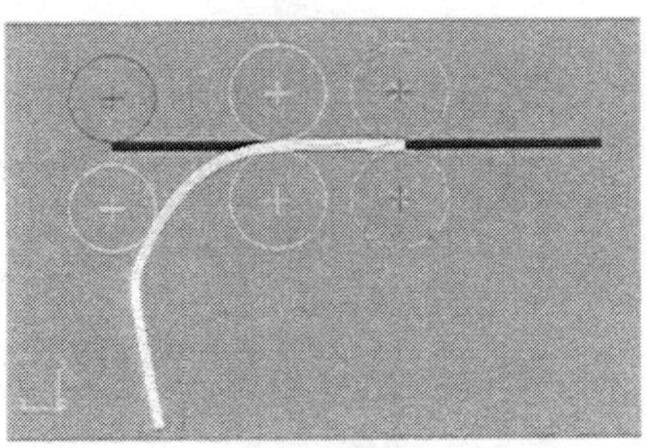

STEP 3 (continuous forming)

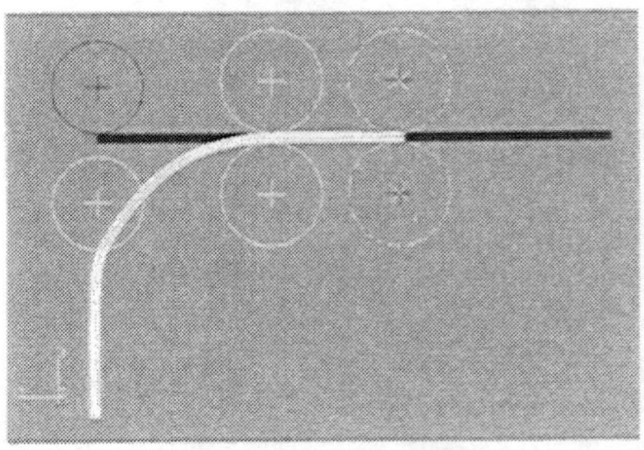

STEP 4 (spring back)

Fig. 10 Stepwise-deformed shape of CICC with downward bending roller travel of 250mm (red: undeformed)

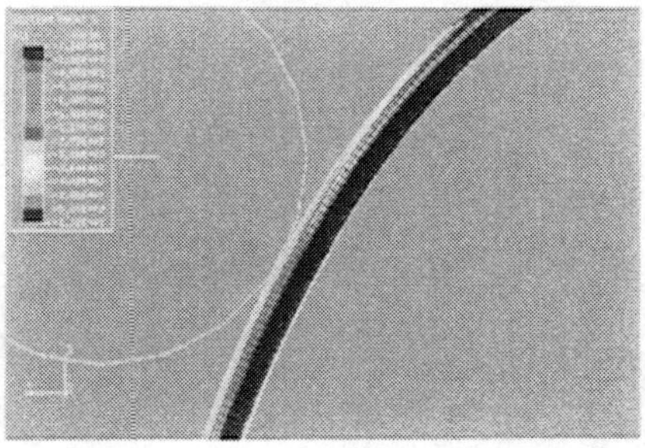

Fig. 11 Contour of longitudinal stress near the bending roller after spring back (step 3)
Material: Incoloy 908; Bending roller travel=250mm

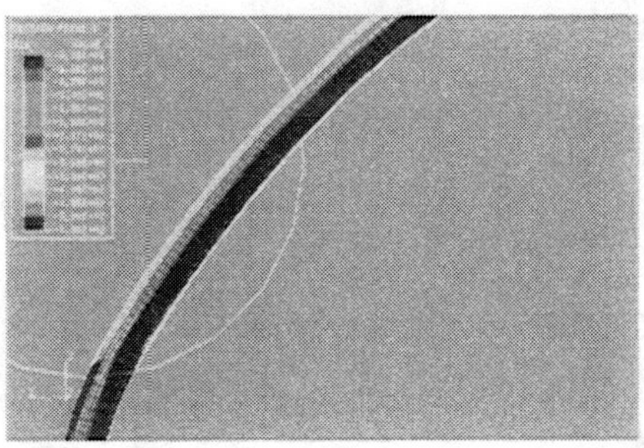

Fig. 12 Contour of longitudinal stress near the bending roller after spring back (step 4)
Material: Incoloy 908; Bending roller travel=250mm

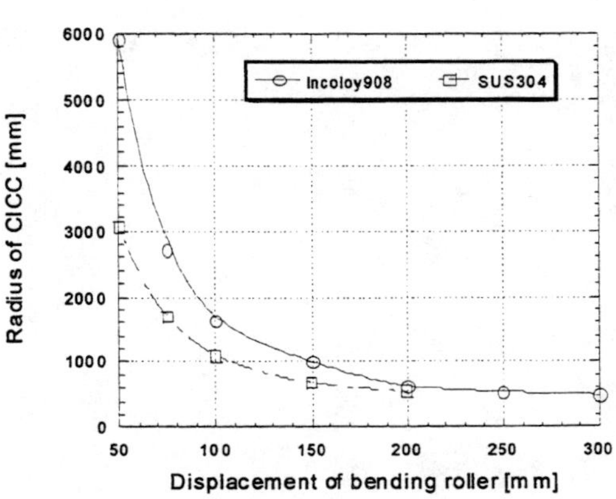

Fig. 13 Predicted radus of formed CICC after spring back vs. distance of the bending roller for both Incoloy 908 and SUS304.
Monitored nodes were located near bending rollers.

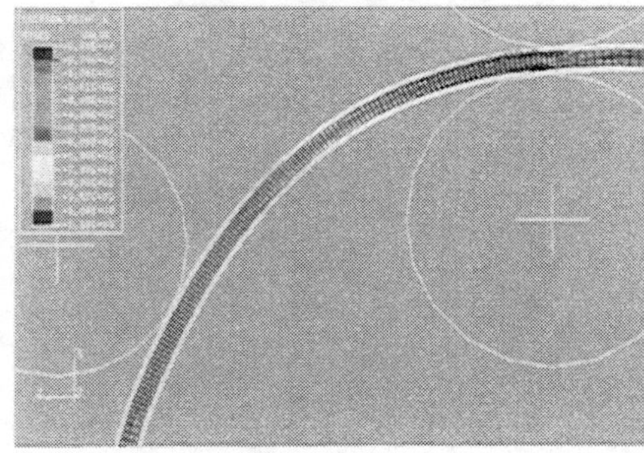

Fig. 14 Contour of von Mises stress when the bending roller traveled by 250mm (Material: Incoloy 908)

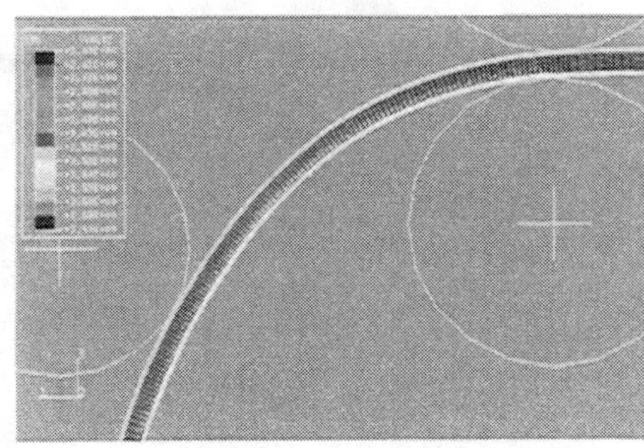

Fig. 15 Contour of thickness when the bending roller traveled by 250mm (Material: Incoloy 908).

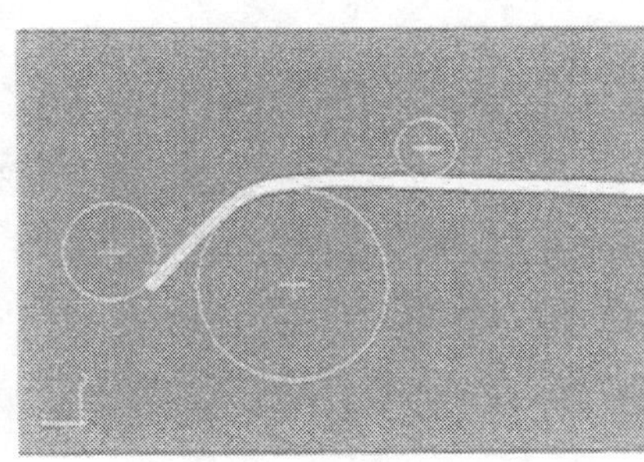

Fig. 16 Modified modeling of 3-roll bending process for the examination of cross-sectional area reduction due to tight radius (197.7mm before spring back) (Travel of the bending roller=250mm)

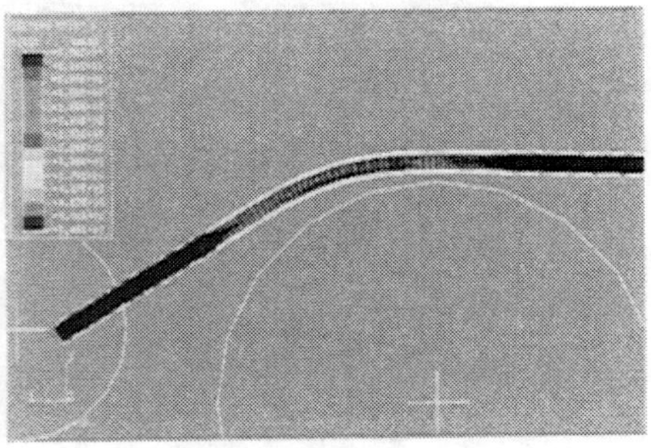

Fig. 17 Contour of von Mises stress near the region with tight radius (234.3mm) after spring back from the modified modeling (Travel of the bending roller=250mm)

(a)

(b)

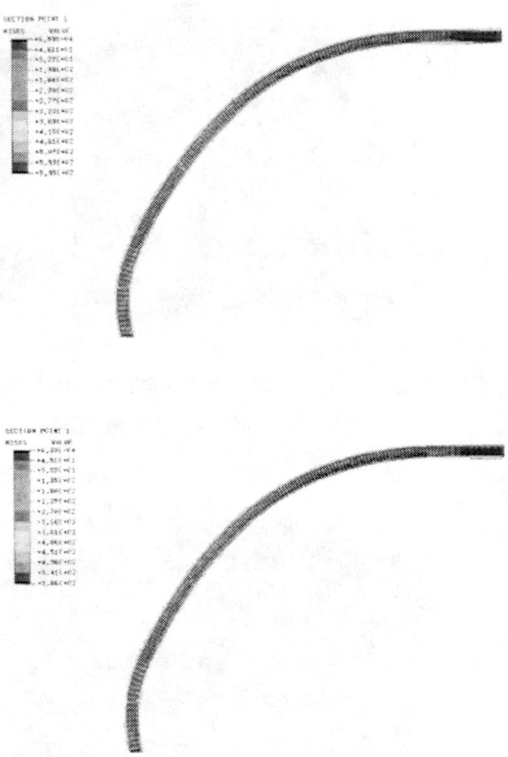

Fig. 18 Comparison of von Mises stress:
(a) when the residual stress was considered as an initial stress and (b) without initial stress after spring back (Travel of the bending roller=300mm)

PVP-Vol. 415, Recent Advances in Solids and Structures — 2000
ASME 2000

YIELD FUNCTION FOR STRUCTURAL POROUS METALS

M. Khairul Alam
Professor
Department of Mechanical Engineering
Ohio University
Athens, OH 45701

Shatil S. Ahmed
Research Scientist
UES Software, Inc.
4401 Dayton-Xenia Road
Dayton, OH 45432

Rex J. Kuriger
Research Associate
Department of Mechanical Engineering
Ohio University
Athens, OH 45701

KEYWORDS: yield function, structural porous metals, powdered metallurgy

ABSTRACT

To reduce weight while maintaining high specific strength, Structural Porous Metals (SPM) are being developed. These materials are alloys containing an inert gas in distributed pores at very high pressures. The pores of gas are incorporated into SPM during processing. Since the pores are closed pores at high pressures, the material can be processed by standard methods such as rolling or forging without collapsing the gas pores. These materials must be treated as a composite material with special deformation behavior under applied stresses. It should be noted that, unlike solids, shear stresses can not be present in a gas medium in static equilibrium. This paper deals with the determination of the yield stress of structural porous metals. The yield function is calculated from the principle of virtual work. In this approach it is assumed that a constant amount of energy needed to deform the material to the yield point. In the case of SPM, the yield function must include the work needed to deform the gas pores as the material yields. The energy of deformation for the gas in the pores is evaluated according to the ideal gas law. The total work done in deforming the gas and the solid metal is taken into account in determination of the yield function of SPM. The result of the analysis is then compared with classic yield functions for monolithic alloys.

INTRODUCTION

Development of advanced materials for structural application in aerospace has been driven by a need to increase the strength-to-weight ratio. The notable recent success has been development and implementation of composite materials in structural application. But some of the recent studies indicate that composites are not the best material to be used for structural application, as they have a tendency to degrade with aging [1, 2]. Also an increased emphasis is being placed on affordability and recyclability of these novel materials.

In the design of an ultra lightweight material, the goal is to achieve increased strength-to-weight ratio offered by composites while maintaining the affordability of conventional metallic components. In search of such material, a partnership led by McDonnell Douglas Aerospace (MDA) is developing an innovative class of materials called Structural Porous Metals (SPM). The basic difference between SPM and conventional composite materials lies in the fact that SPM would be developed as a sandwich material with low density core (LDC) and two solid face sheets as a continuous cross-section within metal mill products.

Current commercial methods of producing porous metallic material result in products that are fragile due to inter-connection and the angular nature of porosity, therefore their use in structural components is restricted. Present commercial methods cannot produce large billets that can be converted to standard mill products by deformation processing. Recent developments in the metal processing technology has made possible the production of SPM. This material has a potential for successfully controlling porosity characteristics in billet stock at cost comparable to standard billet production.

A process that has been successfully implemented is the low density core (LDC) process, where powder metallurgy methods are used to produce a porous sandwich core material bonded to two face-sheets. A primary feature of this process is the introduction of inert gas in the solid billet. Unlike the traditional powder metallurgy process, the trapped inert gas inside the voids is intentionally kept in the LDC and resists compressive stresses during processing.

In a typical process, the raw powder is packed into HIP cans, which are then filled with an inert gas. The cans are HIPed at a high temperature and pressures that are comparable to the yield pressure. The final product is a metal sponge-like material that resists deformation because of entrapped gases at high pressure. This material can be forged into desired shapes. During the forging process, the material behaves very differently from traditional powdered metallurgy and monolithic solid metals. This is due to the fact that such material is not subject to the traditional yield criterion for solid ingot metals (I/M) or powder metallurgy (P/M) systems. The deformation processing involves straining of the material beyond yield point such that plastic flow occurs. To determine the stresses required for the plastic flow, a yield criterion applicable to this material had to be developed.

The yielding of P/M material is more complicated than I/M for the reason that hydrostatic stress plays a role in yielding. For SPM it is even more complicated because in P/M the voids inside do not resist the deformation due to hydrostatic or deviatoric stress components. However the pores in SPM resist deformation, due to the fact that they are enclosed isolated pores with internal gas pressure. Thus a special yield function which takes into account the effect of hydrostatic stress and resistance of pores to deformation is required. In this study it will be assumed that

- SPM is homogeneous and isotropic.
- No permanent change in relative density occurs up to the yield point.
- Trapped gas inside pores obeys the ideal gas law and show ideal fluid behavior.
- Gas pores only resist the hydrostatic stress.
- At yield point, pressure in the gas is equal to hydrostatic stress.

In the development of the yield criterion the principal of virtual work is used. The starting point of this analysis is the yield criterion for conventional P/M processes [3] given by

$$AJ_2' + BJ_1^2 = Y_R^2 \tag{1}$$

where J_2' is the second invariant of deviatoric stress and J_1 is the first invariant of stress tensor, Y_R is the yield stress of P/M materials, A and B are parameters which are functions of relative density R. Many researchers have determined these parameters using different semi-empirical method and experimental results. A more successful approach was developed by Doraivelu et al. [3] by using the principle of virtual work. Following the same approach, the yield criterion for SPM material is developed to include the effect of enclosed pores with internal pressure.

DERIVATION OF YIELD FUNCTION FOR POWDER METALLURGY (P/M) MATERIAL

For partially dense P/M material, it is assumed that material is homogeneous and isotropic. Considering P/M material as continuous, it is proposed that it will yield when the apparent total deformation energy reaches a critical value. Up to the yield point, the behavior of the P/M is considered to be linear, with no permanent change in density. For linear behavior up to the yield point, the elastic strain energy can be written as follows:

$$w_{pm} = \frac{1}{2}\sigma_{ij}\varepsilon_{ij} \tag{2}$$

where w_{pm} is also the work done per unit volume of P/M material, considering the body is in equilibrium so that none of this work goes into kinetic energy, this work is stored as strain energy of deformation in the body. Elastic stress-strain relationship is given by

$$\varepsilon_{ij} = \frac{(1+\upsilon)}{E}\sigma_{ij} - \frac{\upsilon}{E}\sigma_{kk}\delta_{ij} \tag{3}$$

where υ is the Poisson's ratio and E is the Young's modulus of the aggregate. From the definition of deviatoric stress tensor, we can write

$$\sigma_{ij} = \sigma'_{ij} + \frac{\sigma_{kk}}{3}\delta_{ij} \tag{4}$$

Substituting Eq. (3) and (4) in Eq. (2), it follows

$$w_{pm} = \frac{1}{2}\left[\left(\sigma'_{ij} + \frac{\sigma_{kk}}{3}\delta_{ij}\right)\left(\frac{1+\upsilon}{E}\sigma_{ij} - \frac{\upsilon}{E}\sigma_{kk}\delta_{ij}\right)\right] \tag{5}$$

after some algebraic simplification, Eq. (5) can be written as

$$w_{pm} = \frac{1}{2}\left[\frac{1+\upsilon}{E}\sigma'_{ij}\sigma'_{ij} + \frac{1-2\upsilon}{3E}\sigma_{kk}^2\right] \tag{6}$$

In Eq. (6) the equality of $\sigma_{ij}\sigma'_{ij} = \sigma'_{ij}\sigma'_{ij}$ has been used. Now according to this criterion, this energy is a constant at yield, independent of the stress state. Therefore like any other classical yield criterion, the strain energy can be equated to that obtained for the simple tension. For uniaxial tensile state of stress, let $\sigma_1 = Y_R$. Then the stress tensor and the deviatoric stress tensor can be represented as:

$$\sigma_{ij} = \begin{bmatrix} Y_R & 0 & 0 \\ 0 & 0 & 0 \\ 0 & 0 & 0 \end{bmatrix} \quad (7)$$

$$\sigma'_{ij} = \begin{bmatrix} Y_R - \frac{Y_R}{3} & 0 & 0 \\ 0 & \frac{Y_R}{3} & 0 \\ 0 & 0 & \frac{Y_R}{3} \end{bmatrix} \quad (8)$$

So from Eq. (7) and (8) it is calculated as follows

$$\sigma'_{ij}\sigma'_{ij} = \frac{2}{3}Y_R^2 \quad (9)$$

and $$\sigma_{kk} = Y_R \quad (10)$$

Substituting the values from Eq. (9) and (10) into Eq. (6), w_{pm} is expressed as

$$w_{pm} = \frac{1}{2}\left(\frac{3Y_R^2}{E}\right) \quad (11)$$

According to the criterion, equating Eq. (6) and (11) gives

$$3(1+\upsilon)\sigma'_{ij}\sigma'_{ij} + (1-2\upsilon)\sigma_{kk}^2 = 3Y_R^2 \quad (12)$$

But

$$J'_2 = \frac{1}{2}\sigma'_{ij}\sigma'_{ij}$$

and

$$J_1 = \sigma_{kk}$$

So Eq. (12) is rewritten as follows

$$2(1+\upsilon)J'_2 + \left(\frac{1-2\upsilon}{3}\right)J_1^2 = Y_R^2 \quad (13)$$

The above Eq. (13) represents the yield function for P/M material and it is now evident that the yield stress Y_R of the aggregate depends on the deviatoric as well as the hydrostatic stress.

Comparing Eq. (13) with Eq. (1), parameters A and B can be evaluated as

$$A = 2(1+\upsilon) \quad (14)$$

$$B = \frac{1-2\upsilon}{3} \quad (15)$$

The next step is to relate the dependence of yield function on relative density R. A power law dependence of Poisson's ratio υ on relative density R has been proposed by Zhadannovich [4], which states that

$$\upsilon = 0.5R^n \quad (16)$$

Later the value of the exponent n has been experimentally verified as approximately 2 for aluminum and ferrous powders by Kuhn [5]. For modeling purpose $n \cong 2$ is sufficiently accurate for P/M materials. Therefore, the yield function for P/M materials (Eq. 13) can be written in terms of relative density R as

$$(2+R^2)J'_2 + \frac{1-R^2}{3}J_1^2 = Y_R^2 \quad (17)$$

where A and B in terms of relative density R are given as,

$$A = 2 + R^2 \quad (18)$$

$$B = \frac{1-R^2}{3} \quad (19)$$

This completes the review on P/M yield criteria and the way it was derived. This is the basis for the following derivation of a new yield criterion for SPM.

DERIVATION OF YIELD CRITERION FOR SPM

SPM will also be treated as a homogeneous, isotropic, continuum with uniformly dispersed porosity. Additional assumptions are:

- No permanent change in relative density occurs up to yield point.
- Trapped gas inside pores obeys ideal gas law and show ideal fluid behavior.
- Gas pores will only resist the hydrostatic stress.
- At yield point pressure in gas is equal to hydrostatic stress.
- Mechanical operations for SPM are isothermal.

In a P/M material the voids inside cannot resist deformation due to hydrostatic or deviatoric stress components. However, the pores in SPM are closed pores with gas trapped inside, so they will resist deformation due to hydrostatic stress. In the derivation of the yield criterion for SPM, the goal is to extend the P/M yield criterion to incorporate the effect of enclosed pores with internal pressure. Using the principle of virtual work, this idea can be expressed as follows

$$w_{spm} = w_{pm} + w_{gas-comp} \quad (20)$$

where w_{spm} is the total work done per unit volume of aggregate, w_{pm} is the work required per unit volume considering the material to be P/M, and $w_{gas-comp}$ is the additional work required to compress the gas inside the pores per unit volume of SPM. w_{pm} is already derived in the previous section so w_{gas_comp} has to be derived next. Considering the process to be isothermal, the additional work done to compress the gas inside the pores is going to be the integral of the pressure times the small change in volume of pores

$$\overline{w}_{gas-comp} = -\int_{\overline{v}_1}^{\overline{v}} P d\overline{v} \quad (21)$$

where P is the pressure inside the pores and $\bar{v}_1$ and $\bar{v}$ are the initial and current volume of pores per unit mass respectively. Please note that instead of $w_{gas-comp}$, $\bar{w}_{gas-comp}$ is used to distinguish between the work done per unit volume and per unit mass respectively. This is necessary because the mass remains constant for SPM deformation, but the volume changes. The relation between the two can be given as

$$w_{gas-comp} = \rho \bar{w}_{gas-comp} \tag{22}$$

where ρ is the density of the aggregate. Let R_1 and R respectively be the corresponding initial and current relative density of SPM. By definition of relative density it is the ratio between the porous material density to that of fully dense material's density, that is

$$R = \frac{\rho}{\rho_0} \tag{23}$$

where ρ is the density of the porous body and ρ_0 is the density of the same material if it was fully dense. Let $\bar{v}_s$ be the volume of solid per unit total mass, then ρ_0 can be expressed as

$$\rho_0 = \frac{1}{\bar{v}_s} \tag{24}$$

Let $\bar{v}_p$ be the volume of pores per unit total mass, then ρ can be written as

$$\rho = \frac{1}{\bar{v}_s + \bar{v}_p} \tag{25}$$

From Eqs. (23), (24), and (25) the relative density R can be expressed in terms of $\bar{v}_s$ and $\bar{v}_p$ as

$$R = \frac{\bar{v}_s}{\bar{v}_s + \bar{v}_p} \tag{26}$$

At initial state let $\bar{v}_p = \bar{v}_1$ and at current state $\bar{v}_p = \bar{v}$. Further, v_1 and v are denoted as the volume fraction of pores at initial and current state respectively. Again the relation between volume of pores per unit mass and volume fraction of pores can be written as

$$\begin{aligned} v &= \rho \bar{v} \\ v_1 &= \rho_1 \bar{v}_1 \end{aligned} \tag{27}$$

where ρ_1 and ρ are the initial and current density of SPM. Considering the process to be isothermal from Gas Law, it gives

$$P\bar{v} = P_1\bar{v}_1 = const. \tag{28}$$

Substituting the value of P from Eq. (28) in Eq. (21), work done to compress the gas can be calculated as

$$\bar{w}_{gas\_comp} = -P_1\bar{v}_1 \int_{\bar{v}_1}^{\bar{v}} \frac{1}{\bar{v}} d\bar{v} = P_1\bar{v}_1 \ln \frac{\bar{v}_1}{\bar{v}} \tag{29}$$

Substituting $\frac{\bar{v}_1}{\bar{v}} = \frac{P}{P_1}$ from Eq. (28), Eq. (29) changes to

$$\bar{w}_{gas\_comp} = P_1\bar{v}_1 \ln \frac{P}{P_1} \tag{30}$$

Using the relation in Eq. (22) and Eq. (30), w_{gas_comp} can be expressed as

$$w_{gas\_comp} = \rho P_1 \bar{v}_1 \ln \frac{P}{P_1} \tag{31}$$

If it is considered now that up to yield point the density change in SPM is very small i.e., $\rho \approx \rho_1$, then from Eq. (27) and Eq. (31) it can written, that

$$w_{gas\_comp} = P_1 v_1 \ln \frac{P}{P_1} \tag{32}$$

The pressure P at yield can be assumed to be proportionate to the hydrostatic stress, so it may be written that

$$P = -\beta \frac{\sigma_{kk}}{3} \tag{33}$$

where β is an empirical parameter less than or equal to one. This may be assumed to be 1 till it can be verified by experiments, which implies that at yield the pressure inside the pores is assumed to be equal to the mean stress or hydrostatic stress. Substituting P from Eq. (33) in Eq. (32) and assuming $\beta = 1$ gives

$$w_{gas\_comp} = P_1 v_1 \ln\left(-\frac{\sigma_{kk}}{3P_1}\right) \tag{34}$$

Now the value of w_{pm} is substituted from Eq. (6) and w_{gas_comp} from Eq. (34) into Eq. (20) to find the total energy required for SPM to yield.

$$w_{spm} = \frac{1}{2}\left[\frac{1+\upsilon}{E}\sigma'_{ij}\sigma'_{ij} + \frac{1-2\upsilon}{3E}\sigma_{kk}^2\right] + P_1 v_1 \ln\left(-\frac{\sigma_{kk}}{3P_1}\right) \tag{35}$$

According to our postulated yield criterion the above energy w_{spm} should be some constant at the point of yield, let that constant be K_1, so Eq. (35) can be written as

$$\frac{1}{2}\left[\frac{1+\upsilon}{E}\sigma'_{ij}\sigma'_{ij} + \frac{1-2\upsilon}{3E}\sigma_{kk}^2\right] + P_1 v_1 \ln\left(-\frac{\sigma_{kk}}{3P_1}\right) = K_1$$

Therefore,

$$3(1+\upsilon)\sigma'_{ij}\sigma'_{ij} + (1-2\upsilon)\sigma_{kk}^2 + 6EP_1 v_1 \ln\left(-\frac{\sigma_{kk}}{3P_1}\right) = 6EK_1 \tag{36a}$$

As E can be considered to be a constant for a particular material, so the whole right side of Eq. (36a) can be denoted as another constant K, so Eq. (36a) can be rewritten as

$$3(1+\upsilon)\sigma'_{ij}\sigma'_{ij} + (1-2\upsilon)\sigma_{kk}^2 + 6EP_1 v_1 \ln\left(-\frac{\sigma_{kk}}{3P_1}\right) = K \tag{36b}$$

Eq. (36b) is the Yield function for the SPM. Which reduces to the P/M yield function if there is no entrapped gas inside the pores, because in that case the third term in the above equation or w_{gas_comp} will be zero. Again if the material is considered to be 100% dense or incompressible, then $\upsilon = 0.5$ and the Yield function derived will reduce to Von Misses yield criterion. To evaluate K, like any other yield criterion Eq. (36b) is equated for a special case of uniaxial tensile test, where the applied stress component is Y_{spm}, in which case

$$\sigma'_{ij}\sigma'_{ij} = \frac{2}{3}Y^2_{spm} \quad \text{and} \quad \sigma_{kk} = Y_{spm} \qquad \text{(Eqs. 9 and 10)}$$

For uniaxial state of stress Eq. (36) can be written as:

$$3Y^2_{spm} = K \qquad (37)$$

From Eq. (36b) and Eq. (37), it gives

$$3(1+\upsilon)\sigma'_{ij}\sigma'_{ij} + (1-2\upsilon)\sigma^2_{kk} + 6EP_1v_1 \ln\left(-\frac{\sigma_{kk}}{3P_1}\right) = 3Y^2_{spm} \qquad (38)$$

$\sigma'_{ij}\sigma'_{ij} = 2J'_2$ and $\sigma_{kk} = J_1$ are substituted in Eq. (38) to get

$$6(1+\upsilon)J'_2 + (1-2\upsilon)J_1^2 + 6EP_1v_1 \ln\left(-\frac{J_1}{3P_1}\right) = 3Y^2_{spm}$$

Therefore,

$$2(1+\upsilon)J'_2 + \frac{(1-2\upsilon)}{3}J_1^2 + 2EP_1v_1 \ln\left(-\frac{J_1}{3P_1}\right) = Y^2_{spm} \qquad (39)$$

Eq. (39) is the Yield Function of SPM or porous metals with high-pressure gas pores; where the initial pressure and total volume fraction of the gas pores is given by P_1 and v_1. It is generally difficult to determine Y_{spm} without experimental data therefore it is convenient to nondimensionalize Eq. (39) by dividing throughout by Y^2_{spm}. After substituting the values of J_1 and J'_2 in terms of $\bar{\sigma}_1$, $\bar{\sigma}_2$, and $\bar{\sigma}_3$ we get the following nondimensional equation:

$$\bar{\sigma}_1^2 + \bar{\sigma}_2^2 + \bar{\sigma}_3^2 - 2\upsilon(\bar{\sigma}_1\bar{\sigma}_2 + \bar{\sigma}_2\bar{\sigma}_3 + \bar{\sigma}_1\bar{\sigma}_3) + E_v\bar{P}_1 \ln\left(\frac{\bar{\sigma}_1 + \bar{\sigma}_2 + \bar{\sigma}_3}{3\bar{P}_1}\right) - 1 = 0 \qquad (40)$$

where

$$E_v = 2\bar{E}v_1$$

Eq. (40) is difficult to plot because the log term produces a non-real number when $\bar{\sigma}_1 + \bar{\sigma}_2 + \bar{\sigma}_3 < 0$. Figures 1a to 1e show the resulting yield surfaces for a Poisson's ratio $\upsilon = 0.5$, and nondimensionalized initial gas pressure $\bar{P}_1 = 0.8$. Although these parameters may not be representative of the actual processing conditions, it is required to attribute values to these parameters for the purpose of plotting the yield surfaces. To obtain a two-dimensional plot the value of $\bar{\sigma}_3$ was held constant in each of these figures and varied from the maximum of +1 to the minimum of −1. The base-line figure is for a powdered metallurgy material with no enclosed gas pores; this is seen to be an ellipse in all five figures. The parameter that is varied in each figure is the quantity E_v that is varied from a maximum of 100 to a minimum of 0. When $E_v = 0$ the log term in Eq. (40) becomes negligible, thereby coinciding with elliptical yield surface of powdered materials. From Eq. (40) it can be observed that the log term dominates when E_v becomes large; this results in a yield surface corresponding to the curve for $\bar{\sigma}_1 + \bar{\sigma}_2 + \bar{\sigma}_3 \cong const.$ In the two-dimensional figures this produces a straight line with a slope approximately equal to −1.

An interesting feature of these yield surfaces is the conditions for the log term that the argument cannot be negative, therefore the yield surfaces are effectively truncated by the line $\bar{\sigma}_1 + \bar{\sigma}_2 + \bar{\sigma}_3 = 0$. This makes the yield surface curves open, which contrasts it with the closed ellipse for powdered materials. As the value of E_v increases, the ellipse appears to open up and gradually stretch out to a straight line. The ellipse and all other yield surfaces are translated on the $\bar{\sigma}_1$ and $\bar{\sigma}_2$ plane as the value of $\bar{\sigma}_3$ changes. For higher values of $\bar{\sigma}_3$, the SPM yield surface moves closer to the origin, moving towards lower values of $\bar{\sigma}_1$ and $\bar{\sigma}_2$. This occurs because higher values of $\bar{\sigma}_3$ implies a higher initial tensile stress condition; this results in lower stresses ($\bar{\sigma}_1$, $\bar{\sigma}_2$) being required to yield the material.

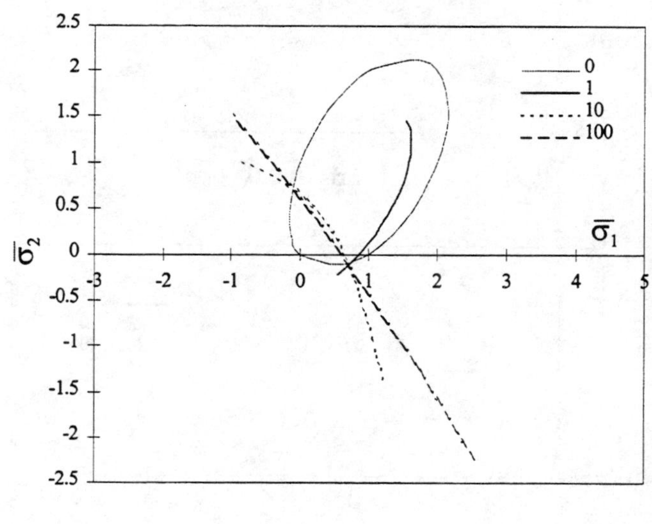

1a. $\bar{\sigma}_3$ = 1.0

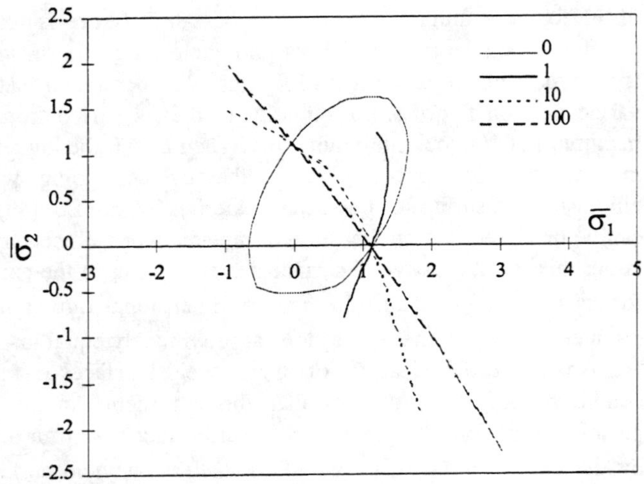

1b. $\overline{\sigma}_3$ **= 0.5**

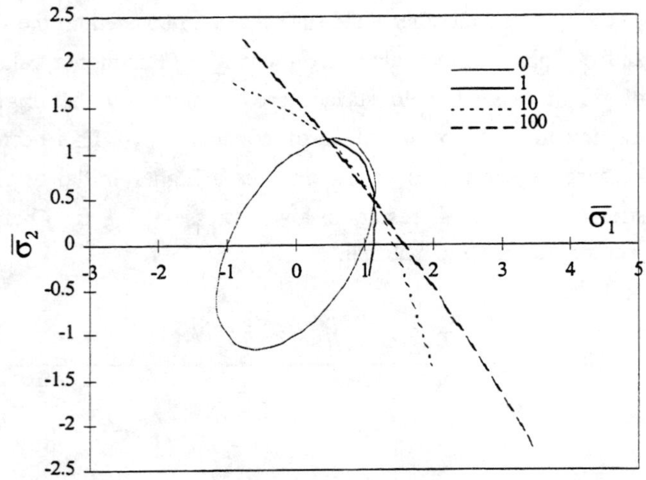

1c. $\overline{\sigma}_3$ **= 0.0**

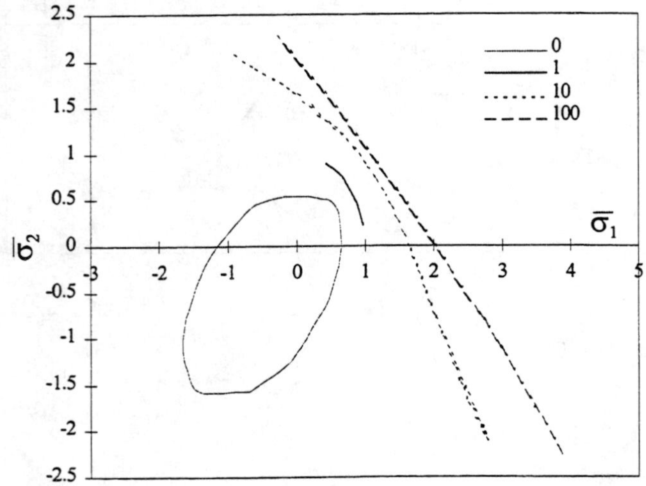

1d. $\overline{\sigma}_3$ **= -0.5**

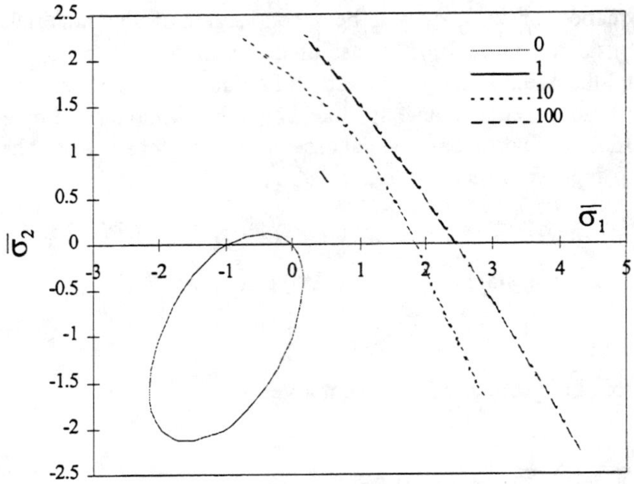

1e. $\overline{\sigma}_3$ **= -1.0**

Figure 1. Two-dimensional yield surfaces with υ=0.5 and $\bar{P}_1$=0.8 for constant values of $\overline{\sigma}_3$ and E_v ranging from 0 to 100.

DERIVATION OF FLOW RULE FOR SPM

We will now derive the general equation for plastic stress-strain relations for any yield criterion based on a unified approach due to Drucker [6,7]. The strain $d\varepsilon_{ij}$ consists of both elastic strain and plastic strain i.e.,

$$d\varepsilon_{ij} = d\varepsilon_{ij}^{e} + d\varepsilon_{ij}^{p} \tag{41}$$

Because of work hardening,

$$d\sigma_{ij} d\varepsilon_{ij} > 0 \quad \text{and} \quad d\sigma_{ij} d\varepsilon_{ij}^{p} \geq 0 \tag{42}$$

Two more basic assumptions are necessary to derive the general stress-strain relations. We assume that at each state of the plastic deformation there exists a yield function $f(\sigma_{ij})$ so that further plastic deformation takes place only for $f(\sigma_{ij})>K$. Both f and K may depend on the existing state of stress and strain history. The relation between infinitesimal change of stress and plastic strain is assumed to be linear, i.e.,

$$d\varepsilon_{ij}^{p} = C_{ijkl} d\sigma_{kl} \tag{43}$$

The coefficient C_{ijkl} may be functions of stress, strain, and history of loading. It can be shown that

$$d\varepsilon_{ij}^{p} = G \frac{\partial f}{\partial \sigma_{ij}} df \tag{44}$$

where G is a scalar which may depend on stress, strain, and history. Similarly the effective plastic strain rate $\dot{\varepsilon}_p$ can be written as

$$\dot{\varepsilon}_{ij}^{p} = \frac{\sigma_e \frac{\partial f}{\partial \sigma_{ij}}}{\sigma_{mn}\left(\frac{\partial f}{\partial \sigma_{mn}}\right)} \dot{\varepsilon}_p \quad (45)$$

where σ_e is the effective stress

We have already defined the yield criterion for SPM in Eq. (40). If we denote our yield function as *f*, then we have

$$\left[2(1+\upsilon)J_2' + \frac{(1-2\upsilon)}{3}J_1^2 + 2EP_1 v_1 \ln\left(-\frac{J_1}{3P_1}\right)\right]^{\frac{1}{2}} = Y_{spm} = f \quad (46)$$

Now to use the derived yield function in Eq. (45) we have to find the partial derivative of *f* with respect to σ_{ij}. Let us denote (similar to P/M):

$$A = 2(1+\upsilon) \quad (47)$$

$$B = \frac{(1-2\upsilon)}{3} \quad (48)$$

then we obtain the following

$$\dot{\varepsilon}_{ij}^{p} = \frac{\sigma_e\left(A\sigma_{ij}' + 2B\delta_{ij}J_1 - \delta_{ij}\frac{2EP_1 v_1}{J_1}\right)}{2\left(AJ_2' + BJ_1^2 - EP_1 v_1\right)} \dot{\varepsilon}_p \quad (49)$$

$$\text{Let} \quad E_g = \frac{EP_1 v_1}{J_1} \quad (50)$$

Note that E_g has the same unit as E, that is of stress. So we can rewrite Eq. (49) as

$$\dot{\varepsilon}_{ij}^{p} = \frac{\sigma_e\left(A\sigma_{ij}' + 2B\delta_{ij}J_1 - 2\delta_{ij}E_g\right)}{2\left(AJ_2' + BJ_1^2 - E_g J_1\right)} \dot{\varepsilon}_p \quad (51)$$

We can define the effective stress σ_e as follows:

$$\sigma_e = \frac{AJ_2' + BJ_1^2 - E_g J_1}{Y_{spm}} \quad (52)$$

Using the above definition of effective stress, Eq. (51) can be written as

$$\dot{\varepsilon}_{ij}^{p} = \frac{\dot{\varepsilon}_p}{2Y_{spm}}\left(A\sigma_{ij}' + 2B\delta_{ij}J_1 - 2\delta_{ij}E_g\right) \quad (53)$$

which is the *flow rule* for SPM materials.

CONCLUSION

The new yield criterion for Structural Porous Metals (SPM) has been developed which took into account the inert gas trapped inside the isotropic homogenous pores. Using the derived yield criterion, the flow rule for SPM has been developed. So far, the yield function and flow rule have not been validated. Experimental and numerical work is in progress to carry out the validation. Preliminary work has been done using SPM material developed for an experimental project, and the flow rule is being incorporated in a finite element code.

REFERENCES

1. Dao, K. C., Shockey, D. A., Seaman, L., Curran, D.R., and Rowcliffe, D. J., May 1979, "Particle Impact Damage in Silicone Nitride"; Contract No. N00014-76-057, Annual Reports, Part III, Office of Naval Research, Washington, DC.
2. Razvan, A., and Reifsnider, K. L., "Fiber Fracture and Strength Degradation in Unidirectional Graphite/Epoxy Composite Materials", *Theoretical and Applied Fracture Mechanics*, **16**, No. 1, Oct 1991, pp. 81.
3. Doraivelu, S. M., Gegel, H. L., Gunasekera, J. S., Malas, J. C., Morgan, M. T., and Thomas J. F. Jr., 1984, "A New Yield Function for Compressible P/M Materials", *Int. J. Mech. Sci.*, **26**, No. 9/10, pp. 527-535.
4. Zhdannovich, G. M., June, 1971, "Theory of compacting of metal powders", Translated from *Teorizc Pressovaniya Metzllichaskikli Poroshkov*, by the Foreign Technology Division, WPAFB, Ohio, U.S.A. (Translation No. FTD-HC-23-775-70), pp. 1-262.
5. Kuhn, H.A., 1978, "Deformation Process of Sintered Powder Metal", Ch. 14, *Powder Metallurgy Processing: New Techniques and Analysis*, Academic Press, New York, pp. 99.
6. Drucker, D. C., 1950, "Some Implications of Work Hardening and Ideal Plasticity", *Quart. Appl. Math.*, **7**, pp. 411-418.
7. Drucker, D. C., 1952, "A More Fundamental Approach to Plastic Stress-Strain Relations", *1st U.S. Congress of Applied Mechanics*, ASME, New York, pp. 487-491.

PVP-Vol. 415, Recent Advances in Solids and Structures — 2000
ASME 2000

REDESIGN OF SPHERICAL ACRYLIC SUBMERSIBLE FOR MANNED OPERATION TO 3000 ft (914.4 m) OCEAN DEPTH

Partha S. Das, Ph.D.
Senior R&D Engineer
Harbor Branch Oceanographic Inst., Inc.
Fort Pierce, Florida 34946, USA
Ph.: (561) 465-2400, Ext. 533
Fax: (561) 464-9094
e-mail : psdas@hboi.edu

ABSTRACT

Harbor Branch Oceanographic Institution (HBOI) is using its spherical acrylic manned submersible for ocean exploration for last 3 decades. However, the development of tiny shear cracks at the interface areas of these submersibles following only few hundred dives require frequent, expensive repairs. To overcome this crack generation problem, a two-phase research program is initiated at HBOI. In the Phase I of this study, a detailed nonlinear 3-D Finite Element Analysis (FEA) is performed at first to increase the understanding of the mechanical behavior at the interface of this submersible and then various analyses are carried out to develop a guidelines for redesigning the spherical acrylic submersible. Complete redesigning of the bottom of acrylic submersible is only presented here in details, as the discussion on the top of acrylic submersible is presented earlier.

Based on the new design guidelines, in the Phase II of this study, a new spherical acrylic submersible is fabricated at HBOI. Brief discussion of the experimental results on the new submersible is also presented here. A significant reduction in peak stresses and a very small relative displacement at the gasket/acrylic interface which are believed to be two of the main causes for crack development at the interface areas clearly indicate a major improvement in the new design of the acrylic submersible, as they are also suggested by the extensive FEA results. This improvement in design is expected to extend the crack free cyclic fatigue life of the acrylic submersible at 3000 ft (914.4 m) ocean depth significantly.

NOMENCLATURE

R_i – inner radius of acrylic sphere
R_o – outer radius of acrylic sphere
OD – outer diamter
ID – inner diameter
dl – radial displacement
dl_{rel} – relative radial displacement
rad_int - radial stress (S_R or SX) in acrylic at acrylic/nylon interface
long_int - longitudinal stress (S_θ or SY) in acrylic at acrylic/nylon interface
hoop_int - hoop stress (S_ϕ or SZ) in acrylic at acrylic/nylon interface
S1, S2 and S3 – first, second and third principal stresses, respectively
normalized stress – ratio of effective stress to applied pressure (S_{eff}/P_{appl})
normalized wall thickness – ratio of the distance from the inner wall to the total wall thickness
P_{appl} – applied pressure
P_{oper} – operational prressure, 1335 psi (9.21 MPa)

1. INTRODUCTION

The spherical acrylic submersible, as shown in Fig. 1, similar to NEMO (Naval Experimental Manned Observatory) type submersible, have served for several years as safe pressure hulls for Johnson-Sea-Link (JSL) submersibles #1 and #2. Although its performance in service to 3000 ft. (914.4 m) ocean depth initially satisfactory, the appearance of tiny shear cracks in the conical bearing surfaces of the penetrations in the spheres following only few hundred dive cycles requires frequent, expensive repairs. Significant savings in the operation of the JSL submersibles can be achieved if the crackfree cyclic fatigue life is extended to several hundred, or even thousand dive cycles. Various factors that contribute to formation of shear cracks have been identified through the present investigation, and accordingly, several major changes are made to the spherical hull assemblies in order to extend with certainty the crackfree cyclic fatigue life of the acrylic bearing surfaces.

Various researchers have conducted extensive hydrostatic tests and Finite Element Analysis (FEA) on scale-down and full size models for the design and improvement of the performance of these manned acrylic submersibles [1-14]. All the previous studies clearly indicate a significant increase in longitudinal strains and stresses at the acrylic bearing interfaces, and these are the areas where the shear cracks are to be noticed within acrylic sphere. Stachiw [7] observed in a 66 in (1.68 m) OD X 58 in (1.47 m) ID acrylic sphere (NEMO MOD 2000) under external hydrostaic pressure of 3000 ft (914.4 m) depth that the interior longitudinal strain at the top polar opening was 100% greater than the interior longitudinal strain at the equator, while the interior hoop strain at the top polar opening was 50% less than the interior hoop strain at the equator. This type of distribution of strains is considered typical for acrylic plastic sphere with a conical penetration sealed by a metallic hatch. This is also obvious from testing of acrylic spheres with t/D_i ratios in the 0.036 to 0.076 range during the previous test programs, which produced similar results [1, 2, 7, 9, 10 and 11]. Even with t/D_i = 0.2 with polycarbonate/nylon gasket for 8000 ft. (2438.4 m) operational depth, the meridional or longitudinal strain at the penetration in the upper hemisphere is found to be approximately 20% higher and the hoop strain at the same location approximately 80% lower than the strains at the equator [13]. As a result, this type of stress distribution, i.e. higher compressive stresses in the longitudinal direction, seems to be one of the main reasons for developing cracks at the interface area within acrylic sphere.

It is also known that the relative radial displacement at the interfaces, caused by the variation of stiffness between different materials [10], is considered to be another major factor in contributing to initiation of cracks in the acrylic bearing surfaces. Snoey and Katona [10] reported the maximum relative movement at the acrylic-steel interface for only 1000 ft (304.8 m) depth in a 66 in (1.68 m) OD X 61 in (1.55 m) ID acrylic submersible to be 0.11 in (0.0028 m). Therefore, if the magnitude of the longitudinal stress within acrylic sphere and relative radial displacement at the acrylic/nylon interface could be reduced at design depth of 3000 ft. (914.4 m), the cyclic fatigue life would improve. To achieve these goals, a two phases research program is initiated at HBOI to redesign the spherical acrylic submersible so that the peak stresses and the relative radial displacement at the edges of the penetrations in the acrylic sphere could be reduced.

The objectives of Phase I of the present research program are to : 1) develop a full scale 3-D non-linear FEA model, similar to the acrylic spherical submersible that HBOI uses for ocean exploration, in order to simulate the structural behavior at the interface and throughout the sphere, 2) develop various models on the basis of the different hatch & plug geometry, shape and thickness of nylon gaskets, openings of the acrylic sphere, placement of rubber padding and attachment of retainer ring to the hatch ring or bottom plug, to investigate their effect on the overall performance of the acrylic spherical submersible, and, 3) redesign and develop the new design guidelines for the acrylic plastic spherical pressure vessel on the basis of the findings from (1) and (2), which will lower the stresses and relative movement in acrylic hull at the interfaces as well as maintain the stresses in other parts within their yield limit.

The objectives in Phase II of the present program are to: 1) fabricate a new acrylic submersible according to the proposed new design guidelines, 2) perform the hydrostatic proof test to satisfy the requirements of PVHO (Pressure Vessels for Human Occupancy) and ABS (American Bureau of Shipping) criteria for manned submersibles, and, 3) discuss the experimental results and their correlation with previously published experimental results, based on NEMO types submersibles as well as the FEA results, as presented in the Phase I of the studies.

PHASE I

2. 3-D FINITE ELEMENT ANALYSIS

Development of the 3-D FEA model and initial analysis to determine the causes for crack generation at the interface of the acrylic submersible is provided in details in Das's earlier work [14]. Redesigning of top of acrylic submersible including various analyses, such as variation of thickness and shape of nylon gasket, a small gap at the inner side of acrylic hull and nylon gasket, variation of hatch material and its shape, redesign of aluminum hatch and hatch ring, are also presented in details in this earlier work. As a continuation, the redesigning of the bottom of acrylic submersible will only be presented here, which basically includes only the redesigning of aluminum bottom plug or penetrator plate. In addition, the effects of the placement of rubber gasket in between acrylic sphere and retainer ring, as well as, attachment of retainer ring with aluminum hatch ring will also be discussed.

2.1 Redesign of Aluminum Bottom Plug or Penetrator Plate

Redesign of the aluminum bottom plug is carried out with the same objective as the top aluminum hatch and hatch ring, i.e. to lower the rigidity of the plug in order to lower the effective peak stresses at the interfaces of the acrylic sphere. Figure 2 shows the 3-D FEA model [16,17,18] with external loading and boundary conditions for the bottom of acrylic submersible. Three parameters are considered for this purpose and they are : i) fillet radii, ii) diameter of the bottom plug, and iii) placement of the holes in the plug, as shown in Fig. 3. Increase in fillet radius at position B from 0.281 in. (0.007 m) to 0.6 in. (0.015 m) reduces the peak stress (Seqv) in the aluminum bottom plug, without increasing the stress in the acrylic sphere, see Table 1(a). The effect of the variation of the diameter of the bottom plug is found similar to the hatch diameter and the peak stresses in the acrylic, as can be seen from Table 1(b), are observed to lower with the smaller bottom plug. However, the presence of the 60 holes in the bottom plug for the electrical and hydraulic connections represents an obstacle for any arbitrary reduction in its size. Therefore, without reducing the number of holes in the bottom plug, an alternative way of reducing the size of the bottom plug is the minimization of the spacing between the holes. Moreover, if the holes in the plug are moved closer to the center of the plug than the present positions, the peak stresses in the plug itself (near hole location) are noticed to decrease, and, this seems to be due to the lower bending moment generated near the hole. The reduction of peak stresses near the hole (considered only Hole # 1) in the bottom plug due to the movement of the hole closer to the center of the plug are shown in Table 1(c). Therefore, the movement of the holes towards the center not only helps to reduce the diameter of the bottom plug, but also helps to increase one of the fillet radii from 0.281 in (0.007 m) to 0.6 in (0.015 m), which helps to drop the effective stresses in the bottom plug significantly. As a whole, the following modifications are made in the new design :

a) Diameter of the bottom plug is reduced from 19.637 in. (0.4988 m) to 18.688 in. (0.475 m)
b) One of the fillet radii in the bottom plug is increased from 0.281 in. (0.007 m) to 0.6 in. (0.015 m) and
c) The holes within the bottom plug are being moved towards the center of the plug and are being rearranged in such a way so that there is enough space for wrench placement and the stress within the penetrator plate remains always under its allowable yield limit (i.e. 28000 psi/193.05 MPa) at operational ocean depth pressure of 3000 ft (914.4 m).

All these modifications also clearly indicate that the bottom plug, as the top aluminum hatch and hatch ring [14], is also over designed in the existing HBOI's acrylic submersible.

2.2 Placement of Rubber Gasket & Attachment of Aluminum Retainer Ring with Aluminum Hatch Ring

Figure 4 indicates the method of retaining the aluminum penetrations in the acrylic hull : retainer rings bolted to the top hatch ring and the bottom plug. A rubber gasket is used as a cushion between the acrylic hull and the retainer ring (at both top and bottom of the sphere). The main purpose of this complete arrangement is to make the acrylic sphere leak proof while it starts going under the water, as well as, to keep all the relative positions of various parts almost same throughout the operations. Since the rubber gasket is basically used for cushioning, it should not exert any additional forces on the acrylic sphere to increase the peak stresses at the interface. Therefore, the effect of the position of the rubber gasket with respect to the acrylic/nylon interface is investigated, based on the following assumptions for the FEA modeling : i) retainer ring is tightly bolted with the hatch ring, i.e. the retainer ring and the hatch ring are a continuous part, and, ii) compressed rubber gasket in between acrylic hull and the retainer ring is modeled with solid elements, SOLID45, with a stiffness of 409E+2 psi (281.995 MPa) and poisson's ratio of 0.4996. Two cases have been studied, with rubber gasket of 0.23 in. (0.0058 m) and 0.84 in. (0.0213 m) away from the acrylic/nylon interface, as shown in Figs. 5(a) & 5(b), respectively. Increase in peak stresses are observed for both the cases, however, the closer the position of rubber gasket to the interface, higher is the value of the peak stresses, as shown in Table 2. Therefore, it seems to be a better choice to place the rubber gasket away from the interface, instead of placing this exactly at the interface, as is done in the existing HBOI's acrylic submersible.

As the retainer ring is tightly bolted with the hatch ring, the acrylic sphere generates load on the retainer ring, while it is under the external pressure. As a result, the bending stress is generated within the retainer rings which acts as tensile and bending loads through the bolts and may elongate these bolts beyond their yield limit and damage permanently. To circumvent this situation, spring loaded thrust washer is introduced to eliminate any extra tensile load on the bolts by allowing further compression of the thrust washer (see Fig. 4). Moreover, in order to avoid any bending loads on the bolts due to bending of the retainer rings, larger slots in the retainer rings for the bolts connections are proposed, as shown in Fig. 6, and, this will allow the retainer ring to bend/rotate without applying any bending loads on the bolts. Since the movement of the retainer ring is not restricted due to the presence of the thrust washer and the larger slots, this will simultaneously help to eliminate any extra load on the acrylic sphere due to the presence of the retainer ring and rubber gasket, which is our main objective. Therefore, the rubber gasket between acrylic sphere and retainer ring is placed 0.5 in (0.0127 m) away from the acrylic/nylon interface, as well as, larger slots in the retainer rings are introduced in the new design to avoid any further increase in peak stresses within the acrylic sphere.

3. NEW DESIGN

3.1 Design Modifications

On the basis of the above analyses, following design modifications (only for bottom of acrylic submersible, top of submersible is already presented in [14]) have been made within PVHO guidelines, which shows a significant reduction in stresses within acrylic submersible. The modifications are in :

i) Penetration, i.e. opening (α) at bottom is increased to 50^0 [14].

ii) Inner diameter of the bottom plug (D_{plug}) is reduced from 19.637 in (0.4988 m) to 18.688 in (0.4747 m) to provide space for thicker nylon gasket.

iii) The fillet radius at position B is increased from 0.281 in (0.00714 m) to 0.6 in (0.01524 m).

iv) The number of through holes for electrical and hydraulic connections are kept same and all the holes are rearranged and moved closer to the center of the plug.

v) The thickness of the bottom nylon gasket is increased from 1.26 in (0.032 m) to 3.025 in (0.0768 m) (measured at the inner side).

vi) The shape of the nylon gasket is changed from straight to spherical at the inner side to maintain the continuity between acrylic hull and nylon gasket.

vii) The outside flange is removed from the nylon gasket, instead, it is flushed with the outside flat area of the acrylic sphere.

viii) Larger slots are provided in the retainer rings for bolts connections, and, the rubber gasket is placed 0.5 in (0.0127 m) away from the nylon/acrylic interface for cushioning.

3.2 Results and Discussions

With the incorporation of the above modifications, the bottom of the acrylic submersible is analyzed in two steps. In the first step, the analysis on the bottom of acrylic submersible is carried out without any holes in the bottom plug. The maximum peak stress (S3) in acrylic at the acrylic/nylon interface is noticed to be –6674 psi (-46.016 MPa), dropping from –7535 psi (-51.952 MPa) in the existing HBOI's submersible (plug with no holes). In the second step, similar analysis is carried out with the inclusions of the holes in the bottom plug. For this case, variations of the different stress distributions through the acrylic wall at the bottom acrylic/nylon interface are shown in Fig. 7(a). The reduction of peak stress (S3) at the bottom acrylic/nylon interface from –7053 psi (-48.628 MPa) in HBOI's existing submersible (including holes in the bottom plug) to -6346 psi (-43.754 MPa) in new submersible is observed due to these modifications and this seems to help to prolong the life of acrylic sphere. Moreover, the normalized radial displacements, as shown in Fig. 7(b), at the bottom interfaces clearly indicate a low relative movement of only 0.011 in (0.00028 m) of acrylic hull with respect to nylon gasket, which is desirable. As a result, the overall stress distributions throughout the bottom of acrylic submersible, as shown in Fig. 7(c), are found as : a) <u>within acrylic hull</u> : S3 = -6346 psi (-43.754 MPa), Seqv = 5286 psi (36.446 MPa) at acrylic/nylon interface near inner side, b) <u>within nylon gasket</u> : S3 = -12782 psi (-88.129 MPa), Seqv = 9467 psi (65.273 MPa) at nylon/plug interface near inner side, and, c) <u>within aluminum plug</u>: S3 = -24978 psi (-172.217 MPa), Seqv = 24292 psi (167.487 MPa) near hole edge. It is to be mentioned that although the peak stresses in the aluminum bottom plug always found lower than its yield limit, however, the peak stresses (S3 and Seqv) in bottom nylon gasket is noticed always to be higher than its yield limit, as the top nylon gasket [14].

3.2.1 Elasto-Plastic Analysis

Since the peak stresses in nylon gaskets are always found to be higher than its yield strength, another analysis is carried out for the top of acrylic submersible considering the material non-linearity for the nylon gasket. In this FEA model, nylon gasket is modeled as elastoplastic material description based on von Mises yield criteria coupled with an isotropic work hardening assumption, known as Multilinear Isotropic Hardening (MISO) option [16]. The behavior of the material also includes large deformation and larger strain criteria. Due to the material non-linearity, the peak stresses (S3 and Seqv) are noticed to drop slightly in nylon gasket, as can be seen in Table 3, locally in a small region at the gasket/hatch ring interface. However, this does not seem to have any change of effect on the acrylic sphere, as the peak stresses in the acrylic sphere remain same. Therefore, although the stress in nylon gasket decreases in the new model and the gasket has never shown any visible cracks even the initial stress is observed to be higher than its yield strength of 12320 psi (84.943 MPa) in the existing submersible, it is recommended to replace the nylon gasket from time to time to avoid any adverse effect on acrylic sphere due to the residual stresses in the nylon gasket, where the time frame is yet to be established. Similar results have also been noticed for the bottom of acrylic submersible.

<u>PHASE II</u>

4. FABRICATION OF NEW ACRYLIC SUBMERSIBLE

Based on the new design guidelines, presented in the Phase I of this paper, a new spherical acrylic submersible is fabricated. The acrylic sphere is fabricated from two hemispherical acrylic castings joined together with a single adhesive bond by Reynolds Polymer Technology, Inc., Colorado. The nylon gaskets (Type 6) and all the aluminum (Type 6061-T6) parts, i.e. top hatch, hatch ring, bottom plug and the retainer rings are machined at HBOI, based on the new design. The acrylic sphere is then fitted with inserts and closures on the top and bottom ports for the hydrostatic tests. A detail cross sectional view of HBOI's new spherical acrylic submersible is shown in Fig. 8.

5. PRESSURE TESTING

The 68.3 in (1.735 m) OD × 57.8 in (1.468 m) ID HBOI's new spherical acrylic submersible is subjected to a series of two hydrostatic tests of 1.25 × 1335 psi, i.e. 1670 psi (3750 ft) or 11.5 MPa (1143 m) proof pressure at Southwest Research Institute (SwRI), San Antanio, TX. The objectives of these pressure tests are threefold:

i) to qualify the new hatch and penetrations bulkhead assembly designs for conformance to American Bureau of Shipping (ABS) for building and classing of underwater vehicles;
ii) to satisfy the acrylic material performance requirements of the American Society of Mechanical Engineers (ASME) PVHO Safety Standard for acrylic windows; and,
iii) to generate experimentally the strains and stresses in acrylic sphere at 3000 ft (914.4 m) operational depth so that they could be used for comparison with previously generated data on NEMO type submersibles as well as to further increase our understanding of the mechanical behavior at the interfaces in order to solve crack generation problem.

The two tests at 25% overpressure is required to satisfy the requirement for classing of metallic pressure hull components by both ABS rules, and ASME PVHO Safety Standard for proof testing of new windows prior to placement in manned service.

To satisfy the first objective, the hatch and penetration bulkhead assembly are instrumented on their low pressure surfaces with 14 triaxial electrical strain gages (CEA-13-125UR-350) located including the areas of peak stresses according to the FEA results, as shown in Fig. 9. To satisfy the second objective, the new acrylic sphere is instrumented on the interior with 10 biaxial strain gages (CEA-13-125UT-350) at locations, such as, equator, edges of penetrations and 5 in (0.127 m) away from penetrations, see Fig. 9. In addition, two linear displacement gages are also placed on the upper hatch assembly (see Fig. 9) to monitor the movement between nylon inserts and the acrylic hull relative to the retainer ring. The static test series consisted of the following two tests conducted at temperature range of 34-40^{0}F in the pressure test facilities of SwRI :

a) First Test – pressurize to 1670 psi (11.5 MPa) at 50 psi/min (0.0057 MPa/s) rate, hold at that pressure for 4 hr, depressurize to 0 psi (0 MPa) at 50 psi/min (0.0057 MPa/s) rate, and relax at that pressure for 12 hr prior to beginning of the second test.
b) Second Test - pressurize to 1670 psi (11.5 MPa) at 50 psi/min (0.0057 MPa/s) rate, hold at that pressure for 1 hr, depressurize to 0 psi (0 MPa) at 50 psi/min (0.0057 MPa/s) rate, and relax at that pressure for 1 hr prior to removal from the vessel.

6. TEST RESULTS AND DISCUSSIONS

The new acrylic submersible withstood successfully the two hydrostatic pressure loadings to 1670 psi (11.5 MPa) without appearance of cracks in the acrylic and nylon gasket bearing surfaces at the polar openings. There is found no leakage, or any permanent deformation of metallic penetration closures. The redesigned HBOI top hatch, hatch ring and bottom penetrator plate/plug assemblies performed at 1.25 design pressure, i.e. at 1670 psi (11.5 MPa), as designed.

6.1 In Aluminum Penetrators

The strains recorded by 14 strain gages that are attached to the top and bottom aluminum penetrators and corresponding stresses are discussed in details in [20,21]. Only the peak values at operational depth pressure of 1335 psi (9.21 MPa) and at proof pressure of 1670 psi (11.5 MPa) are presented here and compared to the predicted values of FEA results. Maximum compressive strain in top penetrator assembly is experienced by gage # 17, which is attached to the top hatch ring, whereas, maximum strain in bottom plug is recorded by the gage # 24, see Fig. 9. The highest magnitudes of recorded principal stresses at 1670 psi (11.5 MPa) are –19707 psi (-135.87 MPa) in the hatch ring and –9403 psi (-64.83 MPa) in the bottom plug. Both these stresses are well below the allowable 28000 psi (-193.05 MPa) of ABS stress limits for 6061-T6 aluminum alloy. Moreover, the strains are observed not to change in aluminum penetrators during 4 hr. or 1 hr. holding of the pressure loading [21], which indicates absence of any creep within aluminum hatch, hatch ring and the bottom plug. Therefore, it can be concluded that the redesigned aluminum penetrators not only safe, but also satisfy all the ABS requirements for manned service.

To compare with the FEA results, strains and stresses at two locations (Gage # 17 & Gage # 24) are presented in Table 4. Although a little variation in the experimental results and the predicted values by FEA are seen here, however, the close resemblance in the values and pattern of strain/stress behavior clearly indicate the accuracy of the FEA calculations. Moreover, the results from FEA calculations, as found to be always higher than the experimental values, also indicates the initial design to be more conservative and safe.

6.2 In Acrylic Sphere

The 68.3 in (1.735 m) OD, 5.25 in (0.133 m) thick acrylic sphere with 50^0 polar penetrations exceeds all of the PVHO requirements for a NEMO window for a design pressure of 1335 psi (9.21 MPa) at 50^{0}F(10^{0}C) design temperature [20]. Moreover, the acrylic sphere has not shown any cracks or crazing at the bearing surfaces, nor any permanent deformation after two proof tests. Detail presentations of strains and stresses through the acrylic sphere at inside surface under hydrostatic tests are given in [20,21]. Here we only concentrate on the experimental results at operational depth pressure and proof pressure, and, discuss their correlation with previously published experimental results, based on NEMO types submersibles as well as the FEA results, as presented in the Phase I of the studies.

The maximum strains at the equator on the interior surface at the beginning of sustained pressurization at proof pressure of 1670 psi (11.5 MPa) are found to be –8550 micro-in/in (-8550 micro m/m) in hoop and –8305 micro-in/in (-8305 micro m/m) in longitudinal direction. After 1 hr of sustained loading, they increased to –9007 and –8710 micro-in/in (-9007 and –8710 micro m/m), respectively [20], which indicates the presence of creep strain within acrylic sphere. The maximum compressive stress at the beginning of the sustained loading is observed to be –5096 psi (-35.14 MPa) in the hoop direction and –5017 psi (-34.59 MPa) in the longitudinal direction, on the basis of $E_{acrylic}$ = 420E+3 psi (2895.8 MPa), & Poisson's ratio = 0.3. However, according to Lame' solution for a continuous thick sphere, the compressive stress at equator in the hoop/longitudinal directions are supposed to be –6359 psi (–43.84 MPa), which suggests for acrylic material with $E_{acrylic}$ = 528E+3 psi (3640.4 MPa), instead of 420E+3 psi (2895.8 MPa). As a result, for direct comparison with FEA results, the FEA models are rerun with only change in acrylic stiffness, and all the experimental stresses are calculated using the new values for acrylic.

Variation of longitudinal and hoop strains through the FEA model of new acrylic sphere at inner surface is shown in Fig. 10 for an operational depth of 3000 ft (914.4 m). The strains at the equator (i.e. Angle = 0 deg.) are very close to the Lame's solution, as expected. However, they change pattern of distribution as they move closer to the opening edges, i.e. near ±65 deg. Angle, where the hoop strain drops about 36% and the longitudinal strain increases about 27%. This is exactly the similar pattern of strain distribution, reported by Stachiw [7] on NEMO MOD 2000. The major difference is in the value of longitudinal strain distribution, which is reported as 100% higher at the top polar opening edge than at the equator, see Table 5. Not only that, in all the previous studies, the maximum strains in acrylic sphere are found at the interior edge of the penetration in longitudinal direction, which is believed to be one of the main reasons for developing cracks at the acrylic interface. It's magnitude is found to be approximately 50-100% higher (depending on the gage locations from the edge) than the measured strain on the interior surface at the equator [7,9,11,12]. In contrast, the strains on the interior surface at the opening edge of the new acrylic submersible, as shown in Table 5, are about 61% lower in longitudinal direction and about 6% lower in hoop direction than their corresponding strains at the equator. As a result, the maximum stresses are found at 5 in (0.127 m) away from the top opening (in the longitudinal direction) as –5727 psi (-39.49 MPa) and –6931 psi (–47.79 MPa) for operational depth and proof test pressures, respectively, whereas, at the edge of the acrylic sphere, the maximum stresses in longitudinal direction are found as only –2365 psi (-16.31 MPa) and –2998 psi (-20.67 MPa), respectively, for the same pressures. Therefore, the peak stresses throughout the acrylic sphere are always found to be lower than acrylic's long term yield strength of -7000 psi (-48.263 MPa). The lower values of strains and corresponding stresses at the opening edges are very encouraging, although the strains predicted through FEA model are found higher than the measured strains in the longitudinal direction of the new acrylic submersible, see Table 5. As a whole, this is a very significant improvement compared to any previous design in reducing the peak stresses at the interface area within acrylic sphere. Therefore, the overall new design is believed to increase the crack free cyclic lives of the acrylic submersible considerably.

6.3 Relative Displacement Between Nylon Gasket and Acrylic Sphere

Variation of normalized relative radial displacements of top nylon gasket and acrylic sphere with respect to upper hatch assembly are shown in Figs. 11(a) and 11(b) for 1st and 2nd test, respectively, during pressurization from 0 to 1670 psi (11.5 MPa). A very small relative displacement of 0.02 in (0.0005 m) is observed between gasket and acrylic sphere during 1st test at 1670 psi (11.5 MPa), whereas, during the 2nd test, a relative displacement of only 0.01 in (0.00025 m) is observed at the same pressure. This clearly shows absence of any significant differential movement between the gaskets and the acrylic sphere at the interface, as also suggested by the FEA results, see Fig. 7(b) . The lower relative displacement between gasket and acrylic sphere indicates that the overall system is more compliant. As a result, this is further believed to help in delaying the development of tiny shear cracks at the interface area.

7. SUMMARY

i) The redesigned hatch, penetrator plate/plug, and gasket assemblies have reduced significantly the magnitude of peak stresses typically encountered on the interior surface of the sphere near penetrations.

ii) Very small or negligible relative movement between gasket and acrylic sphere at the interface also indicates that the system is more compliant, and thus reduce the risk of developing shear cracks in acrylic sphere at the interface.

iii) The materials, configuration, fabrication, and quality of new HBOI acrylic sphere meet all the requirements of ASME PVHO-1 and ABS for design depth of 3000 ft (914.4 m) under 1335 psi (9.21 MPa) of operational pressure.

iv) The structural performance of new acrylic sphere at 1670 psi (11.5 MPa) hydrostatic proof pressure indicates that the sphere is safe for repeated dives of any duration to design depth of 3000 ft (914.4 m).

9. CONCLUSIONS

The changes in structural performance of new acrylic submersible introduced by the redesigned hatch and penetrator plug assemblies are beneficial and is expected to contribute to extend crackfree cyclic fatigue life of the sphere at 3000 ft (914.4 m) operational depth.

10. RECOMMENDATIONS AND FUTURE WORK

The following recommendations are proposed to further improve the performance of the acrylic submersible as well as to increase the understanding of the interface behavior. These are :

i) Increase the opening angle of the acrylic sphere to more than 50^0, i.e. beyond PVHO limitations, to incorporate a thicker nylon gasket.
ii) Replace the aluminum penetrators with penetrators made of nylon or similar material to reduce the structural rigidity between the acrylic hull and the penetrators.
iii) Record and monitor the strain data to determine the change of behavior of acrylic hull and nylon gasket at the interface over several hundred of dives, and,
iv) Perform an extensive and systematic creep analysis of the submersible, since the presence of residual strain in the acrylic sphere due to creep following repeated dives seems to be another underlying reason for the presence of shear cracks within acrylic sphere at the interface.

ACKNOWLEDGEMENTS

The author acknowledges with great appreciation the support for this work from HBOI, Inc., Fort Pierce, FL. The author is deeply indebted to Dr. Jerry Stachiw for his useful discussions and suggestions. The author would also like to thank Rick Herman, Tim Askew, Jerry Neely and all the members of the Sphere Committee at HBOI for their generous support throughout this work.

REFERENCES

1. Stachiw, J. D., "Spherical Acrylic Pressure Hulls for Undersea Exploration," American Society of Mechanical Engineers, Paper No. 70-WA/UnT-3, Journal for Engineering for Industry, Vol. 53, 1971a.
2. Stachiw, J.D., "Acrylic Pressure Hull for Submersible NEMO", ASME Paper 71-Unt-2, 1971b.
3. Staciw, J. D. and Mack, K. L., "The Spherical Acrylic Pressure Hull for Hydrospace Applications, Part 2. Experimental Stress Evaluation of Prototype NEMO Capsule," Naval Civil Engineering Laboratory, Technical Note N-1113, October 1970 (AD 715772).
4. Ottsen, H., "The Spherical Acrylic Pressure Hull for Hydrospace Applications, Part 3. Experimental Stress Evaluation of Prototype NEMO Capsule," Naval Civil Engineering Laboratory, Technical Note N-1094, March 1970 (AD 709914).
5. Staciw, J. D., "The Spherical Acrylic Pressure Hull for Hydrospace Applications, Part 4. Cyclic Fatigue of NEMO Capsule #3," Naval Civil Engineering Laboratory, Technical Note N-1134, October 1970 (AD 715345).
6. Stachiw, J. D., "Acrylic Pressure Hull for Submersible NEMO," American Society of Mechanical Engineers, Paper No. 71-UnT-2, December 1970.
7. Stachiw, J. D., "NEMO Type Acrylic Plastic Spherical Hull for Manned Operation to 3000 ft Depth," American Society of Mechanical Engineers, Paper No. 75-WA/OcE-3, Journal for Engineering for Industry, May 1976, pp. 537-549.
8. Maison, J. R. and Stachiw, J. D., "Acrylic Pressure Hull for Johnson-Sea-Link Submersible," American Society of Mechanical Engineers, Paper No. 71-UnT-2, December 1970.
9. Snoey, M. R. and Katona, M. G., "Structural Analysis of a Full Scale Acrylic Plastic Pressure Hull," Naval Civil Engineering Laboratory, Technical Note R-176, March 1971.
10. Snoey, M. R. and Katona, M. G., "Stress Analysis of a Spherical Acrylic Pressure Hull," American Society of Mechanical Engineers, Paper No. 70-UnT-B, December 1970.
11. Snoey, M. R. and Briggs, E. M., "The NEMO Submersible", ASME Paper 70-Unt-C, 1970.
12. Stachiw, J. D. and Dolan, R. B., "Spherical Acrylic Pressure Hulls with Multiple Penetrations", ASME Journal of Engineering for Industry, Vol. 100, 1978.
13. Stachiw, J. D., Clark, A. and Brenn, C. B., "Acrylic Plastic Spherical Pressure Hull for 2439 m (8000 ft) Design Depth: Phase I", Journal of Energy Resources Technology, Vol. 109, March 1987, pp. 40-47.
14. Das, P. S., "A Parametric Investigation on the Structural Analysis of the Interface Between Metallic Penetrations and the Acrylic Hull of a Spherical Pressure Vessel," PVP-Vol. 331, Pressure Vessels and Piping Design, Analysis, and Severe Accidents, American Society of Mechanical Engineers, 1996.
15. Das, P. S., "Detailed Stress Analysis of a Spherical Acrylic Submersible by 3-D Finite Element Modeling", PVP-Vol. 398, Recent Advances in Solids and Structures, American Society of Mechanical Engineers (ASME), 1999, pp. 153-165.
16. ANSYS User's Manuals, September 1997.
17. ANSYS Elements Reference, Release 5.4, September 1997.
18. ANSYS Modeling and Meshing Guide, Release 5.4, September 1997.
19. Safety Standard for Pressure Vessels for Human Occupancy, ASME PVHO-1-1990 Edition.
20. Das, P. S., "Redesign of Johnson-Sea-Link Acrylic Submersible for Manned Operation to 3000 ft (914.4 m) Ocean Depth", Final Report, HBOI Project No. CM006.1, April, 2000.
21. Ramon, J. I. and Jones, M., "Hydrostatic Pressure Tests on a Manned Submersible Acrylic Spherical Hull for a Johnson-Sea-Link II Submersible", Final Report, Southwest Research Institute Project No. 18.03184, March 2000.

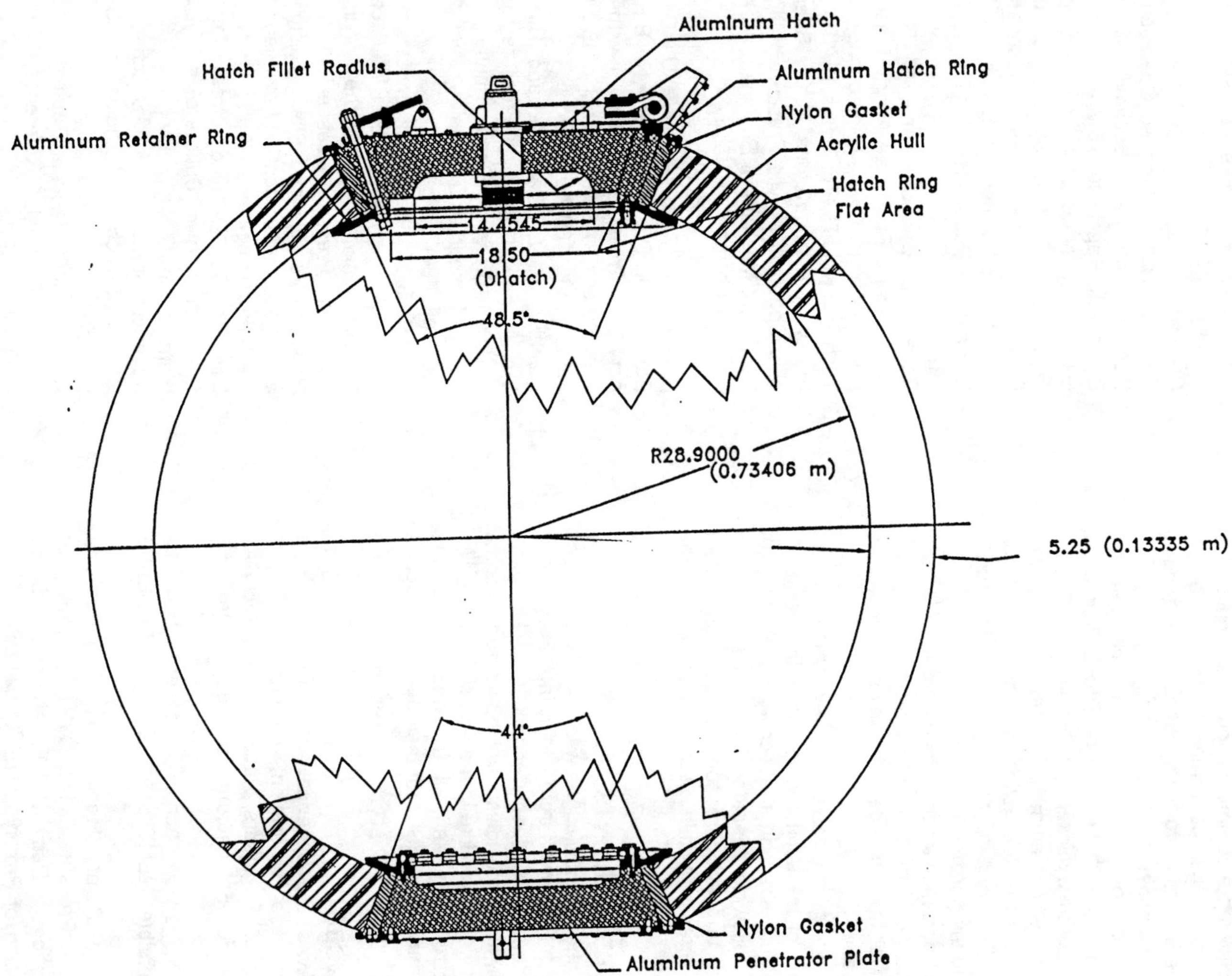

Figure 1 : A Sectional View of HBOI's Acrylic Spherical Submersible

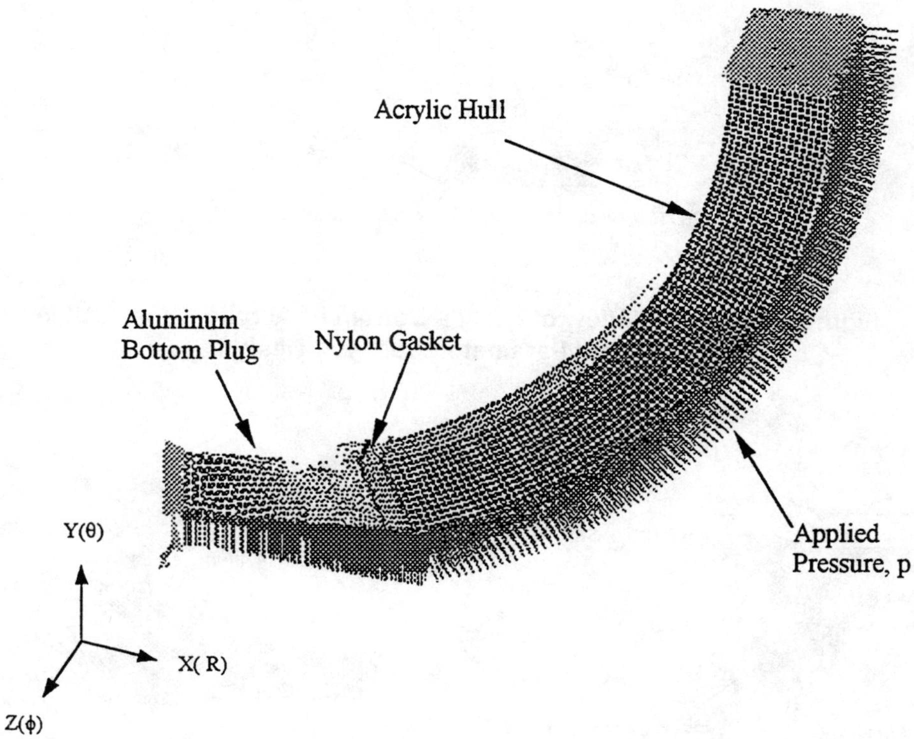

<u>Material Properties:</u>

Materials	**Compressive Modulus of Elasticity psi (MPa)**	**Poisson's Ratio**	**Compressive Yield Stregth psi (MPa)**
Aluminum	9550E+3 (65.845E+3)	0.3	35000 (241.317)
Nylon	535E+3 (36.89E+3)	0.3	12320 (84.943)
Acrylic	420E+3 (28.96E+3)	0.3	16530 (113.970) -Short Term
			7000 (48.263) -Long Term

<u>Properties at contact/sliding interfaces –</u>
Between aluminum hatch ring and nylon gasket
a) μ, i.e. coefficient of friction = 0.27
b) contact stiffness = 9550E+3 psi (65.845E+3 MPa)
Between nylon gasket and acrylic sphere
a) μ, i.e. coefficient of friction = 0.32
b) contact stiffness = 535E+3 psi (36.89E+3 MPa)

Figure 2 : A Discretized FEA Model for Bottom of HBOI's Acrylic Submersible with External Loading and Boundary Conditions, Including Material Properties

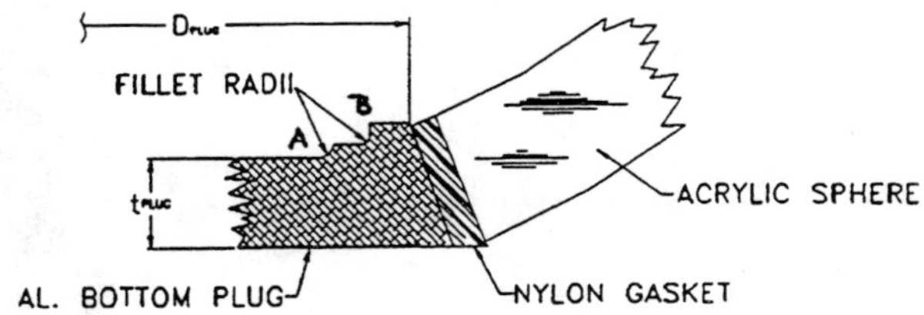

Figure 3 : Sectional View of HBOI's Bottom of Acrylic Submersible with Different Parameters for Plug Design

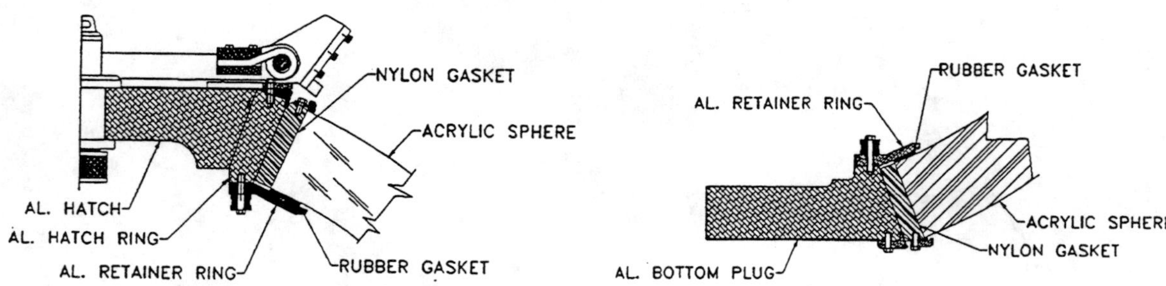

Figure 4 : Top and Bottom of Acrylic Submersible Showing the Method of Retaining the Aluminum Penetrations in the Acrylic Hull

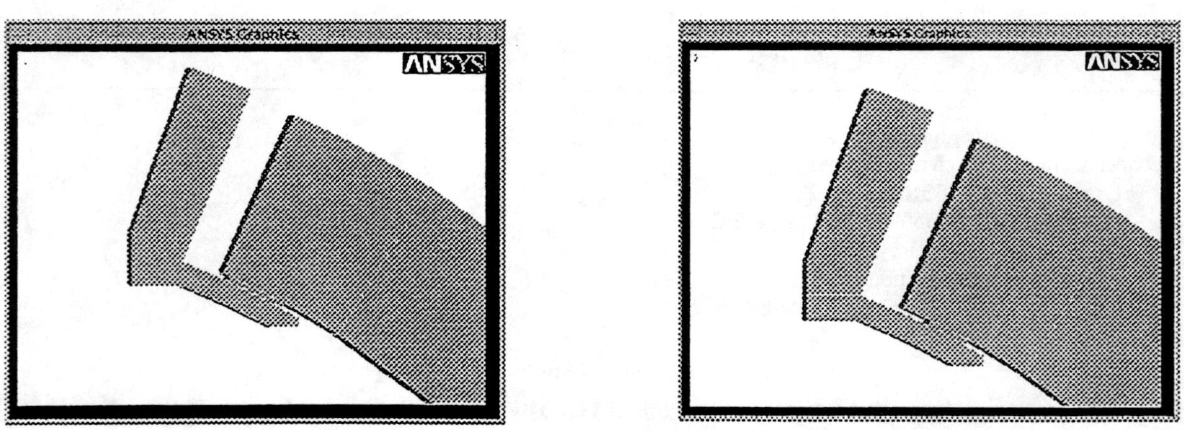

(a) x = 0.23 in (0.0058 m)

(b) x = 0.84 in (0.0213 m)

Figure 5 : Placement of Rubber Gasket

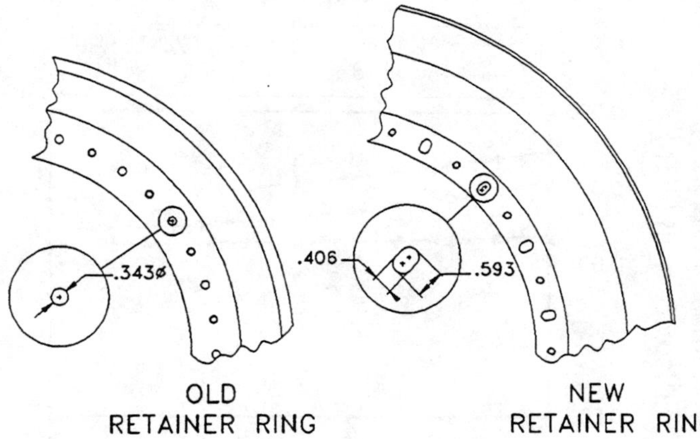

Figure 6 : Variation of the Bolt Holes in the Retainer Rings

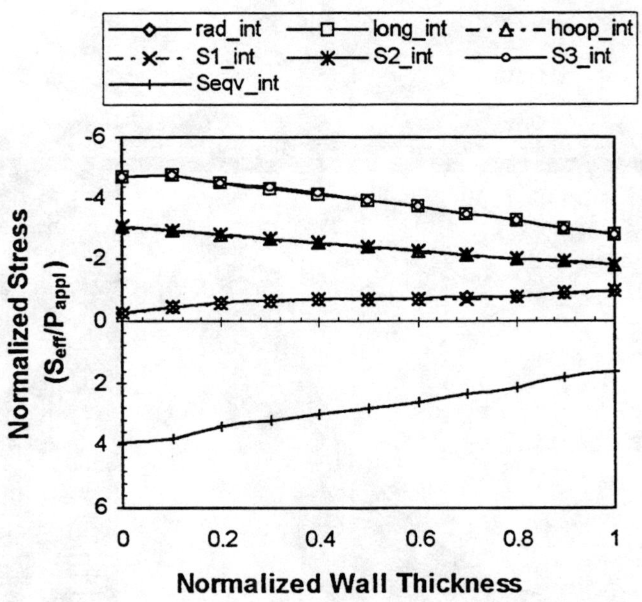

Figure 7(a) : Distribution of Different Stresses (Normal, Principal and Equivalent) at Acrylic/Nylon Interface Through the Bottom Wall of New Acrylic Submersible at 3000 ft (914.4 m) Ocean Depth

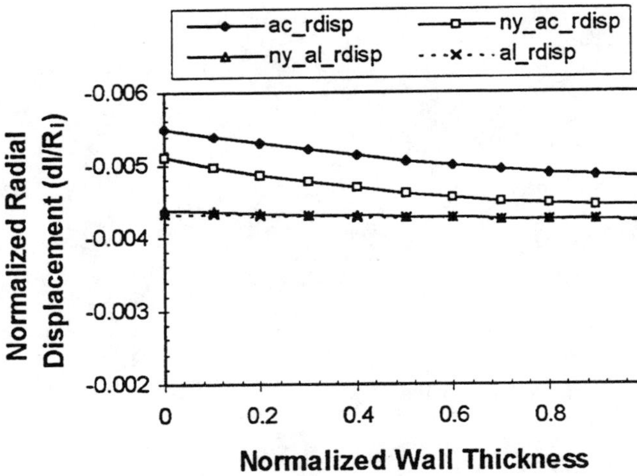

Figure 7(b) : Variation of Normalized Radial Displacement of Acrylic Hull and Nylon Gasket at the Bottom Acrylic/Nylon Interface and of the Nylon Gasket and Aluminum Bottom Plug at the Gasket/Plug Interface Through the Wall of The New Acrylic Submersible at 3000 ft (914.4 m) Ocean Depth

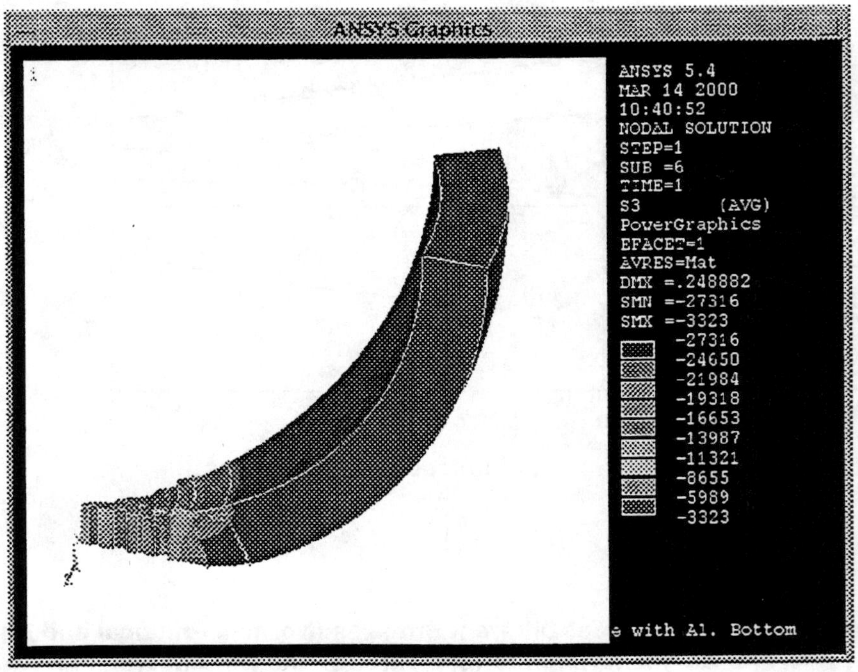

Figure 7(c): Overall Stress (S3) Distribution Through The Bottom Of New Acrylic Submersible (Al. Plug with Holes) At 3000 ft. (914.4 m) Ocean Depth

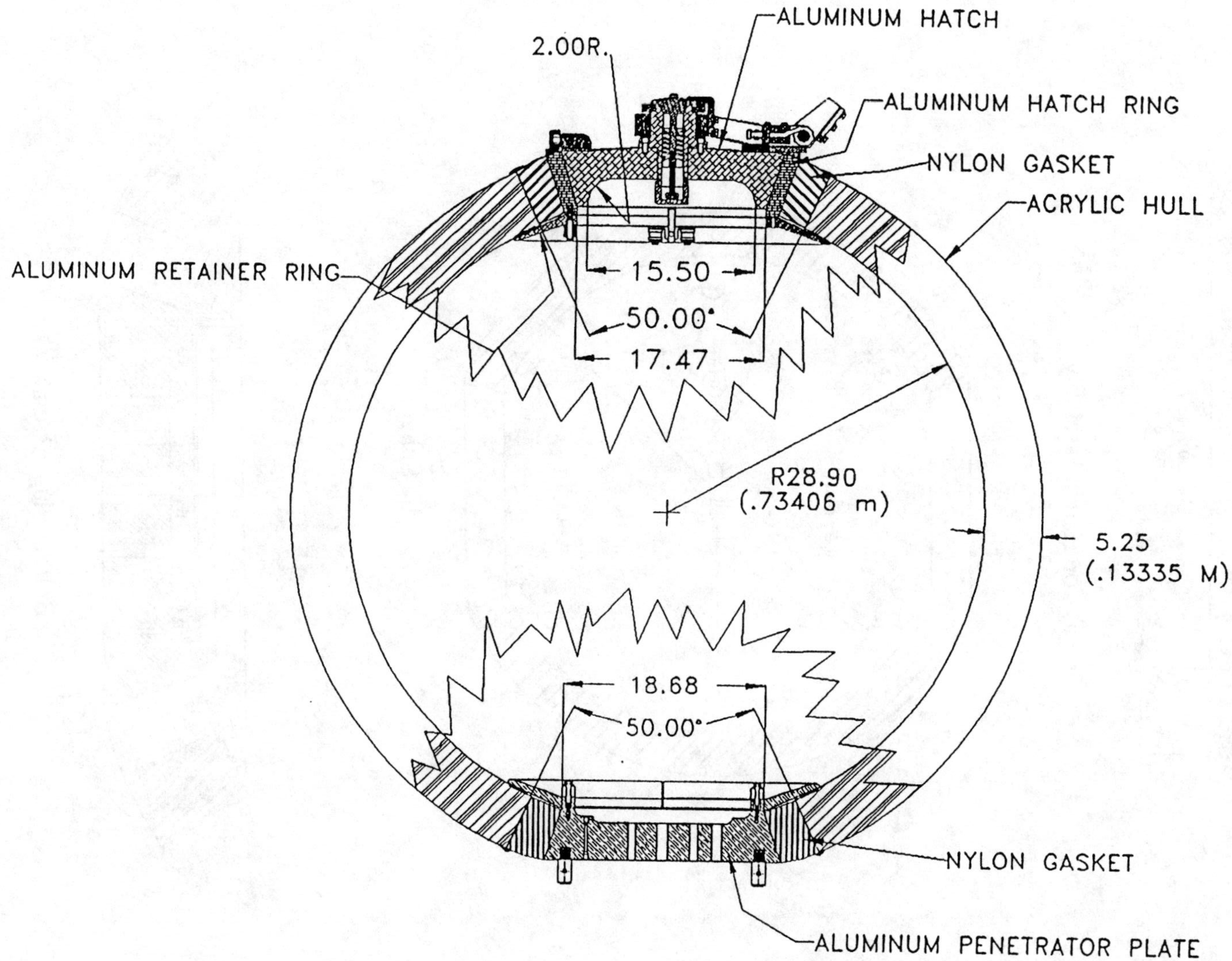

Figure 8 : A Sectional View of HBOI's New Acrylic Submersible

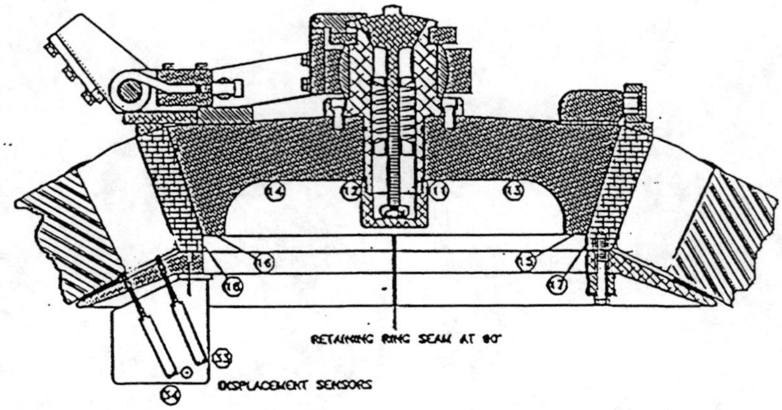

a) Top Hatch Assembly

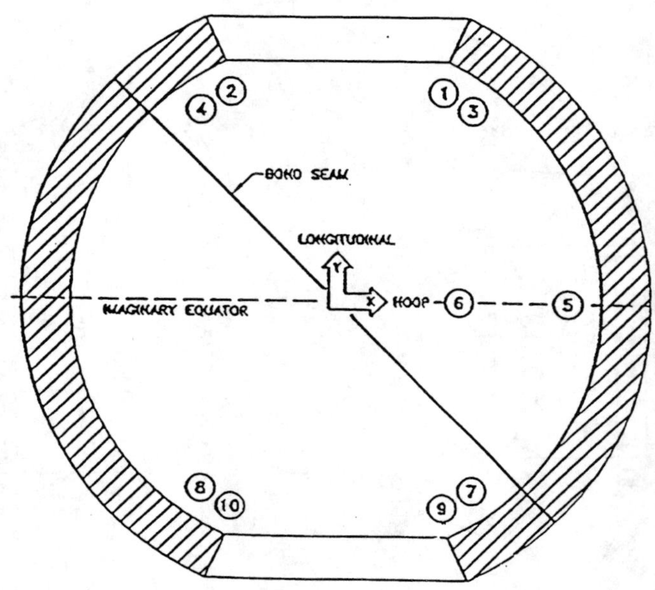

b) Acrylic Sphere

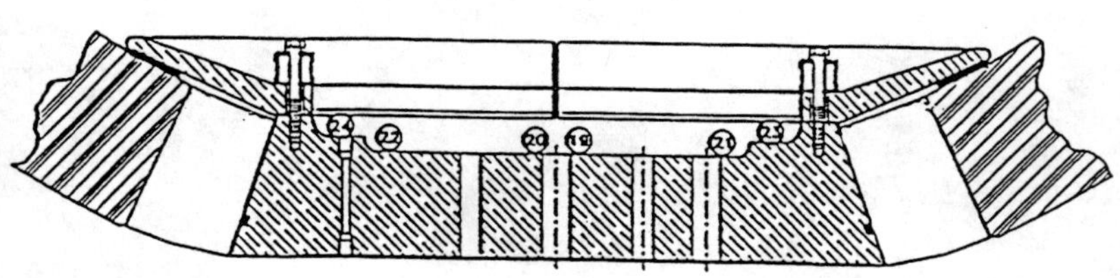

c) Bottom Plug Assembly

Figure 9 : Strain Gage Locations in HBOI's New Acrylic Submersible During Proof Tests

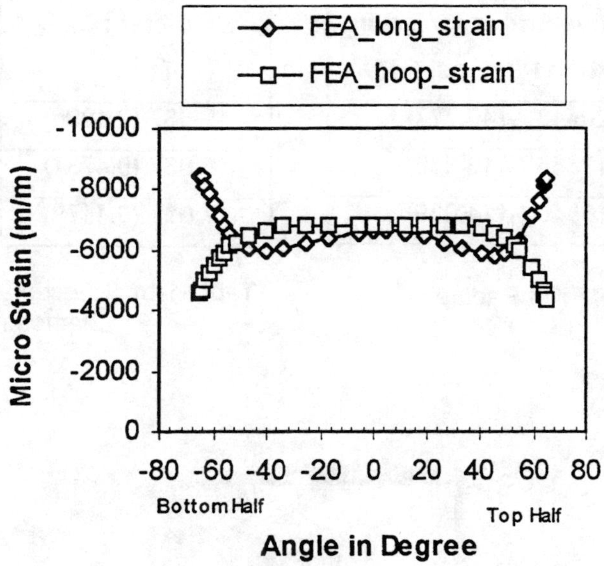

Fiugure 10: Variation of Strains (Longitudinal and Hoop) Through the Acrylic Sphere at Inside Surface at Operational Depth of 3000 ft. (914.4 m)

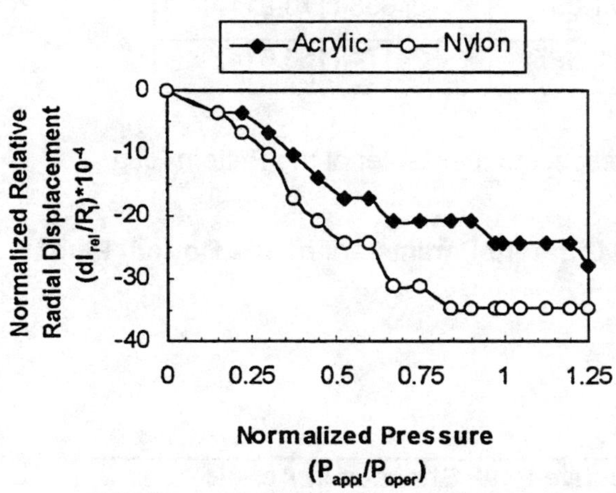

a) First Test

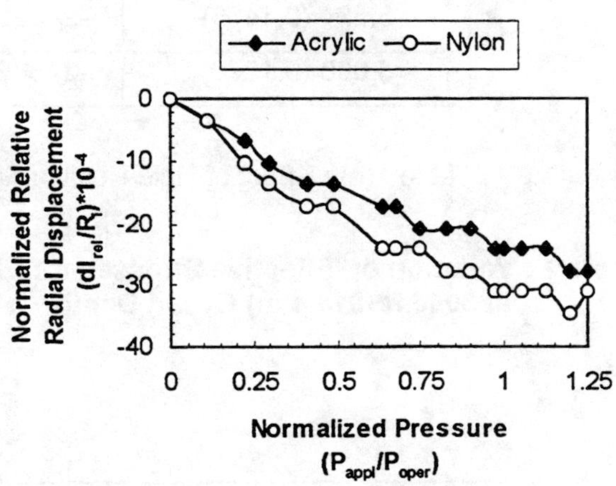

b) Second Test

Figure 11: Normalized Relative Radial Displacements of Top Nylon Gasket and Acrylic Sphere with respect to Aluminum Hatch Assembly During Proof Tests

Fillet Radius Position B, in (m)	Effective Stress (Seqv) in Al Bottom Plug, psi (MPa)
0.281 (0.0071)	-20418 (-140.777)
0.400 (0.0102)	-17388 (-119.886)
0.600 (0.0152)	-16540 (-114.039)

Table 1(a) : Effect of Fillet Radius

Al. Bottom Plug Diameter : D_{plug} in (m)	Effective Stress (S3) In Acrylic, psi (MPa)
19.637 (0.4988)	-7535 (-51.952)
18.637 (0.4734)	-7250 (-49.987)
17.637 (0.4479)	-6870 (-47.367)

Table 1(b) : Effect of Aluminum Bottom Plug Diameter

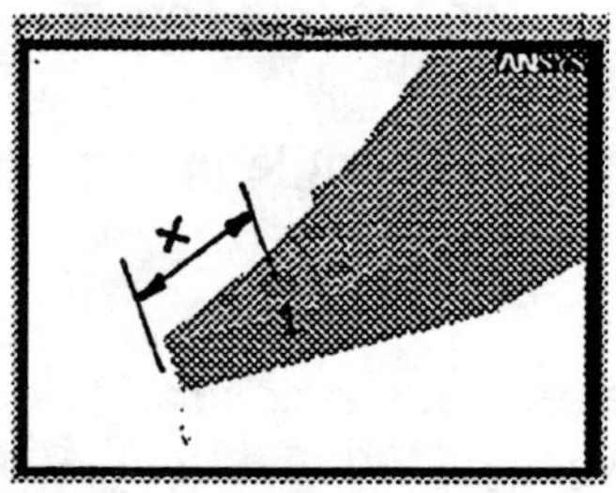

Position of Hole # 1 x in (m)	Effective Stresses Near Hole in psi (Mpa)	
	S3	Seqv
5.534 (0.1406)	-24925 (-171.852)	24658 (170.011)
5.000 (0.1270)	-24569 (-169.397)	24174 (166.674)

Table 1(c) : Effect of Hole Position with respect to the Center of the Bottom Plug

Table 1 : Variation of Effective Stresses for Three Different Parameters of The Bottom Plug at 3000 ft (914.4 m) Ocean Depth

Position of Rubber Gasket (x)	Effective Peak Stresses in Acrylic	
	S3 psi (MPa)	Seqv psi (MPa)
0.23 in (0.0058 m) Away From Interface	-7660 (-52.814)	6042 (41.658)
0.84 in (0.0213 m) Away From Interface	-7329 (-50.532)	6025 (41.541)
Without Gasket	-7253 (-50.008)	6014 (41.465)

Table 2 : Variation of Effective Peak Stresses in Acrylic Hull Due to Different Positions of Rubber Gasket

Components	Elastic Analysis		Elasto-Plastic Analysis	
	S3 psi (MPa)	Seqv psi (MPa)	S3 psi (MPa)	Seqv psi (MPa)
Acrylic Sphere	-6138 (-42.320)	5207 (35.900)	-6138 (-42.320)	5207 (35.900)
Nylon Gasket	-13428 (-92.583)	9931 (68.472)	-12363 (-85.239)	9111 (62.818)
Al. Hatch Ring	-25529 (-176.02)	25462 (175.55)	-25491 (-175.75)	25384 (175.016)
Al. Hatch	-26217 (-180.76)	22275 (153.581)	-26206 (-180.68)	22273 (153.567)

Table 3 : Variation of Peak Stress Distributions Through the Top of Acrylic Submersible at Operational Depth of 3000 ft. (914.4 m) Due to the Inclusion of Material Non-Linearity for Nylon Gasket

Gage No.	Appl. Pr. in psi (MPa)	Micro Strain (in/in or m/m)				Stress in psi (MPa)			
		Experimental Data		FEA		Experimental Data		FEA	
		Long.	Hoop	Long.	Hoop	σ_{max}	σ_{min}	σ_{max}	σ_{min}
17	1335 (9.21)	+335	-1622	+555	-2217	-17201 (-118)	-599 (-4.1)	-21595 (-149)	-1235 (-8.52)
	1670 (11.5)	+518	-1879	+700	-2650	-19707 (-136)	+634 (+4.4)	-25359 (-175)	-1441 (-9.94)
24	1335 (9.21)	-646	+19	-604	-398	-7384 (-50.9)	-1713 (-11.8)	-7781 (-53.6)	-6272 (-43.2)
	1670 (11.5)	-813	+4	-756	-499	-9403 (-64.8)	-2421 (-16.7)	-9741 (-67.2)	-7859 (-54.2)

Table 4: Comparison of the Experimental Results with FEA Results in Aluminum Penetrators

Locations in Acrylic Sphere at Inner Surface	New Model		NEMO MOD 2000		New Acrylic Submersible	
	Micro Strain (FEA)		Micro Strain (Experimental)			
	Hoop	Long.	Hoop	Long.	Hoop	Long.
Edge of Top Opening	-4380	-8380	-4551	-14799	-6423	-2150
5 in (.127 m) Away from Top Opening	-6000	-6310	-	-	-6428	-7942
Equator	-6810	-6610	-8400	-7749	-6775	-6606
5 in (.127 m) Away from Bottom Opening	-5992	-6720	-	-	-6180	-7941
Edge of Bottom Opening	-4560	-8470	-4501	-12299	-6409	-2278

Table 5: Comparison of the FEA Results with Experimental Results of NEMO MOD 2000 and New Acrylic Submersible at Operational Depth of 3000 ft. (914.4 m)

PVP-Vol. 415, Recent Advances in Solids and Structures — 2000
ASME 2000

SIZING & TOPOLOGY OPTIMIZATION OF A CHASSIS FRAME

Murali M.R. Krishna
Andrew Simboli
DANA Corp.

ABSTRACT

The application of sizing and topology optimization to a complete chassis frame has been undertaken to improve its performance within mass/weight constraints. The static frame stiffness (vertical bending and torsion), the first four modal frequencies and point mobility numbers were targeted for improvement. First, sizing optimization was undertaken to meet frame targets. The gage thicknesses of a number of different components, were allowed to be varied. This resulted in an increase in frame weight. Later topology optimization was undertaken to recommend the area where weight can be taken off without affecting the performance. An MSC NASTRAN optimizer was used for all the analyses.

INTRODUCTION

In chassis design processes, advanced optimization techniques using computers are playing an increasingly important role. A case study of an application of sizing and topology optimization techniques to the design of a chassis frame of an automobile has been presented.

The finite element (FE) model of a full chassis frame was used to establish the baseline parameters, the targets were set and the different components, whose gages could be changed, were identified. Also the upper and lower limits of these gage thickness were set. Some of the components, which were designed with crash criterion, like the front cross-member, rear cross-member etc. were excluded from our analysis. A preliminary gage optimization was then performed on the complete frame to find the optimum gages to meet the targets. It was observed that the weight was increased to meet the targets. As a next step, topology optimization was undertaken to suggest ways to identify the most important and least important elements in the frame. This helped us identify the areas where changes could be made in the form of cutouts without affecting the performance.

This study illustrates how the designer can benefit by using advanced optimization techniques. Earlier the designer had to resort to trial and error method of changing the design parameters, which proved to be tedious, costly, time consuming. Independent finite element studies of the chassis frame by varying different parameters one at a time, yield contradictory design directions. On the other hand, optimization effectively deals with these different parameters simultaneously and gives the designer a good direction for change to meet design targets and at the same time maintain the weight limits.

METHODOLOGY

The finite element model of a full chassis frame was used as a starting point. The frame stiffness performance parameters (vertical bending and torsional bending stiffness) were evaluated using the static analysis. A modal analysis was then performed on the unconstrained (free-free) frame and the first four frequencies (vertical bending, torsional, lateral bending and lateral match boxing modes), excluding the rigid-body modes, were evaluated. A modal frequency response was then performed with excitation frequency at five different positions on the frame. The response at 0.5 Hz was used to gage the frame performance.

The finite element Model of the complete chassis frame has 128,705 shell elements and 126,951 nodes. Static stiffness, modal and modal frequency analyses of the baseline model were performed and the baseline stiffnesses, frequencies and point mobility numbers were calculated.

A modal frequency response (MFR) analysis [1] was performed in MSC/NASTRAN. This method uses the mode shapes of the structure to uncouple the equation of motion and also reduce the problem size. The excitation force is assumed to be sinusoidal with frequencies varying from 0 Hz. to 70 Hz. in steps of 0.05 Hz. The designer wanted us to meet the design targets. It was suggested that optimization techniques be used to bring down the responses. The optimization problem was defined as follows –

Objective: Minimize the total weight.

Design Variables: Gage thickness.

The *Direct Linearization* method [2, 3] was used for optimization and the design sensitivities were generated internally in MSC/NASTRAN. The output gage thicknesses were implemented in the finite element model and a complete analysis of performance parameters performed on the modified frame. The responses were then superimposed on the baseline outputs and a percentage difference calculated as shown in Tables 1 and 2 for comparison.

DISCUSSION OF RESULTS

The baseline modal analysis took 00:07:03 hours of CPU time on a mainframe system. The percentage difference in frequencies of the baseline and optimized model are shown in Table 1 below. The MFR analysis took 00:05:01 hours of CPU time. The NASTRAN optimizer took 8 iterations and a CPU time of 09:30:00 hours on a mainframe to converge to an infeasible solution. The overall project took about 80 hours of analyst time not including the time taken for finite element meshing.

It was observed that the frame weight increased by 7.3%. The optimizer was only allowed to change component gages. It is quite possible to make local changes to components and achieve increased performance parameters with marginal weight penalty.

The Mid-Rail gage (Shown in Figure 3) was increased which added 19.2% to its weight. The optimizer wanted to meet the torsional stiffness target of the frame performance. As weight is an important criterion this suggestion is undesirable. An additional reinforcement to the midrail (only on the vertical wall) may be added instead of increasing the gage of the entire component. This solution would reduce weight but increase the number of components and assembly operations. A decision will have to be made based on the cost difference and priority.

The Front Sheet Metal Mount bracket was not included in the optimization run but increased point mobility performance of the bracket was a goal. One way to achieve the target is to increase the gage thickness of the mid-rail and some of the adjoining components. A better way was approach the problem locally and change the bracket design to be stiffer in vertical direction. Changing the mounting legs so that it has a larger bearing area on the side rails, tapering the vertical walls and/or increasing its gage thickness are possible ways of meeting the target without as much weight penalty.

The above argument holds for the Body Mount Bracket #4 (Shown in Figure 4). Shape optimization techniques were used to determine if point mobility could be improved by making local changes. Point mobility was not improved so we reverted back to increasing the gages of the side rails. As weight is an important criterion the addition of a vertical reinforcement in the mid-rail section to enhance the torsional stiffness should be explored.

Topology optimization was then used to effectively identify the least important elements so they could be considered for removal in the form of cutouts. The results have been plotted in Figures 1 – 4 and possible areas identified as candidates for cutouts to further reduce weight. The figures have been plotted with varying densities. A density of 1.0 indicates that the element is most important and a value of 0.0 indicates that it is least important.

CONCLUSIONS

Sizing and topology optimization techniques have been used to optimize the performance of a complete chassis frame. Although the weight of the frame went up topology optimization was used to suggest changes to bring down the weight without compromising the performance. In conclusion, optimization techniques can be very effectively used to give directions for improving a chassis frame of an automobile.

ACKNOWLEDGEMENTS

The authors gratefully acknowledge the technical design data and FE modeling contributions of Ouqi Zhang & others designers of Parish Vehicles Structures Division, Dana Corporation for this case study.

REFERENCES

1. " MSC/NASTRAN Basic Dynamic Analysis User's Guide", Grant Sitton, The MacNeal-Schwendler Corporation, Los Angles, California, USA, 1993.
2. "Design Sensitivity and Optimization in MSC/NASTRAN", Seminar Notes, The MacNeal-Schwendler Corporation, Los Angles, California, USA, March 1987.
3. "Elements of Structural Optimization", Haftka, R.T., and Kamat, M.P., Martinus Nijhoff, Dordrecht., 1985.

Items	% Difference between Baseline & Optimized Designs
1. Weight (kg)	7.259113
2. Frame Stiffness Performance:	
a. Vertical Bending Stiffness(N/mm)	22.73011
b. Torsional Bending Stiffness(kN-m/rad)	16.78111
3. Frame Modal Parameters	
a. First Vertical Bending Mode (Hz.)	7.771326
b. First Torsional Mode (Hz.)	5.38865
c. I Lateral Bending Mode (Hz.)	3.285714
d. I Lateral match Boxing (Hz.)	4.397606
4. Frame Point Mobility 0-70Hz at 0.5Hz.	
a. Front Sheet Metal Mount (N/mm)	16.67925
b. Body Mount #1 Bracket. (N/mm)	24.63809
c. Body Mount #2 Bracket. (N/mm)	28.50064
d. Body Mount #3 Bracket. (N/mm)	25.81979
e. Body Mount #4 Bracket. (N/mm)	14.14648

Table#1: Performance Parameters

Components	% Difference between Baseline & Optimized Designs
1. CM5 Rear	8.62069
2. CM5 Front	14.82759
3. CM5 Sleeve	-19.3548
4. Frame Clearance Hole	6.896552
5. CM4	11.29032
6. CM4 Bracket	5.172414
7. Rear Rail Inner	15.41667
8. Rear Rail Outer	12.5
9. Mid-Rail	19.23077
10. CM1a	1.724138
11. CM1b	-1.6129
12. CM2 Bracket	-1.17647
13. CM3 Top	0
14. CM3 Bottom	1.25
15. Front Shock Bracket	8.461538
16. Rear Shock Bracket	-4.7619
17. CM2 Top	0
18. CM2 Reinforcement	8.888889
19. Tire Bracket	3.448276
20. Tire Bracket Support	-8.27586
21. CM2 Bottom	-3.84615

Table#2: Gages

Note: Unless otherwise noted all % Difference values are positive.

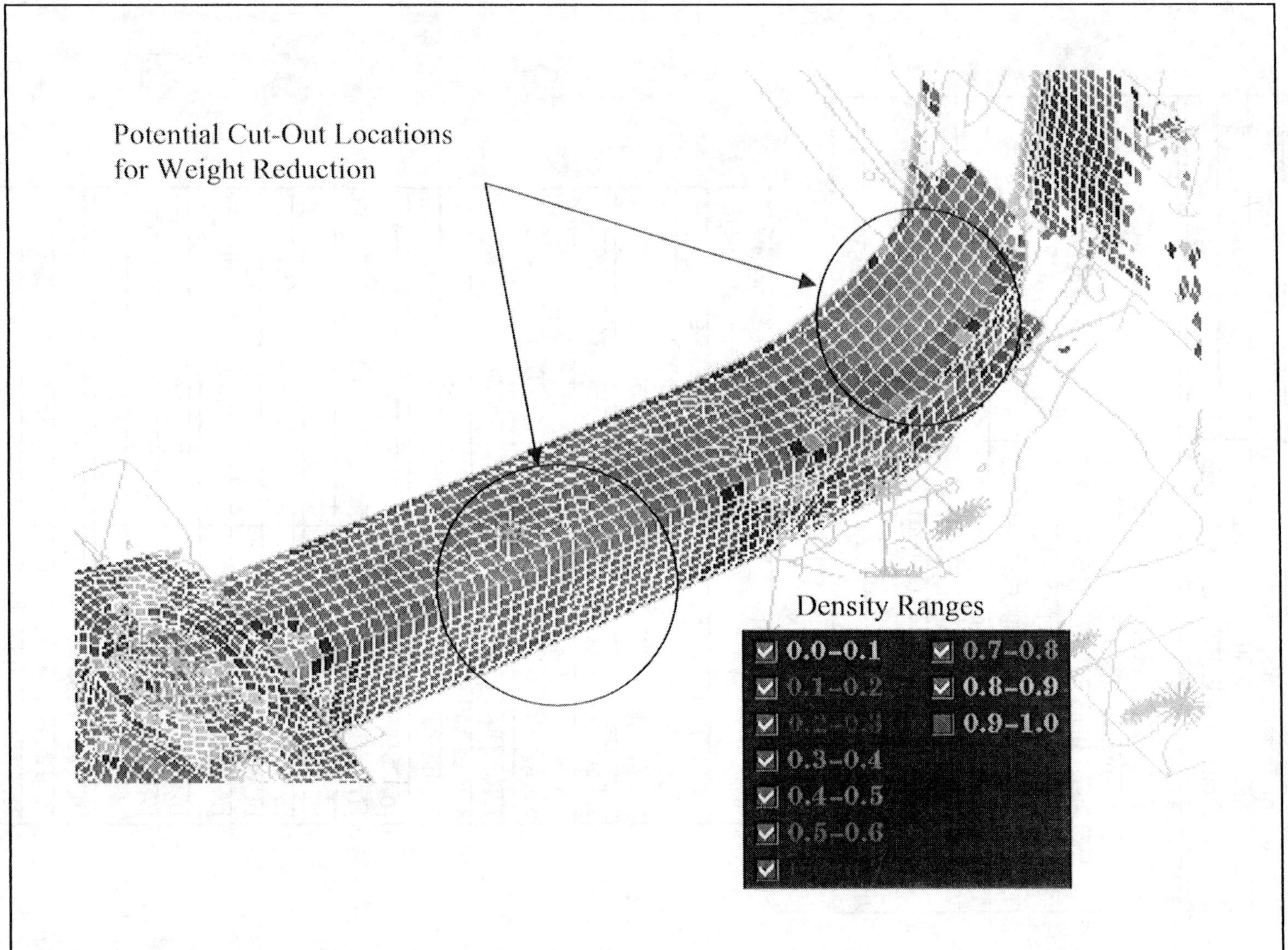

Figure 1: Enlarged View of Forward Crossmember

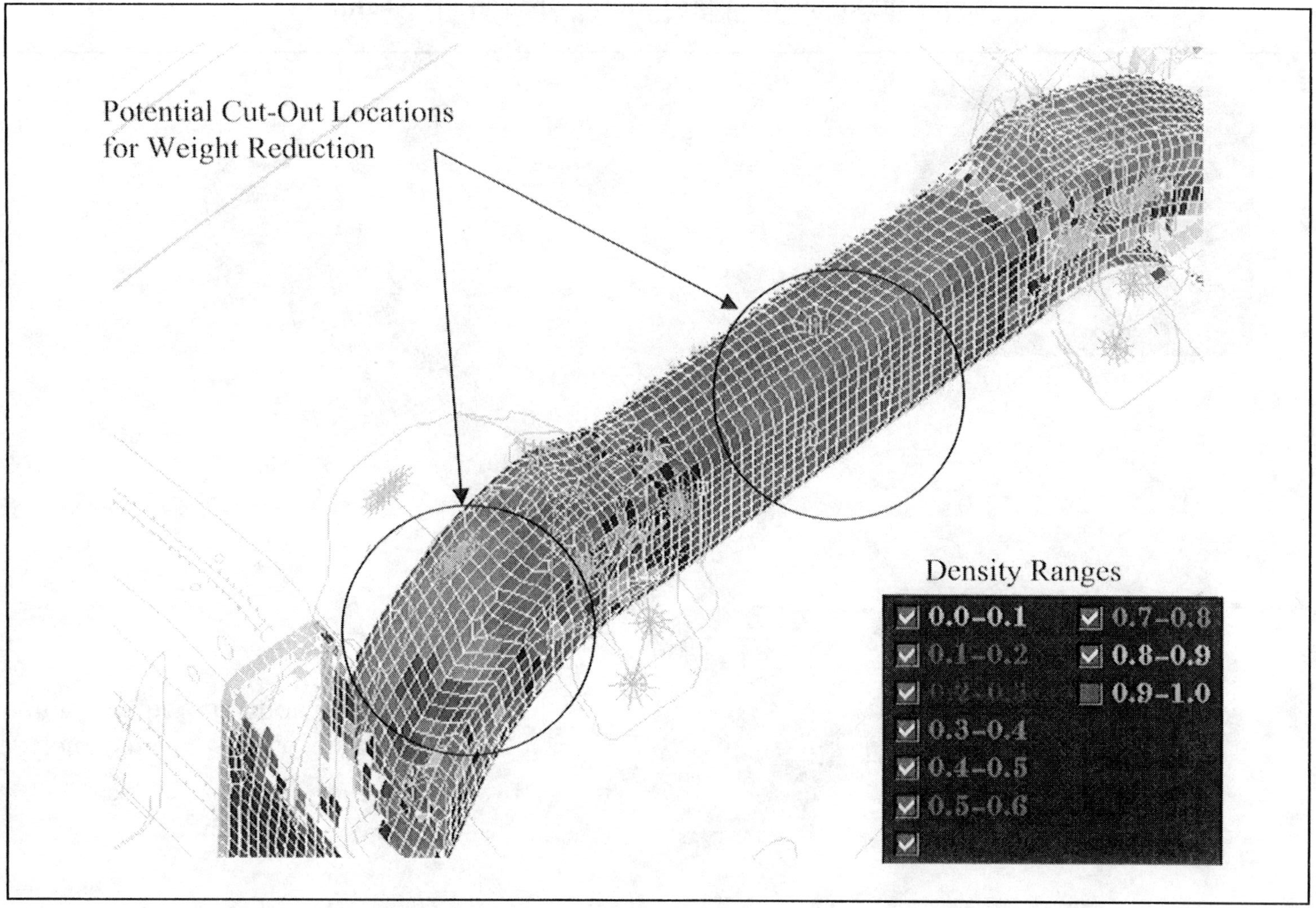

Figure 2: Enlarged View of Forward Crossmember (Different View)

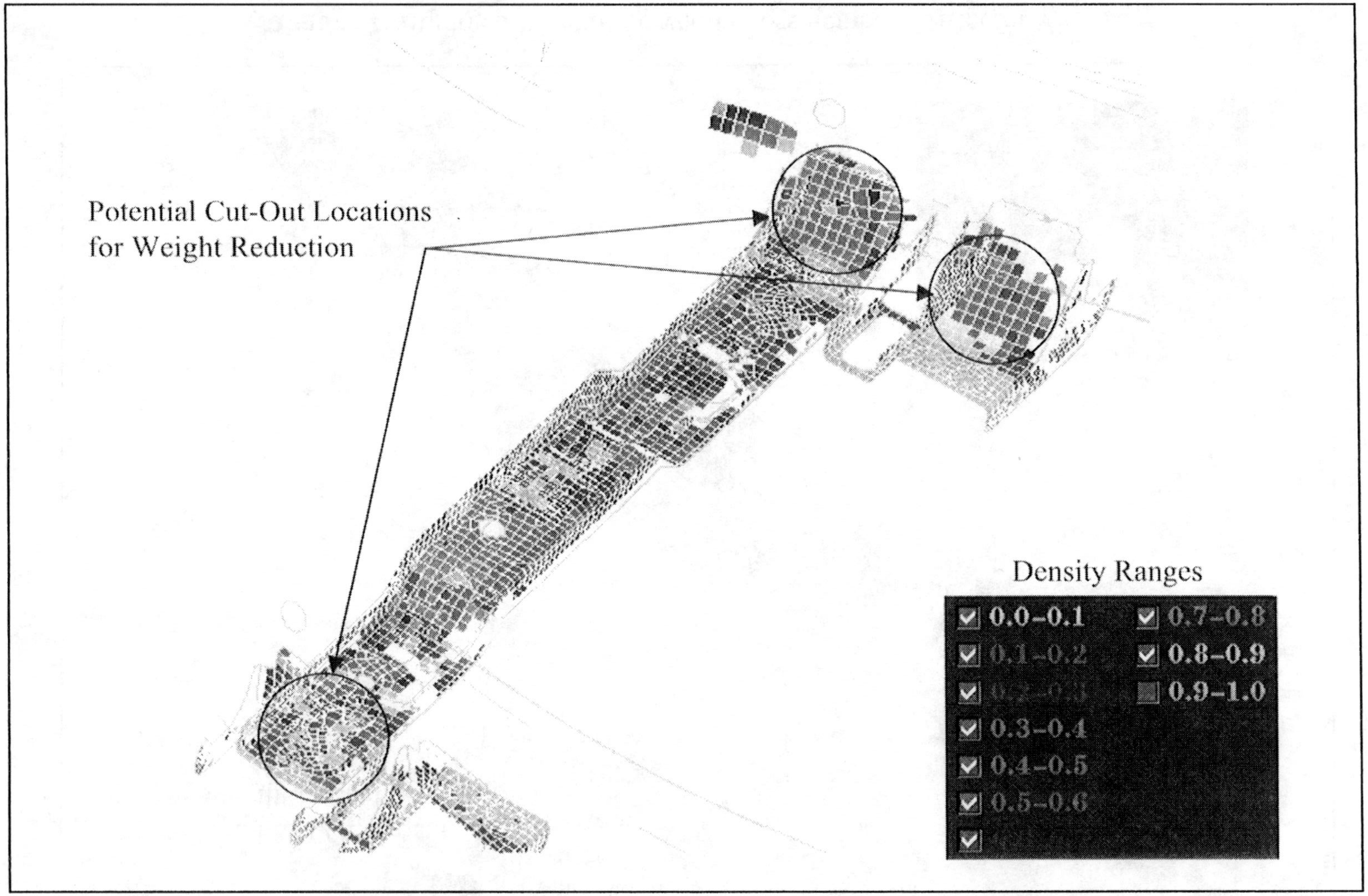

Figure 3: Enlarged View of Mid-Frame Crossmember

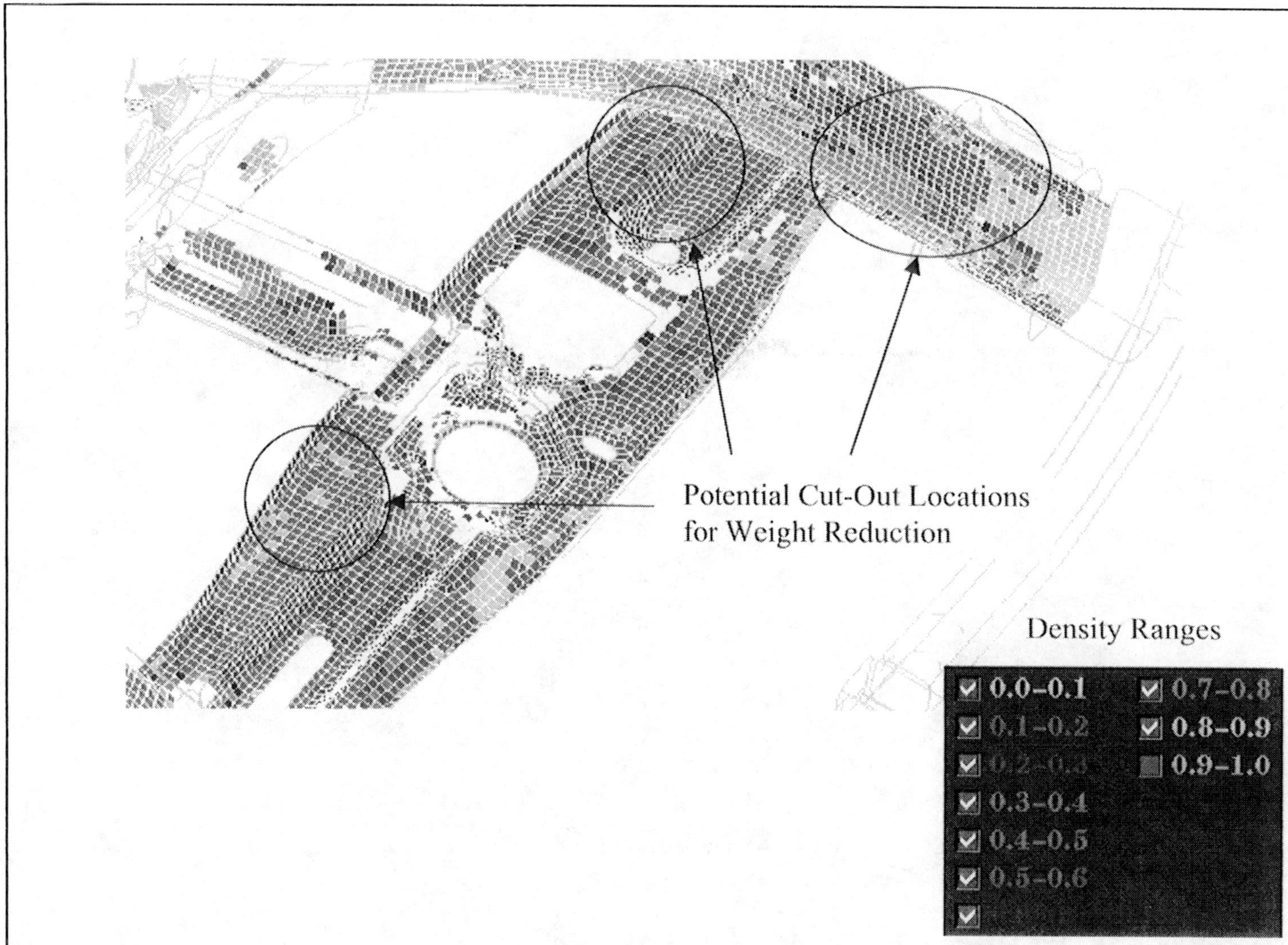

Figure 4: Enlarged View of Rear Crossmember

PVP-Vol. 415, Recent Advances in Solids and Structures — 2000
ASME 2000

BOLT HEAD FILLET STRESS CONCENTRATION FACTORS IN CYLINDRICAL PRESSURE VESSELS

Gowri Srinivasan
i2
5005 W. Royal Lane
Irving, TX 75063

Terry F. Lehnhoff
Department of Mechanical and Aerospace
Engineering and Engineering Mechanics
University of Missouri-Rolla
Rolla, Missouri 65401-0050
Email: lehnhoff@umr.edu

ABSTRACT

Linear three-dimensional finite element analysis (FEA) was performed on bolted pressure vessel joints to determine maximum stresses and stress concentration factors in the bolt head fillet as a result of the prying action. The three-dimensional finite element models consisted of a segment of the flanges containing one bolt, using cyclic symmetry boundary conditions. The flanges were each 20 mm in thickness with 901.7 mm inner diameter. The outer flange diameter was varied from 1021 mm to 1041 mm in steps of 5 mm. The bolt circle diameter was varied from 960.2 mm to 980.2 mm in steps of 5 mm. The bolts used were 16 mm diameter metric bolts with standard head and nut thickness. The threads were not modeled. The internal vessel pressure was 0.6895 MPa (100 psi).

Stress concentration factors in the bolt head fillet were calculated and they ranged from 3.34 to 4.80. The maximum stress in the bolt as well as the stress concentration factors in the bolt head fillet increase with an increase in bolt circle diameter for a given outer flange dimension. Keeping the bolt circle diameter constant, bolt stress and stress concentration factors in the bolt head fillet decrease with increase in outer flange diameter. The maximum stresses in the bolt were also calculated according to the American Society of Mechanical Engineers (ASME) Boiler and Pressure Vessel Code and the Verein Deutscher Ingenieur (VDI) guidelines and compared to the results observed through finite element analysis. The stresses obtained through FEA were larger than those predicted by the ASME and VDI methods by a factor that ranged between 2.96 to 3.41 (ASME) and 2.76 to 3.63 (VDI).

INTRODUCTION

Bolted joint analysis usually assumes axial external loading. Many practical bolted joints, such as pressure vessel flanges, are subjected to eccentric or bending loads, in addition to tension. The objective of this investigation is to analyze the effects on the stresses and stress concentration factors in the bolt head fillet as a result of eccentric external forces. Numerous computer models with varying flange dimensions and bolt positions were considered. The I-DEAS software was used to model a segment of a typical pressure vessel flange containing one bolt. The position of the bolt and the outer flange dimension were varied and the maximum stresses in each case noted. Comparisons were made to the results obtained through the American Society of Mechanical Engineers (ASME) Boiler and Pressure Vessel Code [1] and the Verein Deutscher Ingenieur (VDI) procedure [2].

PRYING ACTION

Prying action is the increase in the amount of tensile and bending stresses produced in the bolt when an external tensile load is applied other than along the bolt centroidal axis, i.e. off to one side of the bolt. Prying occurs any time there is a non-zero distance between the radius of gyration of the joint and the axis of load application. Figure 1 shows a typical eccentrically loaded bolted joint under the effect of prying action. The radius of gyration of the joint is given by

$$r_j = \sqrt{\frac{I_j}{A_j}} \ . \qquad (1)$$

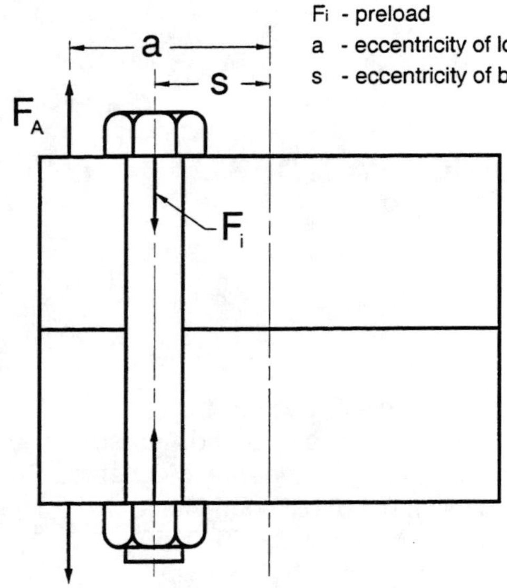

Fig. 1 Typical Eccentrically Loaded Bolted Joint

STRESS CONCENTRATION FACTORS

When irregularities such as notches, holes or grooves exist in the geometry of any structural member, the stress distribution in the neighborhood of the irregularity is altered. Such irregularities are called stress raisers, and the regions in which they occur are called areas of stress concentration. The theoretical or geometric stress concentration factor, which relates the actual maximum stress at the irregularity to the nominal stress, is given by

$$K_t = \frac{\sigma_{act}}{\sigma_{nom}} \quad . \tag{2}$$

In this research, the bolt head fillet is an area of stress concentration, where the actual stress is taken to be the stress in the fillet observed by finite element analysis (FEA). The nominal stress is the total stress applied to the bolt due to the equivalent direct and bending loads, as often calculated by the basic relation

$$\sigma_{nom} = \frac{F_b}{A_b} + \frac{Md}{2I_b} \quad . \tag{3}$$

However, the VDI [2] variation of equation (3) will be used in this research. This method is discussed in the next section of this paper.

METHOD OF ANALYSIS

For a bolt in pure tension, stress concentration factors have been experimentally determined and charts can be found in Pilkey's (Peterson updated) publication [3]. The total stress acting on the bolt, which is the nominal stress, is given as a sum of the bolt prestress and the part of the stress due to the applied axial external load that is transferred to the bolt. In the case of axial tensile loading, the fraction of the load that is transferred to the bolt, or the load factor φ, is determined by the ratio of the stiffnesses and is given by the equation

$$\varphi = \frac{k_b}{k_j + k_b} \tag{4}$$

where the formulas, in the absence of threads, for the bolt and joint stiffness k_b and k_j as given by Shigley and Mischke [4] are

$$k_b = \frac{A_b E}{2t} \tag{5}$$

$$k_j = \frac{0.577 \pi E d}{2 \ln\left(5 \dfrac{0.577 \times 2t + 0.5d}{0.577 \times 2t + 2.5d}\right)} \quad .$$

When there is an eccentric load, prying occurs and the stress concentration factor at the bolt head fillet will be higher than in the axial loading case. In this case, the load factor φ_e depends not only upon the bolt and joint stiffness, but also on other factors such as the eccentricity of load application and the distance of the bolt axis from the radius of gyration of the joint. According to the VDI procedure [2], this equation is

$$\varphi_e = \frac{k_b\left(1 + \dfrac{asA_s}{I_j}\right)}{k_j + k_b\left(1 + \dfrac{s^2 A_s}{I_j}\right)} \quad . \tag{6}$$

The nominal stress is now given by

$$\sigma_{nom} = \varphi_e \frac{F_A}{A_b} + \sigma_i \quad . \tag{7}$$

The stress concentration factor can be calculated from equation (2) or if the nominal stress is known, then the actual stress can be obtained from the product of the factor K_t and the stress from equation (7) according to the VDI method [2].

PREVIOUS RESEARCH

The issue of eccentric loading of bolted joints has been addressed in both the VDI [2] and the ASME Boiler and Pressure Vessel codes [1]. Appendix A of the thesis by Srinivasan [5] contains a FORTRAN program for predicting maximum stresses in the bolt according to Section VIII, Division I of the ASME code [1]. Appendix Y of Division I, in particular, deals with the design of flat face flanges with metal to metal contact.

The VDI approach [1] to calculating stresses in bolts under the action of eccentric loads has also been discussed at length by Bickford [6]. The formula which he provides, for maximum stress, or the actual stress in the bolts, is

$$\sigma_{act} = \left[1 + \left(\frac{1}{\varphi_e} - \frac{s}{a}\right)\frac{1}{l_{eff}}\frac{a\pi d^3}{8I_j}\right]\frac{\varphi_e F_a}{A_b} \tag{8}$$

Appendix B of the above thesis contains the FORTRAN programs for calculation of the eccentricity of loading, radius of gyration of the joint and the stresses in the bolt using the VDI method. Table 1 gives the values of all the parameters mentioned above in each case.

Table 1 Summary of Eccentricity Parameters

Bolt Circle Diameter D_b (mm)	Flange Diameter D_o (mm)	Radius of Gyration R_g (mm)	Eccentricity of Load a (mm)	Eccentricity of Bolt s (mm)
960.2	1021	481.67	27.65	1.57
965.2	1021	481.55	27.53	1.05
970.2	1021	481.43	27.41	3.67
975.2	1021	481.32	27.29	6.28
980.2	1021	481.20	27.17	8.90
970.2	1026	482.83	28.80	2.27
970.2	1031	484.22	30.19	0.88
970.2	1036	485.60	31.58	0.50
970.2	1041	486.99	32.97	1.89

DESIGN CONSIDERATIONS AND PARAMETERS

The pressure vessels considered in this research were assumed to be made of steel with the following properties:

Modulus of Elasticity:	207 GPa
Poisson's Ratio:	0.3
Tensile Strength:	1040 MPa

The bolts were assumed to be grade 10.9 M16 bolts with the minimum 0.6 mm head to shank fillet radius. The flanges were of the loose type with O-ring seals and metal-to-metal contact. The dimensions of the pressure vessels which remained constant were as follows: inner flange diameter equal to 901.7 mm (35.5 in.), wall thickness equal to 6.35 mm (0.25 in.) and flange thickness equal to 20 mm. The thickness of the head of the bolt was 10.75 mm and the nut thickness was 14.8 mm for a standard 16 mm bolt. The standard clearance of 1 mm was used for the 16 mm bolts. The bolt circle diameter was varied from 960.2 mm (37.8 in.) to 980.2 mm (38.59 in.) in steps of 5 mm, for a constant outer flange diameter of 1021 mm (40.2 in.). The outer flange diameter was varied from 1021 mm (40.2 in.) to 1041 mm (40.98 in.) in steps of 5 mm for a constant bolt circle diameter of 970. 2 mm (38.2 in.). The outer flange diameters were chosen in such a way that the bolt head never projected beyond the end of the flange for any position of the bolt. Also, the true effective pressure distribution, which the pressure cone method [4,6] approximates, does not extend beyond the outer boundaries of the flange even for the case with the largest bolt circle diameter and smallest outer flange diameter. The bolt circle diameters were chosen so as to not allow any interference of the bolt with the pressure vessel wall, i.e., there is clearance for wrenching.

The internal pressure was chosen to be 0.6895 MPa (100 psi) based on the pressure vessel material and dimensions. Only the straight cylindrical part of the pressure vessel wall was modeled. This could be achieved by applying the longitudinal stress due to the internal pressure on one end and restraining the other end in the axial direction. The longitudinal stress was calculated from the equation

$$\sigma_A = \frac{Pr}{2w} . \tag{9}$$

Among the various cases considered, the worst case, or the case with the highest stress concentration factor was identified to be the model with the smallest outer flange diameter and the largest bolt circle diameter, since the prying action was maximum in this situation.

The prestress in the bolts was calculated according to the recommendations made by the ASME code. Appendix Y of the Code states, in effect, that the bolt stresses should not exceed the allowable stress at the operating temperature. The allowable stress in this case was one fifth of the tensile strength of the material, hence resulting in a lower value of the prestress than conventional design. Since the prestress was between 10 and 20 percent of the proof strength, as opposed to 75 to 90 percent as recommended by Shigley and Mischke [4], the joint was checked for separation according to the following equation

$$n = \frac{\sigma_i A_b}{F_A\left(1 - \varphi_e\right)} . \tag{10}$$

The load factor, or factor of safety guarding against joint separation was greater than 1 in all cases which also helped to justify using the lowered prestress values.

COMPUTER MODELS

The software chosen for this research was the I-DEAS Master Series 6.0 family of programs. Three-dimensional parabolic tetrahedral elements were employed in modeling both the bolt and the pressure vessel flanges. Both the flanges were identical in all dimensions. Since the flange structure varied for every model, the solid model and mesh had to be

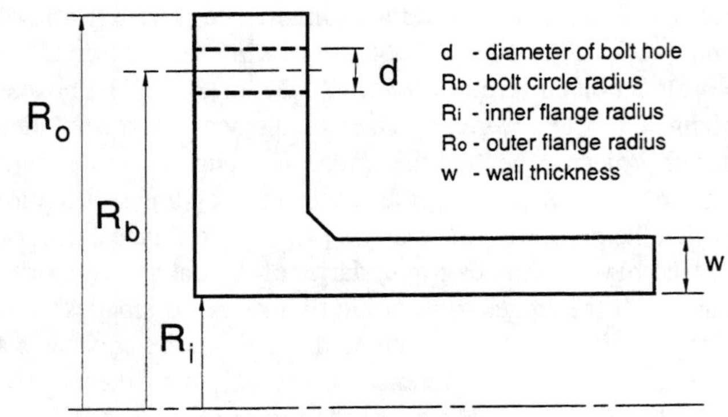

Fig. 2 General Pressure Vessel Flange Dimensions

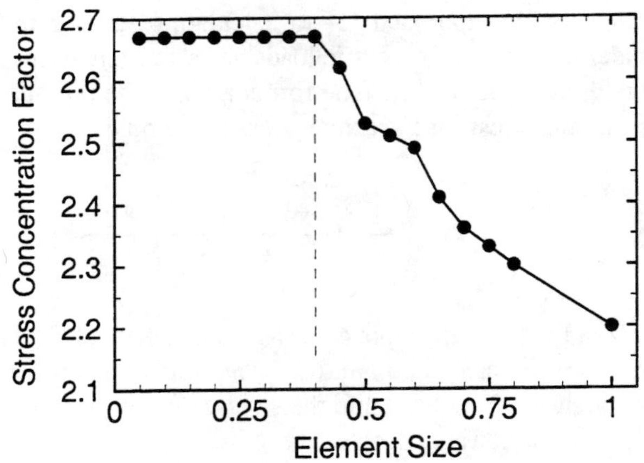

Fig. 3 Stress Concentration Factor versus Element Size for Bolt in Tension

reconstructed for each case. Figure 2 shows the general dimensions of one of the flanges. Each flange was partitioned into three volumes, in order to reduce the size of the problem. The cylindrical volume around the bolt holes had to have the same mesh size as the bolt head and the nut to satisfy contact element requirements. The pressure vessel wall, where stresses were smaller with minimal gradients, had the largest elements.

The bolt-nut system was modeled as a single entity, cut at the mid-surface. As failure often occurs at the bolt heads, the maximum stresses in the bolts, occurring in the fillet area were most relevant in this research. The threads, the other critical area, were not modeled, but some results from axisymmetric studies done by Lehnhoff and Bunyard [7] have been included for reference. The head of the bolt was partitioned in such a way that the critical region was made up of elements small enough to achieve accuracy and yet have a reasonable run time. In order to arrive at the optimum mesh size at the critical region, several models of bolts in pure tension were analyzed, by applying an axial load at the end of the bolt and restraining the top surface of the bolt head in the axial direction, using finite element methods. Based on this analysis, as shown in Figure 3, 0.4 mm was chosen as the element length. The critical region for this study was the lowermost portion of the fillet. The entire fillet region was partitioned so that a 20 degree wedge around the lowermost point consisted of the smallest elements. The mesh size was reduced in steps of 0.05 mm, from 1 mm down to 0.05 mm. The stress concentration factors increased steadily from 2.24 and leveled off at 2.67 when the top surface was fully restrained, thus preventing most bending at the bolt head. The largest element size where the stress concentration factor stabilized, was 0.4 mm. The bolt models with element sizes 0.25 mm to 0.5 mm were also subjected to an axial load as well as a bending moment, by restraining only the circumferential edge of the bolt head surface. This allowed significant head bending and the stress concentration factor in the bolt head fillet was noted to be 4.12. The two above-mentioned restraint cases represent extremes in terms of the head bending configuration and the

Table 2 Theoretical (Geometric) Stress Concentration Factors

Bolt Size (mm)		8	12	16	20	24
Fillet	K_t, [7]*	3.18	3.23	3.63	3.58	3.90
	K_t, [4]°	2.53	2.47	2.47	2.44	2.44
	% diff.	20	24	32	32	37
Thread (min)	K_t, [7]+	4.33	4.32	4.67	4.77	4.82
	K_t, [4]	4.46	4.42	4.37	4.29	4.26
	% diff.	-3	-2	6	10	12
Thread (max)	K_t, [7]#	4.80	4.80	5.12	5.17	5.22
	K_t, [4]	4.46	4.42	4.37	4.29	4.26
	% diff.	7	8	15	17	18

* Factor for the head fillet using the minimum radius
+ Factor for the first thread and the most shallow tolerance limit
\# Factor for the first thread and the deepest tolerance limit
∘ Calculated from fatigue factors for grade 10.9 bolts

actual stress concentration factors should lie between these two numbers. These values were checked with Ref. [7] (see Table 2) which listed the stress concentration factor in the bolt head fillet as 3.63 for the axial loading case corresponding to the cases studied in this research. Since the theoretical stress concentration factors are primarily dependant upon geometry in the case of linear analysis, the bolt model with element size 0.4 mm in the critical area was used in each of the nine pressure vessel models, instead of doing mesh analysis for the final models.

Once the bolt-nut system was aligned with the bolt holes of the flanges, the finite element models of the individual parts were appended together to create one finite element geometry. This geometry is characteristic of full cylindrical pressure vessels, each with slightly different flange and bolt location dimensions. Figure 4 shows a typical completed mesh, and

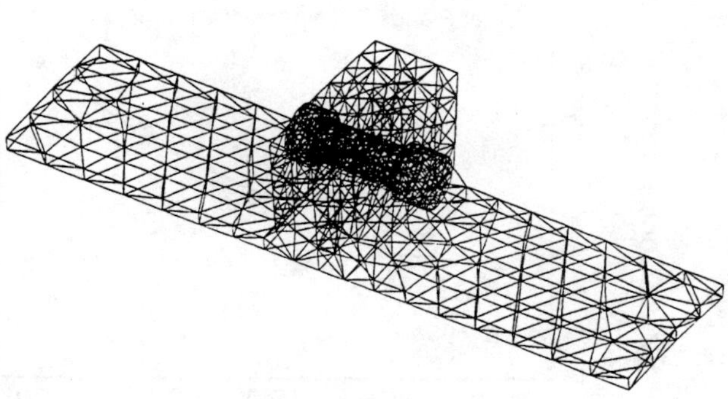

Fig. 4 Typical Finite Element Mesh for Bolted Pressure Vessel Flanges

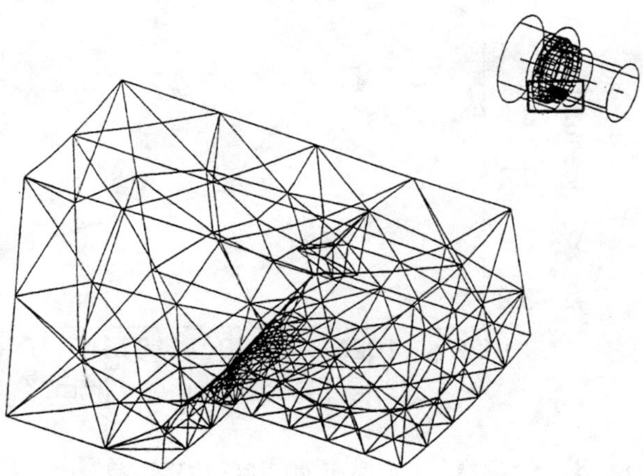

Fig. 5 Close-up of Critical Area of the Bolt Head Fillet

Figure 5 shows a close-up of the lower part of the fillet, which is the critical area. The total number of nodes in the final finite element models ranged from 20282 to 21928 and the number of elements ranged from 13111 to 14178.

The total number of bolts required for the pressure vessel was 40. A 9 degree segment of the flanges containing a single bolt was modeled in order to reduce the size of the problem. This could be achieved by applying cyclic symmetry boundary conditions, or in other words, displacements perpendicular (in the tangential direction) to the cutting plane, were restrained. Only the straight cylindrical portion of the pressure vessel wall was modeled, by restraining one end in the axial direction and applying the longitudinal force due to the internal pressure at the other end. The bolt prestress was applied at both the mid-surface areas. The internal pressure was applied radially along the inner surfaces of the shell. Figure 6 shows the wire frame

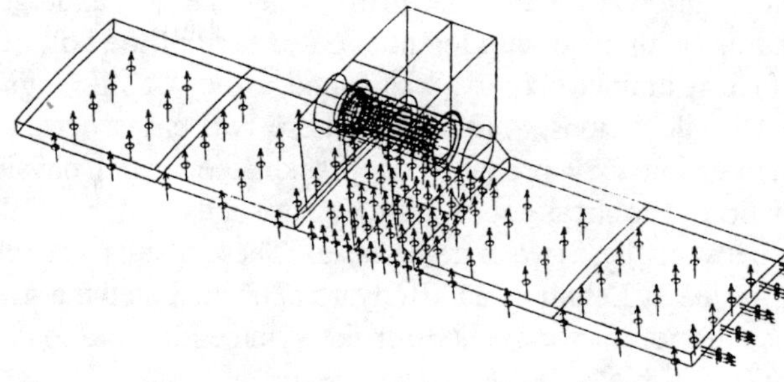

Fig. 6 Wireframe Model of Bolted Pressure Vessel Flanges with Loads

model of a typical pressure vessel flange with all the loads in place.

The various surface pairs that would possibly come into contact were identified. These included the mating surfaces of the flanges, the outer annular portion of the flange around the bolt hole and the inner surface of the bolt head and nut. The contact surfaces were set up as a part of the boundary conditions. The bolt mid-surfaces (the cut on which the prestress was applied) were constrained to remain flat and parallel.

Once the solid model was meshed and suitable boundary conditions and forces were in place, the resulting finite element model was solved for stresses and displacements. The post-processing software identified the maximum stress values and the locations at which they occurred. Stress contours showing regions of varying stress levels were also obtained.

DISCUSSION OF RESULTS

The focus of this research was to evaluate the effect of prying action on the maximum bolt stresses and stress concentration factors at the bolt head fillet. The eccentricity of load application and the distance of the bolt axis from the radius of gyration of the joint were varied by changing either the bolt circle diameter or the outer flange diameter and holding the other constant.

In order to confirm the validity of the computer models, a number of models of the bolt alone under the action of a simple tensile force were tested using the I-DEAS solver. These models were also used to arrive at the required mesh size at the critical region. The approximate experimental stress concentration factor charts for cylindrical stepped shafts in pure tension with a shoulder fillet as provided by Pilkey (Peterson updated) [3] were used for comparison. For the bolts studied here, with the entire bolt head surface restrained along the axial direction, the value was 2.47. The values of stress concentration factors as observed through finite element analysis ranged from 2.24 for 1 mm elements, increased steadily up to 2.67 for 0.4 mm elements and remained constant thereafter for smaller elements. Figure 3 is the graph of element size versus stress concentration factor. The stress concentration at the bolt head fillet when the bolt undergoes bending, with the circumferential outer edge of the bolt head surface restrained axially, was found to be 4.12 through I-DEAS calculations. Neither of these two cases occurs in practice, but they probably do bracket the actual physical situation. The stress concentration factor at the bolt head fillet lies between these two extreme cases. These results were also confirmed by Lehnhoff and Bunyard [7], who listed the stress concentration factor as 3.63 for the symmetric axial loading case (see Table 2).

The stress concentration factors at the bolt head fillet were calculated according to equation (2). The prestress was computed according to the ASME guidelines. In the case of varying the bolt circle diameter for a constant outer flange diameter of 1021 mm, the stress concentration factors at the bolt head fillet ranged from 3.81 to 4.80. These values of the stress concentration factor as a function of the bolt circle are shown in Figure 7. As the bolt circle diameter increased, the lever arm increased, leading to more prying. The stress concentration factors for a constant bolt circle diameter of 970.2 mm, while increasing the outer flange diameter ranged from 4.15 to 3.34. As shown in Figure 8, decreasing the outer flange diameter led to higher stress concentration factors for the same reason as in the previous case. The stress concentration factors as a result of prying, were expected to be higher than the values obtained in the axial loading situation. However, where there is less prying, the values (3.34) are slightly less than the Lehnhoff and Bunyard [7] results (3.63). It is interesting to note that the value of head fillet stress concentration for the 16 mm bolt (3.63), is higher than that for the 12 mm and 20 mm diameter bolts (see Table 2), as was observed by Lehnhoff and Bunyard [7]. The lowest stress concentration factor observed in this research (3.34) is, however comparable to the interpolated value of K_t for the 16 mm bolt (3.41), from the 12 mm and 20 mm bolt diameter results. Where there is more prying, the values (4.80) are considerably greater than the most extreme axial loading case (4.12), where the head bending had been exaggerated by restraining the outer circumference of the bolt head. Appendix C of the thesis lists the FORTRAN program used for the calculation of the stress concentration factors.

The maximum operating FEA stress in the bolt ranged from 472 to 678 MPa. This range was 158 to 236 MPa by the ASME method and 130 to 245 MPa by the VDI method. The VDI calculations were performed according to equation (8). Since the ASME and VDI procedures do not take into account the increased stresses at the bolt head fillet, the values were multiplied by the stress concentration factor for axial loading (2.47), given by Shigley and Mischke [4] and these numbers were considered to be the maximum stresses in the bolt. Figures 9 and 10 depict the maximum stresses at the bolt with varying pressure vessel dimensions in each of the three cases.

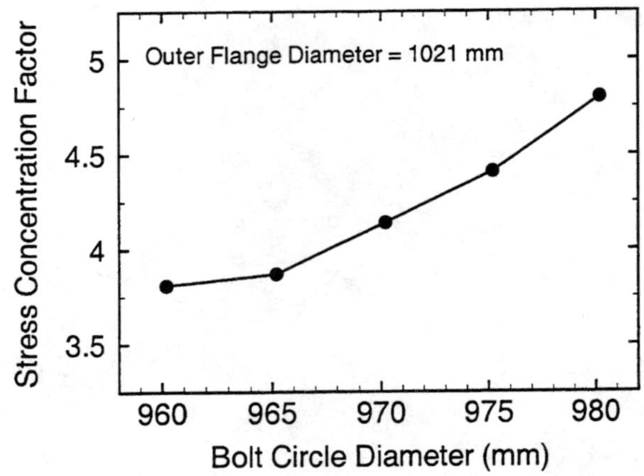

Fig. 7 Stress Concentration Factor versus Bolt Circle Diameter for Outer Flange Diameter = 1021 mm

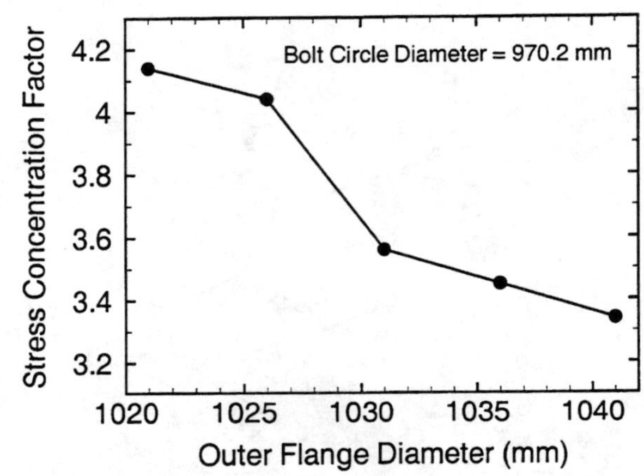

Fig. 8 Stress Concentration Factor versus Outer Flange Diameter for Bolt Circle Diameter = 970.2 mm

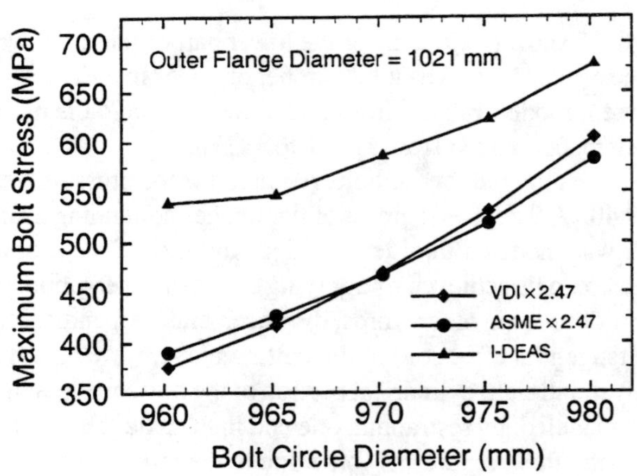

Fig. 9 Maximum Bolt Stress versus Bolt circle Diameter for Outer Flange Diameter = 1021 mm

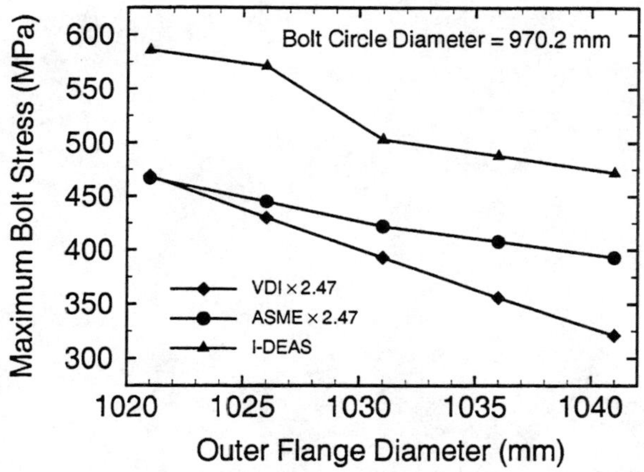

Fig. 10 Maximum Bolt Stress versus Outer Flange Diameter for Bolt Circle Diameter = 970.2 mm

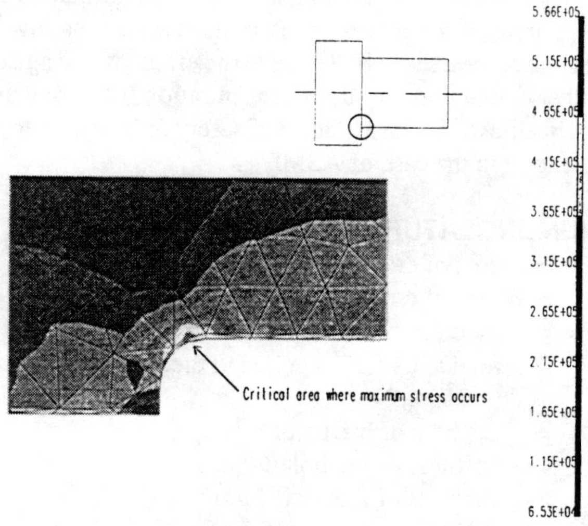

Fig. 11 Typical Stress Contour at Critical Area of the Bolt Head Fillet

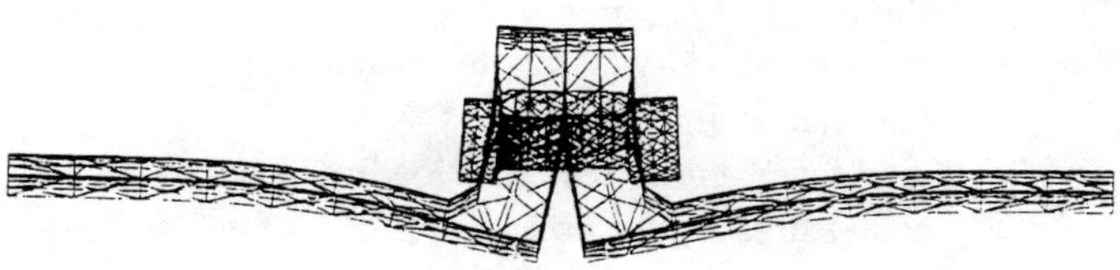

Fig. 12 Deformed Geometry of Bolted Pressure Vessel Flanges

Table 3 Comparison of Stress Values

Bolt Circle Diameter (mm)	Outer Flange Diameter (mm)	Maximum Stress in the Bolt (MPa)			Factor by which I-DEAS values are greater than	
		I-DEAS	ASME	VDI	ASME	VDI
960.2	1021	539	158	152	3.41	3.55
965.2	1021	547	172	169	3.18	3.24
970.2	1021	586	189	190	3.10	3.08
975.2	1021	623	210	215	2.97	2.90
980.2	1021	678	236	245	2.87	2.77
970.2*	1021	586	189	190	3.10	3.08
970.2	1026	571	180	174	3.17	3.28
970.2	1031	503	171	159	2.94	3.16
970.2	1036	488	165	144	2.96	3.39
970.2	1041	472	159	130	2.97	3.63

* This line of data repeated for ease of comparison.

Table 3 gives a comparison of these maximum stresses obtained in the bolt head fillet as observed from finite element analysis to those values predicted by the ASME Boiler and Pressure Vessel Code and the VDI guidelines. The factors by which the ASME and VDI values were less than the FEA results were also tabulated. These factors ranged from 2.96 to 3.41 (ASME) and 2.77 to 3.63 (VDI).

Figures 11 and 12 show the stress contours at the critical area of the bolt head fillet and the typical deformation of the pressure vessel flange, respectively.

CONCLUSIONS

One objective of this research was to show that the stress concentration factors at the bolt head fillet, in the case of eccentric loading, were higher than when the bolt was in pure tension. This somewhat intuitive concept has been quantified for some typical pressure vessel geometries. The stresses in the bolt are also higher in these cases. In examining the results, the values of the stress concentration factors were found to be dependant directly upon the eccentricity of load application and the distance between the radius of gyration of the joint and the bolt centroidal axis. An increase in one or both of these quantities led to higher stress concentration factors. This study showed that the traditional stress concentration factors might not be sufficient in the design of eccentric bolted joints, such as pressure vessel flanges. Table 3 also shows that the maximum stresses in the bolt head fillet were under predicted by the ASME and VDI methods. It is also clear that the stress concentration value cannot be represented by a single number for all the eccentric loading cases, since the factors depend upon the amount of prying involved. In order to avoid computation of these numbers, the

ASME Code prescribes the allowable bolt stress to be 20% of the ultimate strength. The VDI guidelines do not recommend a factor of safety while designing bolted joints due to the effect of prying. In conclusion, it is necessary to allow for the increased stresses in bolts due to the effect of prying action by either computing the stress concentration factor due to prying or, by allowing a large factor of safety (5 based on the tensile strength, in the case of ASME).

NOMENCLATURE

A_b = area of cross section of one bolt, mm^2
A_j = area of cross section of flange, mm^2
E = modulus of elasticity of bolt and flanges, MPa
F_A = total axial force applied to the bolt due to the internal pressure, N
I_j = moment of inertia of the joint, mm^4
k_b = stiffness of the bolt, N/mm
k_j = stiffness of the joint, N/mm
K_t = stress concentration factor
a = eccentricity of load application, mm
d = bolt diameter, mm
l = grip length of the bolt, mm
l_{eff} = effective grip length of the bolt, mm
n = factor of safety guarding against joint separation
s = distance between radius of gyration of the joint and bolt axis, mm
r = inner flange radius, mm
t = flange thickness, mm
w = pressure vessel wall thickness, mm
φ = load factor in tension
φ_e = load factor under eccentric loading conditions
σ_{act} = actual stress in the bolt head fillet, MPa
σ_{nom} = total stress applied to the bolt, MPa
σ_i = bolt prestress, MPa

REFERENCES

[1] ASME Boiler and Pressure Vessel Code, Section VIII, Division 1, Appendix Y, 1998.
[2] VDI 2230, Part 1, "Systematic Calculation of High Duty Bolted Joints: Joints with One Cylindrical Bolt," 1986, VDI, Dusseldorf, Germany.
[3] Pilkey, W.D., 1997, "Peterson's Stress Concentration Factors," John Wiley & Sons, Inc.
[4] Shigley, J.E. and Mischke, C.R., 1989, "Mechanical Engineering Design," McGraw-Hill, 5th Edition
[5] Srinivasan, G., 2000, "Bolt Head Fillet Stress Concentration Factors in Cylindrical Pressure Vessels," MS Thesis-University of Missouri-Rolla.
[6] Bickford, J.H., 1995, "An Introduction to the Design and Behavior of Bolted Joints," Marcel Dekker, Inc., 3rd Edition.
[7] Lehnhoff, T.F. and Bunyard, B.A., 1999, "Bolt Thread and Head Fillet Stress Concentration Factors," PVP–Vol. 398 pp. 103-108 Recent Advances in Solids and Structures, ASME.

PVP-Vol. 415, Recent Advances in Solids and Structures — 2000
ASME 2000

OPTIMIZATION OF MISSION CRITICAL JOINTS IN BOLTED MACHINE TOOL STRUCTURES

Emmanuil Kushnir; Mahendra R. Patel; Terrence M. Sheehan
Hardinge Inc., Elmira, NY

ABSTRACT

In general, precision machine tools consist of a number of structural components, usually castings machined and bolted together to very tight and precise tolerances. Machine bolts are used to prevent the contact surfaces from separating or sliding relative to each other. The issues critical in the design of these precision bolted joints include tensile, compressive, and lateral stiffness, and stability of the joint under different types of load.

In machine tool design, the shape and tightening force in these joints are usually evaluated based on two counter-acting requirements: (1) maintain sufficient stiffness provided by the joint at the cutting (load) point, and (2) allow the joint to slip (breakaway) in the event of a machine crash; in this case, the bolted joint works as a fuse preventing damage and, thus, protect critical/expensive components in the machine and/or avoid extensive repairs.

Field data from machines running production have shown that satisfaction of the two criteria, presented above, using conventional methods of bolted joint design does not always assure that the stability requirements are met. This data shows that in the range of loads that do not exceed the maximum force allowed in the machine, there might occur permanent lateral displacements in the joint. These displacements accumulate during normal operations under repeated loads and the machine looses alignment without obvious instantaneous slippage in the joint.

This paper discusses an approach that gives a qualitative and numerical evaluation of the joint shape, position of the bolts, tightening force, and load that it can withstand without compromising the joint integrity while still providing an effective breakaway for the protection of critical components.

INTRODUCTION

Structural joints in most engineering and industrial applications employ machine bolts to prevent two parts from separating or sliding relative to one another. The issues usually taken in consideration in the design of such bolted joints include (Slocum, 1992):

- Tensile stiffness
- Compressive stiffness

- Lateral stiffness
- Stability, and
- Preload.

Sufficient stiffness of the joint is provided by proper bolt preload. The sliding motion is resisted by frictional forces generated from the normal load on the joint produced by tightening of the bolts and the coefficient of friction between the contact layers of the parts being joined. The evaluation of tightening force is performed usually by well-established methodologies (Standard Handbook of Machine Design, 1996).

For most applications, the primary requirement is joint integrity. The specified bolt preload is large enough to overcome slipping of the joint while maintaining the average bolt stresses within its material elastic limit. However, in precision machine tool design, the shape and tightening force in these joints are usually evaluated based on two counter-acting requirements: (1) maintain sufficient stiffness provided by the joint at the cutting (load) point, and (2) allow the joint to slip (breakaway) in the event of a machine crash; in this case, the bolted joint works as a fuse preventing damage and, thus, protect critical/expensive components in the machine and/or avoid extensive repairs.

Under the influence of the cutting forces from machining operations, the entire joint must remain elastic in order to maintain machine accuracy. The issue in question is stability of the bolted joint in machine tools; the stability from the point of view of keeping the accuracy and alignment under repeated cutting loads while ensuring a positive breakaway under a minimum threshold impact load in the event of a machine crash.

It is well known that lateral displacements in permanently bolted joints have significant influence on the accuracy of the machine tool (Reshetov and Portman, 1988). This is mainly because these joints are usually very far from the point of cutting where the accuracy is specified. In these cases, the stiffness is measured at significant distances from the center of rotation of the joint. Therefore, under the influence of the cutting forces from machining operations, the entire breakaway joint must remain elastic in order to maintain machine accuracy.

Field data from machines running production have shown that satisfaction of the two criteria, presented above, using conventional methods of bolted joint design does not always assure that the stability requirements are met. This data shows that in the range of load that do not exceed the maximum force, allowed in the machine, there might occur permanent lateral displacements in the joint. These displacements accumulate during normal operations under repeated loads and machine looses alignment without obvious instantaneous slippage in the joint.

One of the theories that try to explain this phenomenon assumes that plastic deformations occur in the joint (Levina, 1967). These deformations are very small for each cutting load cycle that exceed some threshold value, but accumulate with repetition of such loading that take place over the course of machining numerous parts. The propagation of the plastic zone results in loss of the joint integrity, eventually leading to poor parts accuracy due to loss of critical alignment of machine components.

The paper describes an approach based on the formulation of plastic ratcheting. As a practical application, the approach is used to evaluate and improve the performance of a joint between the turret body and a slide in a lathe. The developed methodology allowed to optimize the shape of the joint, provide sufficient stiffness in the joint and still keep its ability to work as a protective fuse during a crash.

MODEL OF A BOLTED JOINT

In general, the lateral stiffness of a joint depends on both the surface finish of the contact layers and the preload on the joint. Lateral stiffness is affected by how the surface asperites interlock when the joint is preloaded, and how they subsequently bend and shear when lateral forces are applied to the joint. Attempts at modeling the asperites as random distributions of peaks and valleys have yet to prove satisfactory (Slocum, 1992). Therefore, we will use a simpler approach that allows simulating the deformation in a breakaway bolted joint as deformation in an elastic contact layer. The elastic parameters of the contact layer may be evaluated from tests or taken from literature.

To perform the evaluation, we have to know the shear stress τ_y in the layer at which an irreversible deformation occurs, and also the maximum shear stress τ_m at which slippage occurs. While we do not know the yield point of the artificial material that represents the contact layer, we may assume the value of coefficient of friction "f" in the joint.

By definition, the coefficient of friction $f = \tau_m / \sigma$, where σ is normal pressure in the joint. We now introduce a joint yield coefficient $f_y = \tau_y / \sigma$. This coefficient is similar to coefficient of friction and represents conditions at which plastic (irreversible) deformation may occur in the joint. From available data in literature (Levina, 1967), for most (typically machined cast iron) machine tool joints, $f_y \sim 0.5f$. More accurate data may be obtained from direct testing of particular joint. From technical point of view, it is important to note that if we know the ratio f_y/f, we can evaluate the condition under which plastic deformation may occur in a particular joint. In other words, we can estimate the stability of a joint if we choose the proper coefficient of friction f and the ratio f_y/f. The sensitivity of the results to the chosen values of these parameters is used to demonstrate the validity of the proposed approach.

MODEL OF A JOINT BETWEEN A TURRET AND A SLIDE

The proposed approach is used to evaluate performance of a breakaway joint between the turret body and slide of a lathe. To illustrate the procedure, a simplified model was developed to closely approximate the parts, sizes, and loads in real applications. Both the turret and the slide are machined parts made of cast iron. The solid model (developed for the finite element analysis) of the components is shown in Fig.1.

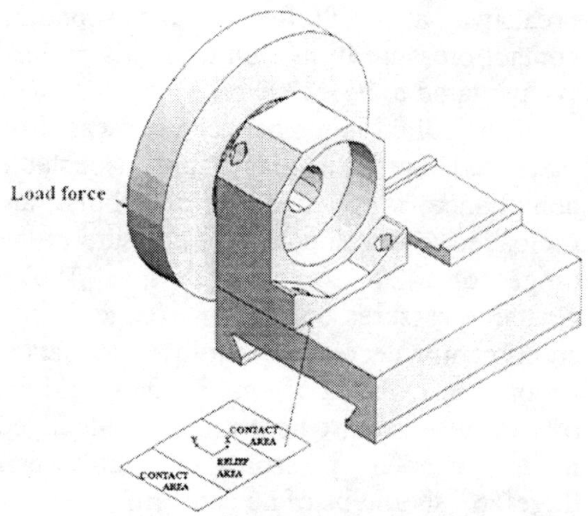

Fig.1. Model of the turret and joint

The FEA model is loaded at the top plate with thrust force in the direction parallel to axis of rotation of the top plate. Again, to illustrate the mathematics involved in the procedure, it is assumed that the specified maximum thrust force during machining does not exceed 2000 lbs.

As shown in Fig.1, the turret body to slide connection is made using four (4) bolts, each 16 mm in diameter. For the problem definition, the joint is required to satisfy following conditions:

1. For the given load point, the joint stiffness is ~1,000,000 lbs/in (typical value for such joints in machine tool structures)
2. The joint maintains its stability (that is no residual displacement) under repeated cutting loading in the range from 0 to 2,000 lbs.
3. The joint must slip (breakaway) under a sudden (impact) load not to exceed ~8,000 lbs in order to limit damage due to a potential crash.

For this example, contact area of the joint is 35 in^2 and the effective joint coefficient of friction f=0.25 (typical test data on cast iron over cast iron machined surfaces for such joints). Using conventional design formulae (Standard Handbook of Machine Design, 1996), each bolt must be torqued to a tensile preload of 19,000 lbs in order for this joint to breakaway at ~8,000 lbs. The corresponding contact pressure in the joint for this preload is calculated to be ~2200 psi.

For bolted joints between two cast iron machined parts, a reasonably acceptable compliance is obtained if the bolts are preloaded to develop contact pressure in the range of 420-520 psi (Levina, 1967). Contact pressures beyond this range do not significantly decrease the joint compliance. From the viewpoint of joint stiffness and the breakaway requirements, the calculated higher contact pressure of 2200 psi, therefore, should be of no concern.

Because of the random interlocking of peaks and valleys of the machined surfaces of the material, the contact layer must be modeled using fictitious (or equivalent) material properties to provide realistic normal and lateral joint compliance under applied load conditions. Based on historically observed data (Levina, 1967), the recommended values for the normal and lateral compliance for machine tool joints are 1.0E-6 in* in^2/lb, and 2.4E-6 in* in^2/lb, respectively. These values were used with finite element analyses to obtain the equivalent elastic modulus (E) and Poisson's ratio values for the contact layer of our particular geometry.

For a joint with symmetrical geometry and loading, the center of rotation coincides with the center of geometry. The joint in our problem is loaded by a force that develops lateral shear and torsion in the joint. As the load is increased and the joint is loaded beyond the elastic limit of the contact layer, the center of rotation will change and must be recomputed for each iterative load increase.

METHODOLOGY OF CALCULATIONS

The friction force at any point of the joint under lateral and torsion load is proportional to its relative displacement. It means, that the local coefficient of friction f_ρ at a given point of the joint is proportional to the distance ρ of this point from the center of rotation, i.e.,

$$f_\rho = f^*\rho/R \qquad (1)$$

where R = distance from center of rotation to the farthest point of the joint.

Therefore, it follows from previous discussion that the plastic deformation at any particular point on a contact layer initiates when

$$f_y = 0.5^*f_\rho \quad \text{or} \quad f_y = 0.5^*f^*\rho/R \qquad (2)$$

Alternatively, a point on the contact layer of the joint will remain elastic as long as

$$(\tau/\sigma)^*R/\rho < 0.5^*f \qquad (3)$$

By introducing a coefficient of local deformation $\phi = (\tau/\sigma)^*R/\rho$, and substituting f=0.25 for our case, Eq. (3) can be rewritten to provide the elastic state if

$$\phi < 0.125 \qquad (3')$$

The point will represent incipient slippage when the coefficient of local deformation ϕ at the point satisfies the condition:

$$\phi = f \text{, or in our case when } \phi = f = 0.25 \qquad (4)$$

Following the initial onset of yielding, the FEA results under gradually increased loading were evaluated to locate the newly developed plastic areas per Eq (3'). The material properties in these areas were changed to parameters of plastic material with hardening prior to next load increment analysis. The process was continued till the load reached the maximum specified limit.

ORIGINAL JOINT EVALUATION

Figure (2) shows the results of the analysis for selected loads.

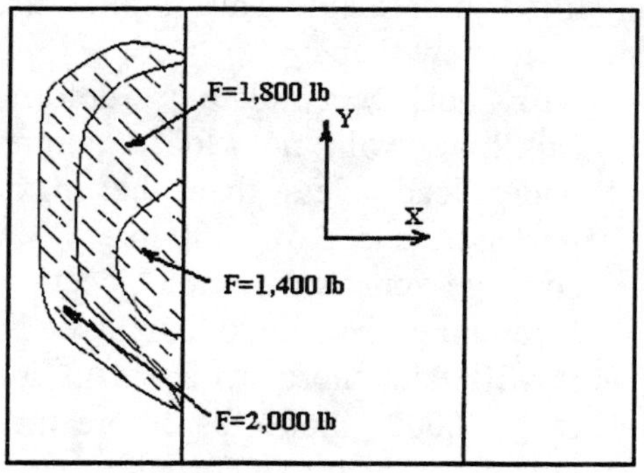

Fig.2. Propagation of plastic zone in the contact layer for applied load on the topplate.

There was no plastic deformation in the contact layer until the load reached 1400 lbs. Under the maximum permissible load of 2000 lbs, the results show that a significant area of the contact layer has undergone plastic deformation, suggesting potential relative motion of the components leading to misalignment and consequent adverse impact on parts accuracy during machining.

The other presentation of the coefficient of local deformation distribution is shown in Fig.3.

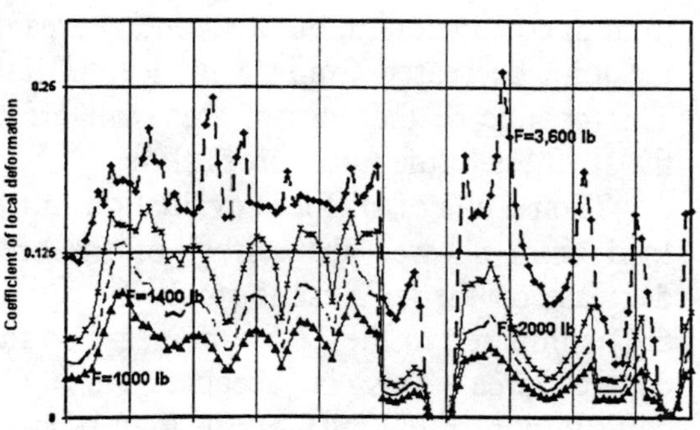

Fig.3. Change in coefficient of local deformation under increase in topplate loading (selected locations on contact layer from FEA model).

It follows from this chart that under the 1,400 lbs load, the coefficient of local deformation $\phi = (\tau / \sigma)*R/\rho$ just begins to exceed the 0.125 value at very few points indicating initial onset of yielding of the contact layer. At 2000 lbs, a large area of the layer has yielded. At 3,600 lbs, ϕ exceeds 0.25 indicating the beginning of slippage in the joint. The calculations were stopped at the onset of slippage.

Based on observed data of actual machine operating histories with analytical correlation such as presented here, one can make the following statements to be used as guidelines for the design of critical joints in precision machine tools:

1. The machine will not loose alignment if the joint under consideration does not have area of plastic deformation under maximum permissible load due to operating cycles.

2. If the operating loads are large enough to induce only localized plastic zones (and no slippage), the machine may eventually lose its alignment due to plastic ratcheting, but only after a sufficiently large number of cycles.

1. The machine will loose alignment with relatively small number of operating cycles, if the operating loads are large enough to develop localized slippage areas in the joint under consideration.

Obviously, the effect of the shear and torsion loading from the operating (cutting loads) cycles can be largely offset and may even be completely negated by increasing the number of bolts and/or using higher bolt preload to induce large, uniform contact pressure in the joint (as is customarily done for most structural joint applications). However, this approach is not normally suitable for critical machine tool joints which typically have limited contact area (limits the number of bolts) and require breakaway.

Experience suggests a more effective approach is to change the shape of the contact area to increase the contact pressure without increasing the breakaway threshold and redistribute the combined shear and pressure stresses to reduce the plastic zone for given cutting loads without materially impacting the joint stiffness. This is illustrated below for the joint in our example.

JOINT WITH NEW SHAPE OF THE CONTACT AREA

Consider changing the shape of the original joint in Fig.2. by taking out the central part of the joint where the plastic deformations occur at first. The area of the revised joint, as shown in Fig. 4, is reduced to 28.5 in^2 from the previous 35 in^2.

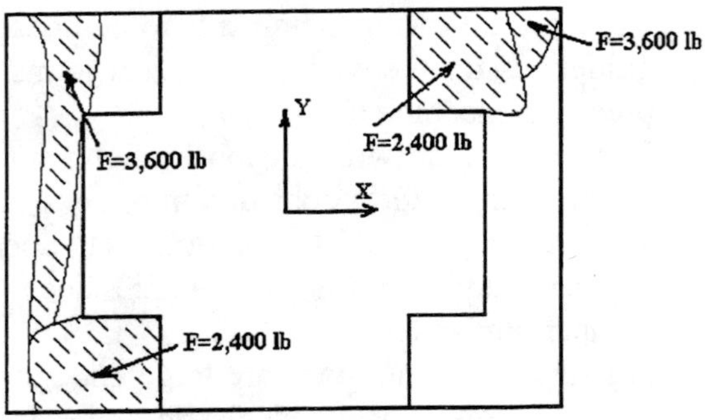

Fig.4. Propagation of plastic zone in the contact layer for applied load at the top plate. Proposed contact area configuration.

Repeating the analyses as before, results show that in the revised joint, plastic deformations begin when the load exceeds 2000 lbs compared to 1400 lbs on the original joint. Figure 4 shows the propagation of plastic zones under 2,400 lbs and 3,600 lbs loads.

The comparison between areas of plastic deformation for the two evaluated shapes of contact area is shown in Fig.5.

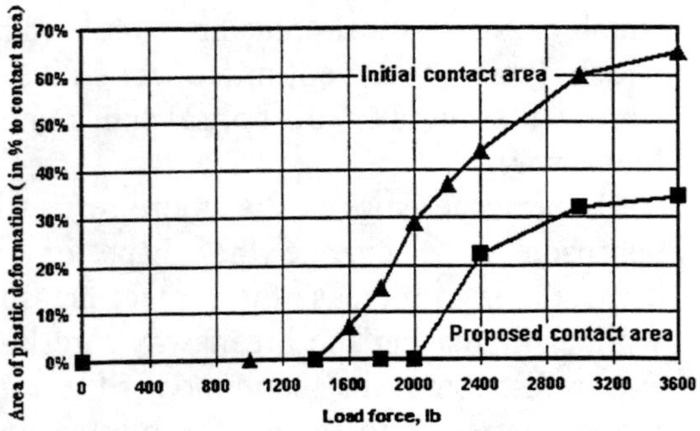

Fig.5. Area of plastic deformation in bolted joint.

In the modified joint, the plastic deformation starts at a higher load. Further, for a given applied load, the percentage area exceeding yield is also smaller for the modified joint shape. This implies that the joint with new shape:

- will not accumulate deformation (machine will not loose alignment) under load less than the maximum permissible load of 2,000 lbs;
- slippage zones in the joint occur under load higher than 3,600 lbs (see Fig.6). It will take more cycles with load in range 2,000 –3,600 lbs before machine will loose its alignment with new shape of the joint, to compare with original one.

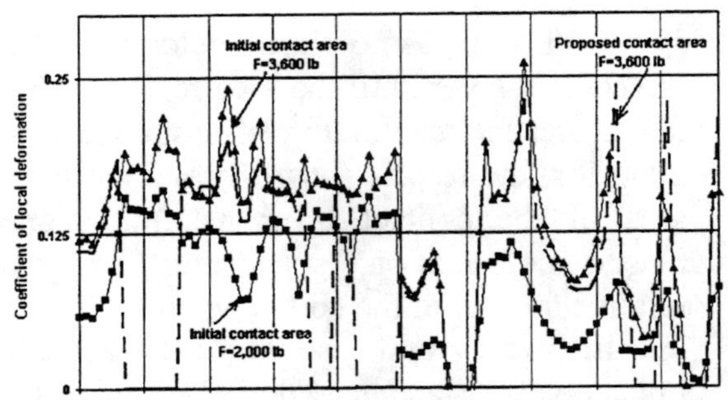

Fig.6. Coefficient of local deformation in contact area for joint with different shape (selected location on contact layer from FEA model).

The elimination of center part of joint helped to solve the stability problem of the joint in consideration, but the contact area of the joint decreased from 35 in^2 to 28.5 in^2. This change of the contact area will effect the stiffness of the turret-cross slide.

The results of stiffness evaluation in the load point showed that effect of the area decrease on the stiffness is small (less than 5%) compared to the 18.5% decrease in the contact area. This is because the main contribution in the stiffness in load point is from the area moment of inertia. This value did not change substantially when the center part of the area was eliminated. If necessary, the small loss in the stiffness may be easily compensated by changes to the outer boundaries of the joint using the same proposed approach.

EFFECT OF FRICTION COEFFICIENT AND CONTACT LAYER MODEL

The presented analyses herein were performed with parameters of the contact layer and coefficient of friction based on reasonable observed historical values. Obviously, actual values based on specific joint characteristics will effect the numerical results. However, the qualitative evaluation of processes in the joint under load presented here will not change.

For example, one may see from Fig.3 that areas, where plastic deformation occurs first, will not change if the coefficient of friction was decreased from 0.25 to 0.15. The plastic deformations will occur in the same areas but at lower load. Similar effect occurs if we change parameters of contact layer or the ratio f/f_y.

It is to be noted that the critical parameter of the joint evaluation is the coefficient of friction. For a specific given case, the coefficient of friction can be obtained from actual testing of the joint. The same is true for the ratio f/f_y.. This means that test data for the specific joint must be used if one wants to have not only qualitative evaluation of the joint, but a more accurate one.

CONCLUSION

Based on illustrated evaluations, the proposed methodology can be used to optimize the design of critical joints of components in precision machine tools to meet the combined requirements of stiffness, joint integrity (no slippage and resulting misalignment) under specified repeated cutting loads, and breakaway in the case of sudden impact loads under potential machine crash conditions. The approach proposed is to modify the contact area of the joint to satisfy stability and crash protection requirements without material impact on the joint stiffness. Reasonable assumptions based on historical information are sufficient for an effective qualitative evaluation. For more accurate, quantitative analysis, coefficients of friction must be obtained based on testing of the specific joint to be evaluated.

REFERENCES

Slocum, A.H. , 1992, "Precision Machine Design", Prentice Hall, NJ

Reshetov D., Portman V., 1988, "Accuracy of Machine Tools", ASME Press, New York

Levina Z. M., " Research on Static Stiffness of Joints in Machine Tools", Proc. of the 8th Int. Machine Tool Design and Research Conf., Sept. 1967, pp. 737-758.

Standard Handbook of Machine Design, 1996, McGraw-Hill, New York

PVP-Vol. 415, Recent Advances in Solids and Structures — 2000
ASME 2000

ON THE EQUIVALENCE OF THE STRESS AND DISPLACEMENT METHODS IN PLANAR ANISOTROPIC ELASTICITY

Xin-Lin Gao
Department of Mechanical Engineering–Engineering Mechanics, Michigan Technological University,
1400 Townsend Drive, Houghton, MI 49931-1295, USA
Robert E. Rowlands
Department of Mechanical Engineering, University of Wisconsin–Madison,
1513 University Avenue, Madison, WI 53706-1572, USA

ABSTRACT

The stress formalism of Lekhnitskii and the displacement formalism of Eshelby, Read and Shockley (ERS) are firstly recapitulated and compared for problems of *planar* anisotropic elasticity. It is *explicitly* demonstrated that for such problems the two formalisms lead to the identical characteristic equation and thus are equivalent. An *alternative* displacement formulation method is then presented for the same problems using an extended version of Green's theorem. This approach introduces a displacement function, which satisfies a fourth-order partial differential equation whose characteristic equation is identical with that obtained using the methods of Lekhnitskii and ERS. The new approach is proved to be *equivalent* to those of Lekhnitskii and ERS for problems of planar anisotropic elasticity. The present displacement function method exhibits similarities with the Airy stress function method.

INTRODUCTION

Plane elasticity theory of *isotropic* materials has been well established. Both the *stress* and *displacement* formulation methods have been successfully applied to solve various problems (see, for example, Boresi and Chong, 1987, pp. 480–484; Gao, 1996). The displacement formulation method was first used by G. Lamé, while the stress method was initiated by E. Beltrami (Teodorescu, 1964). The former leads to the (planar) Navier equations in terms of the two in-plane displacements, while the latter yields the biharmonic equation in terms of the Airy stress function (a *scalar* potential). The similarities and differences between the two formulation methods have also been clearly understood (see, for example, Chou and Pagano, 1967, pp. 89–114; Little, 1973, pp. 83–95). On the other hand, the theory of planar *anisotropic* elasticity, which deals with the *classical plane stress* and *plane strain* problems of anisotropic elastic materials possessing one plane of symmetry at $x_3=0$ (i.e., *monoclinic* elastic materials), is still an active research subject (see, for example, Wang and Shi, 1998; Yang and Ma, 1998).

The familiar *stress* formulation method in the linear elasticity theory of anisotropic materials is due to Lekhnitskii (1963), while the known *displacement* method is originated by Eshelby, Read and Shockley (ERS) (1953). Both Lekhnitskii's and ERS's methods started with the *generalized* plane problems, where all stress, strain and displacement components are independent of the out-of-plane Cartesian coordinate. For a *generalized* plane problem of an *anisotropic* elastic material, the *in-plane* response is coupled with the *out-of-plane* deformation. This necessitates introducing *two* stress functions in Lekhnitskii's formulation, as opposed to a *single* Airy stress function in plane *isotropic* elasticity. Also, the characteristic equation for such a problem is *sextic*, which, in general, cannot be analytically solved (Head, 1979). As a result, the analysis of the generalized plane problem can be very tedious. However, for a *classical* (pure) plane problem of a monoclinic elastic material, where the out-of-plane displacement is a *constant* (often set to zero) and only *in-plane* responses are involved, the formulation is greatly simplified. Since such a problem of *planar* anisotropic elasticity is *strictly* two-dimensional, approaches *similar/parallel* to those in plane *isotropic* elasticity become applicable. Also, the similarities and differences between the stress and displacement methods can be shown more *explicitly* for such problems. In addition, the engineering practice of treating each lamina of a modern composite laminate as a plane stress thin layer of an anisotropic (orthotropic) material also requires a thorough understanding of various methods in planar anisotropic elasticity.

Detailed discussions and general formulas are provided in Lekhnitskii (1963, pp. 134–141) for problems of *planar* anisotropic elasticity. Consequently, most of the existing elasticity solutions for such problems were derived using Lekhnitskii's approach. An illustrative example of this approach is the analysis of the crack problem provided by Sih, Paris and Irwin (1965). Lekhnitskii's *stress* formulation method is very convenient and efficient to use when a problem has only *traction* boundary conditions. For a problem with *displacement* or *mixed* boundary conditions, however, a displacement formulation method is always desired. Unfortunately, *displacement* methods for problems of *planar* anisotropic elasticity have not been thoroughly discussed in the literature. As a result, the relationships between Lekhnitskii's stress method and displacement methods for such problems have not been explicitly shown.

The objective of this paper is to examine the stress and displacement formulation methods for problems of *planar* anisotropic elasticity and to *explicitly* prove their equivalence using *simple/familiar* approaches. Aiming at self-consistency, the constitutive equations for both the classical plane stress and plane strain problems of a monoclinic elastic material are firstly recorded (and compared). This is followed by a recapitulation of Lekhnitskii's method, and then an examination of ERS's method, both in the context of planar anisotropic elasticity, where only *six* independent elastic constants are involved (Lekhnitskii, 1963; Yang and Ma, 1998). The recapitulations of the two familiar methods in the same context are included here for completeness and, more importantly, for comparison. It is *explicitly* shown, by a direct comparison, that the characteristic equation is *identical* for both the Lekhnitskii and ERS methods in the context of planar anisotropic elasticity. Finally, an alternative displacement method and thus a second proof of the equivalence of stress and displacement methods are presented for problems of planar anisotropic elasticity. This approach is based on an extended version of Green's theorem and exhibits similarities with Airy's stress function method. The paper concludes with a summary.

CONSTITUTIVE RELATIONS

From linear elasticity theory, for any homogeneous elastic material undergoing infinitesimal three-dimensional deformations, the generalized Hooke's law gives

$$\sigma_{ij} = C_{ijkl}\varepsilon_{kl}, \tag{1}$$

where σ_{ij} are the stress components, ε_{kl} are the strain components, and C_{ijkl} are the components of the elasticity (*elastic stiffness*) tensor, which are material constants and satisfy the following symmetry relations:

$$C_{ijkl} = C_{klij} = C_{jikl} = C_{ijlk}. \tag{2}$$

Equation (1) can also be rewritten as

$$\varepsilon_{ij} = C_{ijkl}^{-1}\sigma_{kl} \equiv S_{ijkl}\sigma_{kl}, \tag{3}$$

where S_{ijkl}, as defined, are the components of the *elastic compliance* tensor, which also satisfy the full symmetry relations:

$$S_{ijkl} = S_{klij} = S_{jikl} = S_{ijlk}. \tag{4}$$

For a *classical plane strain* case (of a very long *cylindrical* body) with the displacements $u_1=u_1(x_1,x_2)$, $u_2=u_2(x_1,x_2)$, $u_3=0$ and thus $\varepsilon_{31}=\varepsilon_{32}=\varepsilon_{33}=0$ (see, for example, Teodorescu, 1964), it follows from eqns (1) and (2) that the *in-plane* stress components are

$$\begin{aligned}
\sigma_{11} &= C_{1111}\varepsilon_{11} + 2C_{1112}\varepsilon_{12} + C_{1122}\varepsilon_{22} \equiv c_{11}\varepsilon_{11} + c_{12}\varepsilon_{22} + c_{16}\gamma_{12},\\
\sigma_{22} &= C_{2211}\varepsilon_{11} + 2C_{2212}\varepsilon_{12} + C_{2222}\varepsilon_{22} \equiv c_{12}\varepsilon_{11} + c_{22}\varepsilon_{22} + c_{26}\gamma_{12},\\
\sigma_{12} &= C_{1211}\varepsilon_{11} + 2C_{1212}\varepsilon_{12} + C_{1222}\varepsilon_{22} \equiv c_{16}\varepsilon_{11} + c_{26}\varepsilon_{22} + c_{66}\gamma_{12},
\end{aligned} \tag{5}$$

where $\gamma_{12}\equiv 2\varepsilon_{12}$ is the engineering shear strain, and c_{ij}, as defined, are the *stiffness constants* (with $c_{ij}=c_{ji}$). This system of equations can be inverted to obtain

$$\begin{aligned}
\varepsilon_{11} &= b_{11}\sigma_{11} + b_{12}\sigma_{22} + b_{16}\sigma_{12},\\
\varepsilon_{22} &= b_{12}\sigma_{11} + b_{22}\sigma_{22} + b_{26}\sigma_{12},\\
\gamma_{12} &= b_{16}\sigma_{11} + b_{26}\sigma_{22} + b_{66}\sigma_{12},
\end{aligned} \tag{6}$$

where b_{ij}, called the *reduced compliance constants*, are given by

$$\begin{aligned}
b_{11} &= \frac{c_{22}c_{66} - c_{26}^2}{\Delta}, \quad b_{12} = \frac{c_{16}c_{26} - c_{12}c_{66}}{\Delta},\\
b_{16} &= \frac{c_{12}c_{26} - c_{22}c_{16}}{\Delta}, \quad b_{22} = \frac{c_{11}c_{66} - c_{16}^2}{\Delta},\\
b_{26} &= \frac{c_{12}c_{16} - c_{11}c_{26}}{\Delta}, \quad b_{66} = \frac{c_{11}c_{22} - c_{12}^2}{\Delta},
\end{aligned} \tag{7}$$

with

$$\Delta \equiv \begin{vmatrix} c_{11} & c_{12} & c_{16} \\ c_{12} & c_{22} & c_{26} \\ c_{16} & c_{26} & c_{66} \end{vmatrix} > 0 \tag{8}$$

from the *positive definiteness* of the elasticity tensor. By definition, a *monoclinic* elastic material has *one* plane of *symmetry*. Let $x_3=0$ be this plane of symmetry. Then, it follows that (see, for example, Green and Zerna, 1968, p. 154)

$$C_{1123} = C_{1113} = C_{2223} = C_{2213} = C_{3323} = C_{3313} = C_{2312} = C_{1312} \equiv 0. \tag{9}$$

Using eqns (9) and (2) in eqn (1) gives

$$\begin{aligned}
\sigma_{13} &= 2C_{1313}\varepsilon_{13} + 2C_{1323}\varepsilon_{23},\\
\sigma_{23} &= 2C_{2313}\varepsilon_{13} + 2C_{2323}\varepsilon_{23},\\
\sigma_{33} &= C_{3311}\varepsilon_{11} + 2C_{3312}\varepsilon_{12} + C_{3322}\varepsilon_{22} + C_{3333}\varepsilon_{33}.
\end{aligned} \tag{10}$$

With $\varepsilon_{31}=\varepsilon_{32}=\varepsilon_{33}=0$, eqn (10) yields

$$\sigma_{13} \equiv 0,\ \sigma_{23} \equiv 0,\ \sigma_{33} \equiv C_{3311}\varepsilon_{11} + 2C_{3312}\varepsilon_{12} + C_{3322}\varepsilon_{22} \tag{11}$$

as the *out-of-plane* stress components in the *plane strain* case considered. Note that all non-zero stress and strain components in this case are independent of the out-of-plane coordinate x_3.

For a *classical plane stress* case (of a very *thin* plate) with $\sigma_{31}=\sigma_{32}=\sigma_{33}=0$ (see, for example, Teodorescu, 1964), it follows from eqns (3) and (4) that the *in-plane* strain components are

$$\begin{aligned}
\varepsilon_{11} &= S_{1111}\sigma_{11} + 2S_{1112}\sigma_{12} + S_{1122}\sigma_{22} \equiv a_{11}\sigma_{11} + a_{12}\sigma_{22} + a_{16}\sigma_{12},\\
\varepsilon_{22} &= S_{2211}\sigma_{11} + 2S_{2212}\sigma_{12} + S_{2222}\sigma_{22} \equiv a_{12}\sigma_{11} + a_{22}\sigma_{22} + a_{26}\sigma_{12},\\
\gamma_{12} &= 2S_{1211}\sigma_{11} + 4S_{1212}\sigma_{12} + 2S_{1222}\sigma_{22} \equiv a_{16}\sigma_{11} + a_{26}\sigma_{22} + a_{66}\sigma_{12},
\end{aligned} \tag{12}$$

where a_{ij}, as defined, are the *compliance constants* (with $a_{ij}=a_{ji}$). Similarly, one can label the plane of symmetry of the *monoclinic* material as $x_3=0$. Since S_{ijkl} are the components of a fourth-rank tensor and transform according to the *same* rule as C_{ijkl} do, it follows from eqn (9) that

$$S_{1123} = S_{1113} = S_{2223} = S_{2213} = S_{3323} = S_{3313} = S_{2312} = S_{1312} \equiv 0. \tag{13}$$

Using eqns (13) and (4) in eqn (3) gives

$$\begin{aligned}
\varepsilon_{13} &= 2S_{1313}\sigma_{13} + 2S_{1323}\sigma_{23},\\
\varepsilon_{23} &= 2S_{2313}\sigma_{13} + 2S_{2323}\sigma_{23},\\
\varepsilon_{33} &= S_{3311}\sigma_{11} + 2S_{3312}\sigma_{12} + S_{3322}\sigma_{22} + S_{3333}\sigma_{33}.
\end{aligned} \tag{14}$$

With $\sigma_{31}=\sigma_{32}=\sigma_{33}=0$, eqn (14) yields

$$\varepsilon_{13} \equiv 0,\ \varepsilon_{23} \equiv 0,\ \varepsilon_{33} \equiv S_{3311}\sigma_{11} + 2S_{3312}\sigma_{12} + S_{3322}\sigma_{22} \tag{15}$$

as the *out-of-plane* strain components in the plane stress case considered. Again, note that all non-zero stress and strain components in this case are independent of x_3.

Since the non-zero *out-of-plane* stress/strain component is determinable from the *in-plane* strain/stress components, as shown in eqns (11) and (15), we will consider *only* the *in-plane* components in our formulations to follow.

It should be noted that the compliance constants (for the plane stress case), a_{ij}, involved in eqn (12) are related to the reduced compliance constants (for the plane strain case), b_{ij}, introduced in eqn (6). It can be shown that

$$b_{ij} \equiv a_{ij} - \frac{a_{i3}a_{j3}}{a_{33}},\ j = \{1,2,6\},\ i = \{1,2,6\}. \tag{16}$$

Equations (6), (12) and (16) establish the *formal* equivalence between the constitutive relations of the classical plane stress and plane strain cases of the monoclinic elastic material, which enables one to formulate problems of planar anisotropic elasticity using *either* the plane stress *or* the plane strain constitutive relations presented above. This is *parallel* to that of replacing Young's modulus E and Poisson's ratio ν in the plane stress equations by $E/(1-\nu^2)$ and $\nu/(1-\nu)$, respectively, to form the plane strain equations in *plane isotropic elasticity*.

It should also be noticed from eqns (6) and (12) that only *six* independent material constants are needed to describe the *in-plane* response of an anisotropic *thin/long cylindrical* elastic body with $x_3=0$ as the only plane of symmetry, as opposed to *twenty-one* independent constants (i.e., C_{ijkl} defined by eqns (1) and (2) or S_{ijkl} defined by eqns (3) and (4)) needed to represent the deformation of a *three-dimensional* elastic object with arbitrary anisotropy. Any further reduction in the number of independent material constants presupposes some additional plane(s) of material symmetry.

STRESS FORMULATION METHOD: LEKHNITSKII'S FORMALISM

Consider a *plane stress* body (i.e., a very *thin* plate) of homogeneous, anisotropic, elastic material undergoing infinitesimal deformations and having one plane of symmetry at x_3=0. Then, the governing equations, in a *stress formulation* using the usual Cartesian coordinates $\{x_1, x_2, x_3\}$, are the *equilibrium* equations (in the *absence* of body forces):

$$\frac{\partial\sigma_{11}}{\partial x_1}+\frac{\partial\sigma_{12}}{\partial x_2}=0,\quad \frac{\partial\sigma_{12}}{\partial x_1}+\frac{\partial\sigma_{22}}{\partial x_2}=0, \tag{17}$$

the *compatibility* equation:

$$\frac{\partial^2\varepsilon_{11}}{\partial x_2^2}+\frac{\partial^2\varepsilon_{22}}{\partial x_1^2}=\frac{\partial^2\gamma_{12}}{\partial x_1\partial x_2}, \tag{18}$$

and the *constitutive* equations given by eqn (12). Note that the conditions $\sigma_{31}=\sigma_{32}=\sigma_{33}=0$ have been used in reaching eqn (17), and eqn (15) and the assumption of very small thickness have been invoked in obtaining eqn (18) (see, for example, Timoshenko and Goodier, 1970, pp. 274–277) from the equilibrium and compatibility equations of three-dimensional elasticity.

To derive a *general* solution of this equation system, one can introduce the *Airy stress function* $\psi(x_1, x_2)$ by letting

$$\sigma_{11}=\frac{\partial^2\psi}{\partial x_2^2},\quad \sigma_{22}=\frac{\partial^2\psi}{\partial x_1^2},\quad \sigma_{12}=-\frac{\partial^2\psi}{\partial x_1\partial x_2} \tag{19}$$

such that the equilibrium equations given by eqn (17) are identically satisfied. Now using eqns (12) and (19) in eqn (18) and carrying out the algebra will yield

$$a_{11}\frac{\partial^4\psi}{\partial x_2^4}-2a_{16}\frac{\partial^4\psi}{\partial x_1\partial x_2^3}+(2a_{12}+a_{66})\frac{\partial^4\psi}{\partial x_1^2\partial x_2^2}-2a_{26}\frac{\partial^4\psi}{\partial x_2\partial x_1^3}+a_{22}\frac{\partial^4\psi}{\partial x_1^4}=0. \tag{20}$$

This is the *basic* governing equation of the system, which contains all *six* independent material constants and applies to monoclinic thin elastic bodies (with one plane of symmetry at x_3=0). For orthotropic materials with $a_{16}=0=a_{26}$ (i.e., x_1=0 and x_2=0 are two *additional* planes of symmetry such that the *in-plane* tensions and shear are decoupled), eqn (20) reduces to

$$a_{11}\frac{\partial^4\psi}{\partial x_2^4}+(2a_{12}+a_{66})\frac{\partial^4\psi}{\partial x_1^2\partial x_2^2}+a_{22}\frac{\partial^4\psi}{\partial x_1^4}=0, \tag{21}$$

where

$$a_{11}\equiv\frac{1}{E_{11}},\quad a_{22}\equiv\frac{1}{E_{22}},\quad a_{66}\equiv\frac{1}{G_{12}},\quad a_{12}\equiv-\frac{\nu_{12}}{E_{11}}=-\frac{\nu_{21}}{E_{22}}, \tag{22}$$

with E_{11}, E_{22}, G_{12}, ν_{12} (or ν_{21}) being the *four* independent (in-plane) material constants. Furthermore, for *isotropic* materials there exist

$$a_{11}=a_{22}=\frac{1}{E},\quad a_{66}=\frac{1}{G}=\frac{2(1+\nu)}{E},\quad a_{12}=-\frac{\nu}{E}, \tag{23}$$

where E and ν (or G) are the only *two* independent material constants known as Young's modulus and Poisson's ratio (or shear modulus), respectively. Substituting eqn (23) into eqn (21) then leads to

$$\frac{\partial^4\psi}{\partial x_2^4}+2\frac{\partial^4\psi}{\partial x_1^2\partial x_2^2}+\frac{\partial^4\psi}{\partial x_1^4}=0, \tag{24}$$

which is the well-known *biharmonic* equation in two-dimensional *isotropic* elasticity.

The solution of eqn (20) can be represented by (Lekhnitskii, 1963)

$$\psi(x_1,x_2)\equiv\Psi(z),\quad z\equiv x_1+\mu x_2,\quad \mu\equiv\alpha+i\beta, \tag{25}$$

where $\Psi(z)$ is an analytic function of the complex variable z yet to be determined, $i^2=-1$, and α and β are real constants. Substituting eqn (25) into eqn (20) yields, with the help of the *chain rule*,

$$\Psi^{(IV)}(z)[a_{11}\mu^4-2a_{16}\mu^3+(2a_{12}+a_{66})\mu^2-2a_{26}\mu+a_{22}]=0. \tag{26}$$

Since $\Psi(z)$ can be an arbitrary function such that its fourth-order derivative $\Psi^{(IV)}(z)\not\equiv0$, eqn (26) implies that

$$a_{11}\mu^4-2a_{16}\mu^3+(2a_{12}+a_{66})\mu^2-2a_{26}\mu+a_{22}\equiv0. \tag{27}$$

This is known as the *characteristic equation* of eqn (20). Clearly, the solution of eqn (20), in the form of eqn (25), depends on the roots of eqn (27). It can be shown (Lekhnitskii, 1963, p. 122) that this *fourth-degree* polynomial equation has *only* complex roots, μ_i, which consist of two conjugate pairs of the form:

$$\mu_1=\alpha_1+i\beta_1,\ \overline{\mu_1}=\alpha_1-i\beta_1,\ \mu_2=\alpha_2+i\beta_2,\ \overline{\mu_2}=\alpha_2-i\beta_2, \tag{28}$$

where α_i and β_i (i={1,2}) are real constants depending on the values of a_{ij}. If these roots are *distinctive* (with the understanding that non-distinctive roots may be regarded as suitable limits of distinctive roots), then the general solution of eqn (20) is given, via the principle of *superposition*, by

$$\Psi=\Psi_1(z_1)+\Psi_3(\overline{z_1})+\Psi_2(z_2)+\Psi_4(\overline{z_2}), \tag{29}$$

where Ψ_i (i={1,2,3,4}) are four analytic functions of their arguments. Since $\Psi\equiv\psi(x_1,x_2)$ is real, eqn (29) must have the form

$$\psi=2\mathrm{Re}[\Psi_1(z_1)+\Psi_2(z_2)], \tag{30}$$

where Re stands for the *real* part and

$$z_1=x_1+\mu_1x_2,\quad z_2=x_1+\mu_2x_2. \tag{31}$$

Using eqn (30) in eqns (19) and (12) and carrying out the differentiation and integration involved will result in

$$\begin{aligned}\sigma_{11}&=2\mathrm{Re}[\mu_1^2\Psi_1''(z_1)+\mu_2^2\Psi_2''(z_2)],\\ \sigma_{22}&=2\mathrm{Re}[\Psi_1''(z_1)+\Psi_2''(z_2)],\\ \sigma_{12}&=-2\mathrm{Re}[\mu_1\Psi_1''(z_1)+\mu_2\Psi_2''(z_2)],\\ u_1&=2\mathrm{Re}[p_1\Psi_1'(z_1)+p_2\Psi_2'(z_2)],\\ u_2&=2\mathrm{Re}[q_1\Psi_1'(z_1)+q_2\Psi_2'(z_2)],\end{aligned} \tag{32}$$

where

$$p_i\equiv a_{11}\mu_i^2+a_{12}-a_{16}\mu_i,\quad q_i\equiv a_{12}\mu_i+\frac{a_{22}}{\mu_i}-a_{26};\quad i=\{1,2\}, \tag{33}$$

and μ_1, μ_2 are the roots of eqn (27), whose forms are given by eqn (28). Equation (32) gives the stress and displacement components in the *plane stress* body considered. The two analytic functions, $\Psi_1(z_1)$ and $\Psi_2(z_2)$, involved here are to be determined from the boundary conditions or measured data (Lin and Rowlands, 1995) of a specific problem. Since the values of μ_1 and μ_2 depend on a_{ij}, which are material constants, eqn (32) demonstrates that the *stress* components are *dependent* on the *material* properties, *unlike* that in plane *isotropic* elasticity (see, for example, Gao, 1996).

It should be noticed that the solution of eqn (24), the basic governing equation for plane *isotropic* materials, given by the well-known Goursat formula is *not* reducible from eqn (30), the solution of the basic governing equation for *monoclinic* materials given by eqn (20), although eqn (24) itself is reduced from eqn (20). The reason for this is that *isotropic* materials are *degenerate* and, as a result, eqn (24) is even harder to solve than eqn (20).

As remarked near the end of the preceding section, for the classical *plane strain* case all of the above formulas *still* apply *after* replacing a_{ij} by b_{ij}.

Note that the main results presented in this section were first reported in Lekhnitskii (1963, pp. 134–141) (without providing very detailed derivations).

DISPLACEMENT METHOD: ESHELBY, READ AND SHOCKLEY'S FORMALISM

Consider a *plane strain* body (i.e., a very long *cylindrical* object) of homogeneous, anisotropic, elastic material undergoing infinitesimal deformations and having one plane of symmetry at x_3=0. Then, the governing equations, in a *displacement formulation*, are the *equilibrium* equations given by eqn (17), which, in this case, are reduced from the three equilibrium equations of three-dimensional elasticity by using eqn (11), the *geometrical* equations:

$$\varepsilon_{11}=\frac{\partial u_1}{\partial x_1},\quad \varepsilon_{22}=\frac{\partial u_2}{\partial x_2},\quad \gamma_{12}=\frac{\partial u_1}{\partial x_2}+\frac{\partial u_2}{\partial x_1}, \tag{34}$$

and the *constitutive* equations given by eqn (5).

In a displacement formulation, $u_1(x_1,x_2)$ and $u_2(x_1,x_2)$ are regarded as the *basic unknowns* and are to be determined *first*. Using eqns (5) and (34) in eqn (17) and carrying out the algebra will yield

$$\begin{aligned}
&c_{11}\frac{\partial^2 u_1}{\partial x_1^2}+2c_{16}\frac{\partial^2 u_1}{\partial x_1\partial x_2}+c_{66}\frac{\partial^2 u_1}{\partial x_2^2}+c_{16}\frac{\partial^2 u_2}{\partial x_1^2}\\
&\qquad+(c_{12}+c_{66})\frac{\partial^2 u_2}{\partial x_1\partial x_2}+c_{26}\frac{\partial^2 u_2}{\partial x_2^2}=0,\\
&c_{16}\frac{\partial^2 u_1}{\partial x_1^2}+(c_{12}+c_{66})\frac{\partial^2 u_1}{\partial x_1\partial x_2}+c_{26}\frac{\partial^2 u_1}{\partial x_2^2}+c_{66}\frac{\partial^2 u_2}{\partial x_1^2}\\
&\qquad+2c_{26}\frac{\partial^2 u_2}{\partial x_1\partial x_2}+c_{22}\frac{\partial^2 u_2}{\partial x_2^2}=0.
\end{aligned} \tag{35a,b}$$

These are the *basic* governing equations of the problem in terms of displacements u_1 and u_2. Note that all *six* independent material constants appear in the *system* of eqns (35a,b), but not all of them in a *single* equation (i.e., eqn (35a) or eqn (35b)). Physically, this implies, in addition to the mathematical dictations, that the *two* equations given by eqns (35a,b) *together* define the solution of the problem. The situation here appears to differ from that in the *stress formulation* presented earlier. However, it will become clear later (in the next section) that a *displacement formulation* similar to the stress function method can also be developed for this problem. For *isotropic* materials, with

$$\begin{aligned}
&c_{11}=c_{22}=c_{33}=2G+\lambda=\frac{E(1-\nu)}{(1+\nu)(1-2\nu)},\\
&c_{44}=c_{55}=c_{66}=G=\frac{E}{2(1+\nu)},\\
&c_{12}=c_{13}=c_{23}=c_{21}=c_{31}=c_{32}=\lambda=\frac{E\nu}{(1+\nu)(1-2\nu)},\\
&\textit{all other } c_{ij}=0,
\end{aligned} \tag{36}$$

eqns (35a,b) reduce to

$$\begin{aligned}
&(\lambda+G)\left(\frac{\partial^2 u_1}{\partial x_1^2}+\frac{\partial^2 u_2}{\partial x_1\partial x_2}\right)+G\left(\frac{\partial^2 u_1}{\partial x_1^2}+\frac{\partial^2 u_1}{\partial x_2^2}\right)=0,\\
&(\lambda+G)\left(\frac{\partial^2 u_1}{\partial x_1\partial x_2}+\frac{\partial^2 u_2}{\partial x_2^2}\right)+G\left(\frac{\partial^2 u_2}{\partial x_1^2}+\frac{\partial^2 u_2}{\partial x_2^2}\right)=0.
\end{aligned} \tag{37}$$

These are exactly the Navier equations for plane strain case in isotropic elasticity (see, for example, Chou and Pagano, 1967, p. 73).

The *general solution* of eqns (35a,b), a homogeneous, linear, second-order partial differential equation system, can be represented by, following Eshelby et al. (1953) for the *generalized* plane strain case,

$$u_1=k_1 f(z)=k_1 f(x_1+\mu x_2),\quad u_2=k_2 f(z)=k_2 f(x_1+\mu x_2), \tag{38}$$

where $f(z)$ is an analytic function of the complex variable z yet to be determined, and k_1 and k_2 are yet-unknown constants. Using eqn (38) in eqns (35a,b) gives

$$\begin{aligned}
&(c_{11}+2c_{16}\mu+c_{66}\mu^2)k_1+[c_{16}+(c_{12}+c_{66})\mu+c_{26}\mu^2]k_2=0,\\
&[c_{16}+(c_{12}+c_{66})\mu+c_{26}\mu^2]k_1+(c_{66}+2c_{26}\mu+c_{22}\mu^2)k_2=0
\end{aligned} \tag{39}$$

for $f''(z)\neq 0$ (since $f(z)$ can be an arbitrary function of z). This defines an eigenvalue problem. A non-trivial solution for u_1 and u_2 requires that $k_1\neq 0$ and $k_2\neq 0$, which is true *if and only if*

$$\begin{vmatrix} c_{11}+2c_{16}\mu+c_{66}\mu^2 & c_{16}+(c_{12}+c_{66})\mu+c_{26}\mu^2 \\ c_{16}+(c_{12}+c_{66})\mu+c_{26}\mu^2 & c_{66}+2c_{26}\mu+c_{22}\mu^2 \end{vmatrix}=0. \tag{40}$$

Expanding and rearranging eqn (40) will give

$$\begin{aligned}
&(c_{22}c_{66}-c_{26}^2)\mu^4+2(c_{16}c_{22}-c_{12}c_{26})\mu^3+(c_{11}c_{22}+2c_{16}c_{26}-c_{12}^2\\
&-2c_{12}c_{66})\mu^2+2(c_{11}c_{26}-c_{12}c_{16})\mu+(c_{11}c_{66}-c_{16}^2)=0.
\end{aligned} \tag{41}$$

This is the *characteristic equation* of the system listed in eqns (35a,b) in terms of the stiffness constants c_{ij}. Using eqn (7) in eqn (41) results in, with $\Delta>0$ according to eqn (8),

$$b_{11}\mu^4-2b_{16}\mu^3+(2b_{12}+b_{66})\mu^2-2b_{26}\mu+b_{22}=0 \tag{42}$$

as the characteristic equation in terms of the reduced compliance constants b_{ij}.

For the classical *plane stress* case, one replaces b_{ij} by a_{ij} and eqn (42) becomes identical with eqn (27). This proves that the Lekhnitskii and ERS formalisms are *equivalent* in generating the characteristic equation for problems of planar anisotropic elasticity.

From this equivalence, it follows immediately that eqn (42) only has conjugate-pair complex roots of the form:

$$\mu_1=\xi_1+i\eta_1,\quad \overline{\mu_1}=\xi_1-i\eta_1,\quad \mu_2=\xi_2+i\eta_2,\quad \overline{\mu_2}=\xi_2-i\eta_2, \tag{43}$$

where ξ_i and η_i (i={1,2}) are real constants depending on the values of b_{ij}. These roots are called *eigenvalues*. For each of the four eigenvalues, solving eqn (39) will give k_1 and k_2 as the components of the corresponding eigenvector. If these roots are *distinctive*, the general solution of eqns (35a,b) is then given, via the principle of *superposition*, by

$$u_1 = 2\mathrm{Re}\{k_1^{(1)} f_1[x_1 + (\xi_1 + i\eta_1)x_2] + k_1^{(2)} f_2[x_1 + (\xi_2 + i\eta_2)x_2]\},$$
$$u_2 = 2\mathrm{Re}\{k_2^{(1)} f_1[x_1 + (\xi_1 + i\eta_1)x_2] + k_2^{(2)} f_2[x_1 + (\xi_2 + i\eta_2)x_2]\}, \quad (44)$$

where $k_i^{(1)}$ and $k_i^{(2)}$ (i={1,2}) are the components of the eigenvectors associated with μ_1 and μ_2, respectively. Using eqn (44) in eqn (34) yields

$$\varepsilon_{11} = 2\mathrm{Re}[k_1^{(1)} f_1'(z_1) + k_1^{(2)} f_2'(z_2)],$$
$$\varepsilon_{22} = 2\mathrm{Re}[k_2^{(1)}\mu_1 f_1'(z_1) + k_2^{(2)}\mu_2 f_2'(z_2)], \quad (45)$$
$$\gamma_{12} = 2\mathrm{Re}[(k_1^{(1)}\mu_1 + k_2^{(1)}) f_1'(z_1) + (k_1^{(2)}\mu_2 + k_2^{(2)}) f_2'(z_2)],$$

where $z_1 \equiv x_1 + \mu_1 x_2$ and $z_2 \equiv x_1 + \mu_2 x_2$ are the two complex variables, and μ_1 and μ_2 are given by eqns (42) and (43). Substituting eqn (45) into eqn (5) will give the expressions of stress components. Thus, all stress, strain and displacement components have been determined for the *plane strain* body considered, which are in terms of the two analytic functions, $f_1(z_1)$ and $f_2(z_2)$, to be determined from the boundary conditions of a given problem.

Again, it should be noted that the solution for *monoclinic* materials given by eqn (44) cannot be reduced to the solution of eqn (37), Navier's equations for *isotropic* materials, which are *degenerate*.

For the classical *plane stress* case, all of the above formulas *still* apply *after* replacing b_{ij} by a_{ij}.

DISPLACEMENT METHOD: AN ALTERNATIVE APPROACH (GAO AND ROWLANDS, 2000)

In this section, an *alternative* displacement method (*different* from that of ERS discussed in the preceding section) is presented first. It is then demonstrated that this alternative method can also lead to the *same* characteristic equation as that obtained above (in the previous two sections).

In *any* displacement method (for the classical plane strain case of the monoclinic material), one only needs to solve the *two* partial differential equations listed in eqns (35a,b). This can be done by solving one equation *first* and *then* enforcing the other (see, for example, Gao, 1998, 1999a).

Note that eqn (35a) may, in the spirit of Wang (1985), be rewritten as

$$\frac{\partial}{\partial x_1}(c_{11}\frac{\partial u_1}{\partial x_1} + 2c_{16}\frac{\partial u_1}{\partial x_2} + c_{16}\frac{\partial u_2}{\partial x_1})$$
$$= -\frac{\partial}{\partial x_2}[c_{66}\frac{\partial u_1}{\partial x_2} + (c_{12} + c_{66})\frac{\partial u_2}{\partial x_1} + c_{26}\frac{\partial u_2}{\partial x_2}]. \quad (46)$$

Then, by an extended version of Green's theorem (Chou and Pagano, 1967, p. 114; Gao, 1999b) there exists a function $\Lambda(x_1, x_2)$ such that

$$c_{11}\frac{\partial u_1}{\partial x_1} + 2c_{16}\frac{\partial u_1}{\partial x_2} + c_{16}\frac{\partial u_2}{\partial x_1} = \frac{\partial \Lambda}{\partial x_2},$$
$$-[c_{66}\frac{\partial u_1}{\partial x_2} + (c_{12} + c_{66})\frac{\partial u_2}{\partial x_1} + c_{26}\frac{\partial u_2}{\partial x_2}] = \frac{\partial \Lambda}{\partial x_1}. \quad (47)$$

These two equations can also be written as

$$\frac{\partial}{\partial x_1}(c_{11}u_1 + c_{16}u_2) = \frac{\partial}{\partial x_2}(\Lambda - 2c_{16}u_1),$$
$$\frac{\partial}{\partial x_1}[\Lambda + (c_{12} + c_{66})u_2] = \frac{\partial}{\partial x_2}(-c_{66}u_1 - c_{26}u_2). \quad (48a,b)$$

Again, by the extended Green's theorem, eqn (48a) implies that there is a function $\Gamma(x_1, x_2)$ such that

$$c_{11}u_1 + c_{16}u_2 = \frac{\partial \Gamma}{\partial x_2}, \quad \Lambda - 2c_{16}u_1 = \frac{\partial \Gamma}{\partial x_1}, \quad (49a,b)$$

and eqn (48b) implies that there exists a function $\Omega(x_1, x_2)$ such that

$$\Lambda + (c_{12} + c_{66})u_2 = \frac{\partial \Omega}{\partial x_2}, \quad -c_{66}u_1 - c_{26}u_2 = \frac{\partial \Omega}{\partial x_1}. \quad (50a,b)$$

Now solving eqns (49a) and (50b) for u_1 and u_2 gives

$$u_1 = \frac{1}{c_{11}c_{26} - c_{66}c_{16}}(c_{26}\frac{\partial \Gamma}{\partial x_2} + c_{16}\frac{\partial \Omega}{\partial x_1}),$$
$$u_2 = \frac{-1}{c_{11}c_{26} - c_{66}c_{16}}(c_{66}\frac{\partial \Gamma}{\partial x_2} + c_{11}\frac{\partial \Omega}{\partial x_1}) \quad (51)$$

for the general case with $c_{11}c_{26} - c_{66}c_{16} \neq 0$. The special cases of orthotropic and isotropic materials with $c_{26} \equiv 0 \equiv c_{16}$ (and thus $c_{11}c_{26} - c_{66}c_{16} = 0$) have been analyzed separately (see Gao, 1999b, 2000), the findings of which complement the present derivation for the general case.

Next, combining eqns (49b) and (50a) yields

$$2c_{16}u_1 + (c_{12} + c_{66})u_2 = \frac{\partial \Omega}{\partial x_2} - \frac{\partial \Gamma}{\partial x_1}. \quad (52)$$

Using eqn (51) in eqn (52) leads to

$$\frac{\partial}{\partial x_1}[\Gamma + \frac{2c_{16}^2 - c_{11}(c_{12} + c_{66})}{c_{11}c_{26} - c_{66}c_{16}}\Omega]$$
$$= \frac{\partial}{\partial x_2}[\Omega + \frac{c_{66}(c_{12} + c_{66}) - 2c_{16}c_{26}}{c_{11}c_{26} - c_{66}c_{16}}\Gamma]. \quad (53)$$

This implies, once again by the extended Green's theorem, that there exists a function $F(x_1, x_2)$ such that

$$\Gamma + \frac{2c_{16}^2 - c_{11}(c_{12} + c_{66})}{c_{11}c_{26} - c_{66}c_{16}}\Omega = \frac{\partial F}{\partial x_2},$$
$$\Omega + \frac{c_{66}(c_{12} + c_{66}) - 2c_{16}c_{26}}{c_{11}c_{26} - c_{66}c_{16}}\Gamma = \frac{\partial F}{\partial x_1}. \quad (54)$$

Solving eqn (54) then gives

$$\Omega = \frac{1}{A}[\frac{\partial F}{\partial x_1} - \frac{c_{66}(c_{12} + c_{66}) - 2c_{16}c_{26}}{c_{11}c_{26} - c_{66}c_{16}}\frac{\partial F}{\partial x_2}],$$
$$\Gamma = \frac{1}{A}[-\frac{2c_{16}^2 - c_{11}(c_{12} + c_{66})}{c_{11}c_{26} - c_{66}c_{16}}\frac{\partial F}{\partial x_1} + \frac{\partial F}{\partial x_2}], \quad (55)$$

where the constant A ($\neq 0$) is defined as

$$A \equiv 1 - \frac{[c_{66}(c_{12} + c_{66}) - 2c_{16}c_{26}][2c_{16}^2 - c_{11}(c_{12} + c_{66})]}{(c_{11}c_{26} - c_{66}c_{16})^2}. \quad (56)$$

Substituting eqn (55) into eqn (51) results in

$$u_1 = \frac{1}{A(c_{11}c_{26} - c_{66}c_{16})}[c_{16}\frac{\partial^2 F}{\partial x_1^2} + (c_{12} + c_{66})\frac{\partial^2 F}{\partial x_1 \partial x_2} + c_{26}\frac{\partial^2 F}{\partial x_2^2}],$$
$$u_2 = \frac{-1}{A(c_{11}c_{26} - c_{66}c_{16})}[c_{11}\frac{\partial^2 F}{\partial x_1^2} + 2c_{16}\frac{\partial^2 F}{\partial x_1 \partial x_2} + c_{66}\frac{\partial^2 F}{\partial x_2^2}]. \quad (57)$$

These expressions determine the displacement components *solely* in terms of $F(x_1, x_2)$. Hence, $F(x_1, x_2)$ may be called a *displacement function, parallel* to the *Airy stress function* introduced in Lekhnitskii's stress method discussed earlier. A displacement function method for plane *isotropic* elasticity problems was presented by Von K. Marguerre in 1933 (see, for example, Teodorescu, 1964; Little, 1973, pp. 90–92; Gao, 1999b).

Clearly, since the two expressions in eqn (57) are derived from eqn (35a) *alone*, they must be substituted into the other basic governing equation, eqn (35b), to obtain the *defining* equation for $F(x_1,x_2)$. Substituting eqn (57) into eqn (35b) gives

$$
\begin{aligned}
&(c_{11}c_{66}-c_{16}^2)\frac{\partial^4 F}{\partial x_1^4}+2(c_{26}c_{11}-c_{12}c_{16})\frac{\partial^4 F}{\partial x_1^3\partial x_2} \\
&+(c_{11}c_{22}+2c_{16}c_{26}-c_{12}^2-2c_{12}c_{66})\frac{\partial^4 F}{\partial x_1^2\partial x_2^2} \\
&+2(c_{22}c_{16}-c_{12}c_{26})\frac{\partial^4 F}{\partial x_1\partial x_2^3}+(c_{22}c_{66}-c_{26}^2)\frac{\partial^4 F}{\partial x_2^4}=0.
\end{aligned}
\tag{58}
$$

With the help of eqns (7) and (8), this equation becomes

$$
\begin{aligned}
&b_{22}\frac{\partial^4 F}{\partial x_1^4}-2b_{26}\frac{\partial^4 F}{\partial x_1^3\partial x_2}+(2b_{12}+b_{66})\frac{\partial^4 F}{\partial x_1^2\partial x_2^2} \\
&-2b_{16}\frac{\partial^4 F}{\partial x_1\partial x_2^3}+b_{11}\frac{\partial^4 F}{\partial x_2^4}=0.
\end{aligned}
\tag{59}
$$

This is the *basic* governing equation to solve for $F(x_1,x_2)$ for the classical *plane strain* problem considered. By replacing b_{ij} in eqn (59) by a_{ij}, the governing equation for the classical *plane stress* problem of the monoclinic material is obtained, which is identical in the form to eqn (20) except that $\psi(x_1,x_2)$ there is now replaced by $F(x_1,x_2)$. Therefore, we have proved that the displacement method presented herein for problems of planar anisotropic elasticity, which is different from that based on ERS's displacement method discussed in the preceding section, results in a *characteristic* equation *equivalent* to that obtained using the Lekhnitskii (stress) method. This conclusion also applies to the special case of in-plane deformations of *orthotropic* materials, whose solution has been derived by Gao (2000) using a modified approach. For the other special case of *isotropic* materials, it has been known from plane isotropic elasticity that there is no such equivalence to be discussed due to the material/ mathematical degeneracy involved.

Having established this equivalence, we can then follow the *same* procedures as those used in Lekhnitskii's stress method discussed earlier to obtain the characteristic equation (identical to eqn (42)) and consequently the displacement function as

$$
F(x_1,x_2)=2\,\mathrm{Re}[G_1(x_1+\mu_1x_2)+G_2(x_1+\mu_2x_2)],
\tag{60}
$$

where G_1 and G_2 are two yet-unknown analytic functions of their arguments, and μ_1 and μ_2 are the eigenvalues given by eqns (42) and (43). Using eqn (60) in eqn (57) gives the displacement components as

$$
\begin{aligned}
u_1&=\frac{2}{A(c_{11}c_{26}-c_{66}c_{16})}\mathrm{Re}\{[c_{16}+(c_{12}+c_{66})\mu_1+c_{26}\mu_1^2] \\
&\quad\times G_1''(x_1+\mu_1x_2)+[c_{16}+(c_{12}+c_{66})\mu_2+c_{26}\mu_2^2]G_2''(x_1+\mu_2x_2)\}, \\
u_2&=\frac{-2}{A(c_{11}c_{26}-c_{66}c_{16})}\mathrm{Re}[(c_{11}+2c_{16}\mu_1+c_{66}\mu_1^2)G_1''(x_1+\mu_1x_2) \\
&\quad+(c_{11}+2c_{16}\mu_2+c_{66}\mu_2^2)G_2''(x_1+\mu_2x_2)].
\end{aligned}
\tag{61}
$$

It then follows from eqns (34) and (61) that the strain components are

$$
\begin{aligned}
\varepsilon_{11}&=\frac{2}{A(c_{11}c_{26}-c_{66}c_{16})}\mathrm{Re}\{[c_{16}+(c_{12}+c_{66})\mu_1+c_{26}\mu_1^2] \\
&\quad\times G_1'''(x_1+\mu_1x_2)+[c_{16}+(c_{12}+c_{66})\mu_2+c_{26}\mu_2^2]G_2'''(x_1+\mu_2x_2)\}, \\
\varepsilon_{22}&=\frac{-2}{A(c_{11}c_{26}-c_{66}c_{16})}\mathrm{Re}[\mu_1(c_{11}+2c_{16}\mu_1+c_{66}\mu_1^2)G_1'''(x_1+\mu_1x_2) \\
&\quad+\mu_2(c_{11}+2c_{16}\mu_2+c_{66}\mu_2^2)G_2'''(x_1+\mu_2x_2)], \\
\gamma_{12}&=\frac{2}{A(c_{11}c_{26}-c_{66}c_{16})}\mathrm{Re}\{[c_{16}\mu_1+(c_{12}+c_{66})\mu_1^2+c_{26}\mu_1^3 \\
&\quad-(c_{11}+2c_{16}\mu_1+c_{66}\mu_1^2)]G_1'''(x_1+\mu_1x_2)+[c_{16}\mu_2+(c_{12}+c_{66})\mu_2^2 \\
&\quad+c_{26}\mu_2^3-(c_{11}+2c_{16}\mu_2+c_{66}\mu_2^2)]G_2'''(x_1+\mu_2x_2)\}.
\end{aligned}
\tag{62}
$$

Substituting eqn (62) into eqn (5) will give the stress components. Hence, we have obtained the third *general* solution for the classical *plane strain* problem of the monoclinic material (with $c_{11}c_{26}-c_{66}c_{16}\neq 0$ and $A\neq 0$), with G_1 and G_2 to be determined from the boundary conditions (or measured data) of a given specific problem.

Again, we remark that the solution for isotropic materials cannot be reduced from eqn (61) because of the material degeneracy involved.

For the classical *plane stress* problem, all formulas derived above *still* apply *after* replacing b_{ij} by a_{ij}.

SUMMARY

Both the Lekhnitskii and ERS formalisms are examined and compared for problems of planar anisotropic elasticity in a self-consistent manner. It is *explicitly* demonstrated, by a direct comparison, that for such problems the two formalisms lead to the *identical* characteristic equation, a *fourth-degree* polynomial equation having *only* complex roots.

An alternative displacement method based on an extended version of Green's theorem is presented for the same problems of planar anisotropic elasticity. This formulation introduces a *displacement function*, which satisfies a fourth-order partial differential equation whose characteristic equation is *identical* to that obtained using the Lekhnitskii and ERS methods. The new displacement approach is proved to be *equivalent* to the other two methods for problems of planar anisotropic elasticity. Moreover, this method parallels the traditional Airy stress function method and thus the Lekhnitskii method for pure plane problems. Both the new approach and the Airy stress function method start with the equilibrium equations and use the same extended version of Green's theorem to derive the expressions for stress or displacement components in terms of a potential (stress or displacement) function (see also Gao, 1999b, 2000). It is therefore anticipated that the displacement function involved in this new method could also be evaluated from measured data, as done by Lin and Rowlands (1995) and Freire and Cavaco (1990) to determine the Airy stress function experimentally.

Three general solutions for problems of planar anisotropic elasticity follow from the three different formulation methods. All of the *three* solutions depend on the complex roots of the *same* characteristic equation, though they have different forms. This justifies their *equivalence*.

Since *isotropic* materials are *degenerate*, their solutions cannot be reduced from the three general solutions for *monoclinic* materials by simply specifying material constants, despite the fact that the *governing equations* for isotropic materials themselves are directly *reducible* from those for monoclinic materials. Fortunately, isotropy can be regarded as a limiting case of orthotropy such that the general solutions derived for monoclinic materials (and orthotropic materials) can still be applied to isotropic materials (via a suitable *limiting* process) to yield very good *numerical* results (Gao, 2000; Gerhardt, 1984).

REFERENCES

Boresi, A. P., and Chong, K. P., 1987, *Elasticity in Engineering Mechanics*, Elsevier, New York.

Chou, P. C., and Pagano, N. J., 1967, *Elasticity: Tensor, Dyadic, and Engineering Approaches*, Van Nostrand, Princeton, New Jersey.

Eshelby, J. D., Read, W. T., and Shockley, W., 1953, "Anisotropic Elasticity with Applications to Dislocation Theory," *Acta Metall.*, Vol. 1, pp. 251–259.

Freire, J. L. F., and Cavaco, M. A. M., 1990, "Analysis of photoelasticity data through stress function fitting," *Proc. of the 1990 SEM Spring Conf. on Exp. Mech.*, Albuquerque, New Mexico, pp. 213–217.

Gao, X.-L., 1996, "A general solution of an infinite elastic plate with an elliptic hole under biaxial loading," *Int. J. Pres. Ves. Piping*, Vol. 67, pp. 95–104.

Gao, X.-L., 1998, "A mathematical analysis for the plane stress problem of a power-law material," *IMA J. Appl. Math.*, Vol. 60, pp. 139–149.

Gao, X.-L., 1999a, "An exact elasto-plastic solution for the plane wedge problem of an elastic linear-hardening material," *Math. Mech. Solids*, Vol. 4, pp. 289–306.

Gao, X.-L., 1999b, "Alternative derivation of Marguerre's displacement solution in plane isotropic elasticity," *ASME J. Appl. Mech.* (in press).

Gao, X.-L., 2000, "Two displacement methods for in-plane deformations of orthotropic linear elastic materials," (manuscript submitted for publication).

Gao, X.-L., and Rowlands, R. E., 2000, "On displacement methods in planar anisotropic elasticity," (manuscript submitted for publication).

Gerhardt, T. D., 1984, "A hybrid/finite element approach for stress analysis of notched anisotropic materials," *ASME J. Appl. Mech.*, Vol. 51, pp. 804–810.

Green, A. E., and Zerna, W., 1968, *Theoretical Elasticity*, Oxford University Press, Oxford.

Head, A. K., 1979, "The Galois unsolvability of the sextic equation of anisotropic elasticity," *J. Elasticity*, Vol. 9, pp. 9–20.

Lekhnitskii, S. G., 1963, *Theory of Elasticity of an Anisotropic Elastic Body*, Holden-Day, San Francisco.

Lin, S. T., and Rowlands, R. E., 1995, "Thermoelastic stress analysis of orthotropic composites," *Exp. Mech.*, Vol. 35, pp. 257–265.

Little, R. W., 1973, *Elasticity*, Prentice-Hall, Englewood Cliffs, New Jersey.

Sih, G. C., Paris, P. C., and Irwin, G. R., 1965, "On cracks in rectilinearly anisotropic bodies," *Int. J. Fract. Mech.*, Vol. 1, pp. 189–203.

Teodorescu, P. P., 1964, "One hundred years of investigations in the plane problem of the theory of elasticity," *Appl. Mech. Rev.*, Vol. 17, pp. 175–186.

Timoshenko, S. P., and Goodier, J. N., 1970, *Theory of Elasticity*, McGraw-Hill, New York.

Wang, L., 1985, " On general solution of problem of elastic shallow thin shells," *Acta Mech. Sinica*, Vol. 17, pp. 64–71.

Wang, W., and Shi, M. X., 1998, "On the general solutions of transversely isotropic elasticity," *Int. J. Solids Struct.*, Vol. 35, pp. 3283–3297.

Yang, W., and Ma, C.-C., 1998, "Orthotropic transform for planar anisotropic elasticity and reduced dependence of elastic constants," *Proc. R. Soc. Lond.*, Vol. A454, pp. 1843–1855.

PVP-Vol. 415, Recent Advances in Solids and Structures — 2000
ASME 2000

Predicting the Large Deflection Path of End-Loaded Tapered Cantilever Beams

Matthew B. Parkinson
University of Michigan
Ann Arbor, MI

Gregory M. Roach **Larry L. Howell**
Department of Mechanical Engineering
Brigham Young University
Provo, UT

ABSTRACT

A simple (quadratic) mathematical model for predicting the deflection path of both non-tapered and continuously tapered cantilever beams loaded with a vertical end force is presented. It is based on the proposition that the path is a function of the ratio of the endpoints' moments of inertia. The model is valid for both small and large (the tip makes a 70 degree angle with the horizontal) deflections. This was verified through physical testing, comparison to solution of the Bernoulli-Euler equation, and results obtained through nonlinear finite element analysis. Predicted endpoint deflections were found to be accurate within 1.8% of the actual deflection path for moment of inertia ratios varying from 1:1 to 1000:1.

INTRODUCTION

The deflection path of a cantilever beam (Figure 1) is described by the Bernoulli-Euler equation, which states that the curvature of a deflected object is proportional to the applied moment. This equation is typically simplified and used for small deflection analysis. When the small deflection assumption is inappropriate, nonlinear analysis must be performed. For tapered cantilevers, the solution is even more problematic since the resulting differential equation contains variable coefficients.

The elliptic integral solutions for cantilever beams are cumbersome, time consuming, and only usable for simple loading cases. Numerical methods such as nonlinear finite element analysis may also be used to determine the deflection path, but these programs are costly and there is often some difficulty in setting up the proper model. This paper introduces a quick, simplified method for approximating the deflection path.

Incremental solutions to this problem exist. Bisshopp and Drucker (1945) posed the non-tapered elliptic integral problem and solved for the deflection using a combination of integral tables and numerical techniques. Flodin (1957) used a modified-graphical method to solve for the deflection path of tapered cantilever beams undergoing small, linear deflections. The large deflection problem was solved numerically by Kemper (1968) for a variety of cases. However, the Runge-Kutta formulation would not converge for some cases and at times required trial and error. Numerical methods were also used by Spolek and Jefferies (1982), who used finite difference to solve for the deflection path of a tapered cantilever with a static transverse end-load. Ohnishi and Matsuzaki (1985) were able to solve for the deflection path of a tapered cantilever beam (a graphite fishing rod) with an applied vertical end-load using non-linear finite element analysis.

In his small-deflection study, Flodin (1957) used the moments of inertia at various points along the beam to scale the deflection values. Similarly, Baumeister and Sebrosky (1972) scaled the moment of inertia at the fixed-end with a parameter that is a function of the ratio of the beam's widths and heights to each other. The method proposed here is a combination of the two. The deflection path is described as an ellipse, with the parameters defining the ellipse determined by beam length and the ratio of the moments of inertia.

DEFLECTION PATHS

Data sets of deflection paths for beams of varying taper were obtained using a finite element model. These sets were fit with a specific class of second order curve, the ellipse. The coefficients of the ellipses were then made functions of key parameters of the beams (the ratio of the moments of inertia of the ends).

Finite Element Model

Nonlinear finite element analysis was performed for four cases distinguished by their degree of taper, which was quantified using a ratio of the moments of inertia of the two ends.

$$I_{ratio} = \frac{I_{small}}{I_{big}} \tag{1}$$

The maximum value of the ratio is one (a beam with no taper) and the minimum value approaches zero (a beam that tapers to a point). For practical purposes, a ratio of 1000 to 1, or 0.001 was deemed sufficient for this work. The four cases upon which the predictive model is based were spaced logarithmically: 1, 0.1, 0.01, and 0.001.

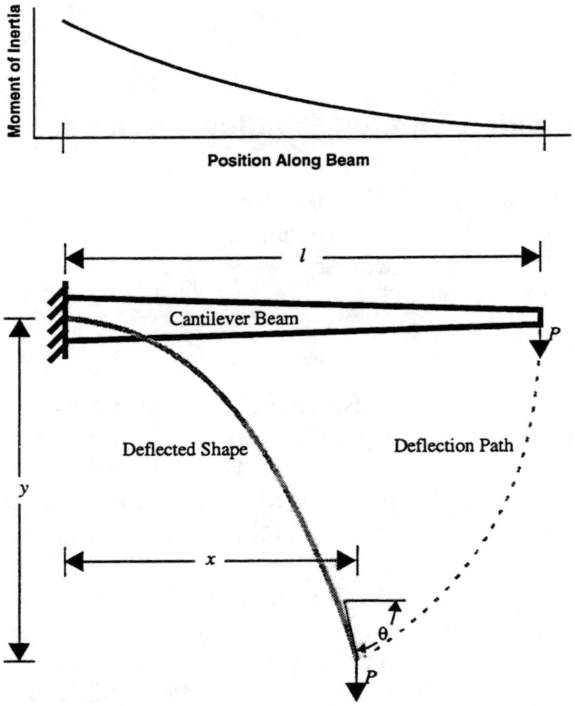

Figure 1: A diagram detailing the loading and deflection range of the finite element model. Although the beam is linearly tapered, the moment of inertia varies quadratically along the beam.

A finite element model was constructed using a tapered beam element. One end of the beam was fixed in all degrees of freedom. A vertical displacement was applied to the free end. The beam was deflected until the tip reached an angle of 70° (Figure 1). The *x* and *y* displacements were output for 100 load steps between 0° and 70° of tip rotation.

CREATING THE MODEL

Examination of the actual deflection paths of several tapered beams showed that the path is nearly elliptical. This realization is critical because it allows the deflection path to be fit with a second-order curve with just a single degree of freedom. Generally, the coefficients *a*, *b*, and *c* must be determined when a second-order curve (i.e. $ax^2 + bx + c$) is fit to data. As demonstrated below, assuming that the path forms an ellipse reduces the problem to the determination of a single coefficient.

The equation for an ellipse is given as

$$\frac{(x-x_0)^2}{a^2} + \frac{(y-y_0)^2}{b^2} = 1 \qquad (2)$$

If the center of the ellipse is located at the fixed end of the beam (0, 0) the equation simplifies to

$$\frac{x^2}{a^2} + \frac{y^2}{b^2} = 1 . \qquad (3)$$

With the x-intercept *a* allowed to be the overall length of the beam, *l*, the equation further simplifies to

$$\frac{x^2}{l^2} + \frac{y^2}{b^2} = 1 . \qquad (4)$$

The equation can be rewritten for *y* as a function of *x* and *b*

$$y = b\sqrt{1 - \frac{x^2}{l^2}} \qquad (5)$$

If the value of the y-intercept *b* is known, the *x* and *y* points of deflection can be calculated. The proper value for *b* that produces the ellipse approximating the deflection path must now be determined.

Calculating "*b*"

The nonlinear finite element analysis produced four sets of data, or deflection paths, which will be discussed later in Figure 3. An ellipse was fit to each of these. This was done by using an optimization routine to calculate the "*b*" value for each ellipse. The objective function, which minimizes the sum of the square of the differences between the finite element data points and those predicted by Equation (5), is then

$$\text{minimize} \sum_{i=1}^{100} \left(\hat{y}_i - b\sqrt{1 - \frac{x_i^2}{l^2}} \right)^2 \qquad (6)$$

The percentage of error was calculated using the following equation for each of the data points in the set.

$$\text{error \%} = \left| \frac{\hat{y} - y}{\hat{y}} \right| \qquad (7)$$

The maximum percentage error for each path is summarized in Table 1. The closeness of the approximation corroborates the decision to use the ellipse as the basis for the model.

The ellipse parameter *b* was normalized by dividing it by the undeflected length of the beam. This normalized value was then correlated with the ratios of the moments of inertia for the

Table 1: Individual Model Results

Ratio I_{small}/I_{big}	1.0	0.1	0.01	0.001
Max Error %	0.70	0.38	0.63	1.73
Avg. Error %	0.41	0.19	0.33	1.09

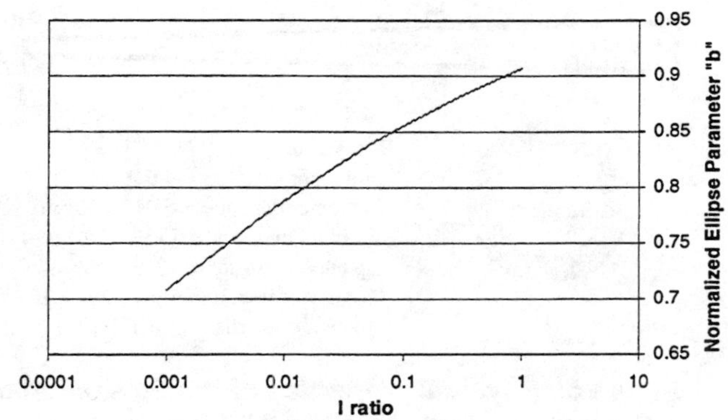

Figure 2: A log plot of the ratio I_{small}/I_{big} and the normalized ellipse parameter "b"

four cases. Figure 2 which presents this data on a log plot, shows the near linearity of this relationship.

The relationship between the natural log of the I_{ratio} and the parameter b is almost linear, and suggests that the previous prediction equation, Equation (5), can be modified by replacing b with the equation for a line. This yields Equation (8).

$$y = (c_1 \ln I_{ratio} + c_2) l \sqrt{1 - \frac{x^2}{l^2}} \tag{8}$$

The optimization problem used previously to determine the b values can now be modified to find the coefficients of the linear fit. In an effort to increase the accuracy of the model, the scope of the objective function is also modified to include all four data sets simultaneously rather than individually. To give equal weight to each of the data sets, the error must be measured in a new way. Rather than minimizing the sum of the square of the differences, the objective is to minimize the maximum percentage of error within the model. These modifications result in the new formulation of the unconstrained optimization problem

minimize

$$\max_{i=1\ldots4}\left\{\max_{j=1\ldots100}\left\{\left|\frac{\hat{y}_{i,j} - (c_1 \ln(I_{ratio_j} + c_2))\, l_i \sqrt{1 - \frac{x_{i,j}^2}{l_i^2}}}{\hat{y}_{i,j}}\right|\right\}\right\} \tag{9}$$

As expected, the results (summarized in Table 2) of this second optimization are not as accurate as those obtained in the first, when each data set is fit with its own ellipse. The results are, however, accurate enough for practical use and they justify the assumption of a linear relationship between I_{small}/I_{big} and

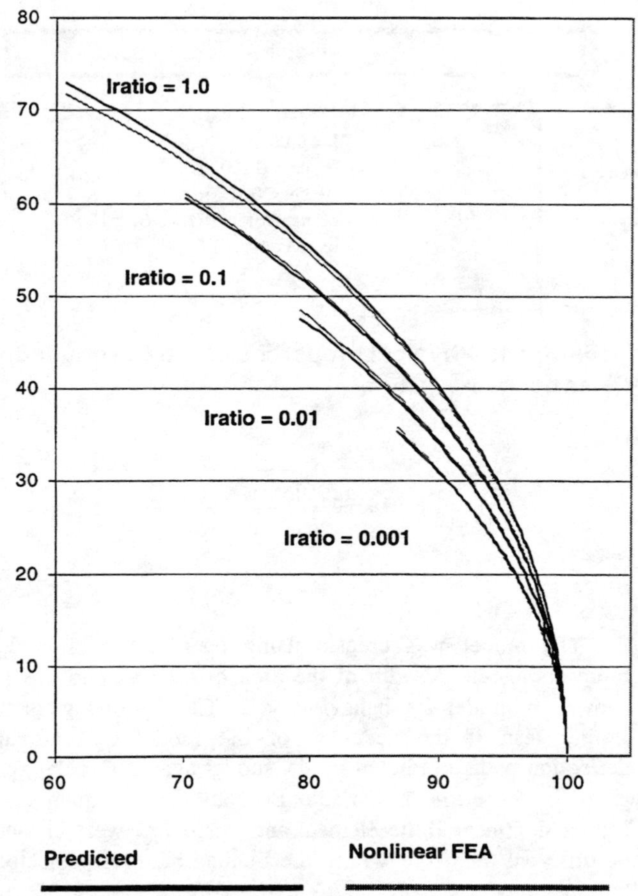

Figure 3: The actual and predicted deflection paths for the four developmental cases

the ellipse parameter b. Figure 3 shows the actual and predicted deflection paths for the four cases.

The final prediction equation, which gives the deflection as a function of the beam length and the ratio of the moments of inertia of the two ends, is then

$$y = (0.030432 \ln I_{ratio} + 0.91897) l \sqrt{1 - \frac{x^2}{l^2}} \tag{10}$$

For the special case of straight beam, the equation simplifies to

Table 2: Global Model Results

$c_1 = 0.030432,\ c_2 = 0.91897$

Ratio	1.0	0.1	0.01	0.001
Max Error %	1.78	1.17	1.78	1.78
Avg. Error %	1.06	0.94	0.93	1.12

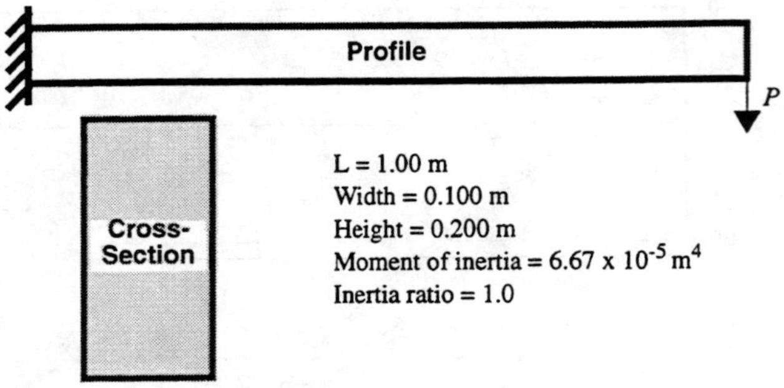

Figure 4: Physical properties of the Bernoulli-Euler beam

$$0.91897l\sqrt{1-\frac{x^2}{l^2}} \tag{11}$$

VALIDATION

The model was created using data sets of four I_{ratios} representing the breadth of the area of interest and has been shown to model its behavior well. The following sections further test of the accuracy of the model by comparing deflection paths predicted by the model against "actual" paths given by experimentation, the Bernoulli-Euler equation, and further nonlinear finite element analysis. I_{ratios} were chosen to be different than, but within the bounds of, those selected to build the model.

Bernoulli-Euler Validation

The results were compared against those obtained by the numerical solution of the Bernoulli-Euler equation. This example investigates one extreme of the model's range: a non-tapered beam (I_{ratio} = 1.0). The profile, cross-section, and physical properties are summarized in Figure 4. Substituting into Equation (10) gives

$$y = 0.9190\sqrt{1-x^2} \tag{12}$$

To use the Bernoulli-Euler equation, material and geometric properties and loading conditions are required. The cantilever beam was loaded only in the y-direction. The loads were incrementally increased until the tip angle measured 50°. Path coordinates at each of the positions were recorded. The results from the Bernoulli-Euler equation and those predicted by Equation (12) are in Table 3.

Physical Test

The second validation test examines the other extreme case: taper to a near-point. It involves the physical deflection of a flyrod blank. These are made of graphite and have a continuous taper from butt to tip. Typically, the I_{ratio} is ~0.003, which is within the scope of this model. A profile and cross-section of the blank used for this test as well as specific material properties are in Figure 5. Fifteen rods were tested. The results of one of the tests (typical of the results for the others) is outlined below.

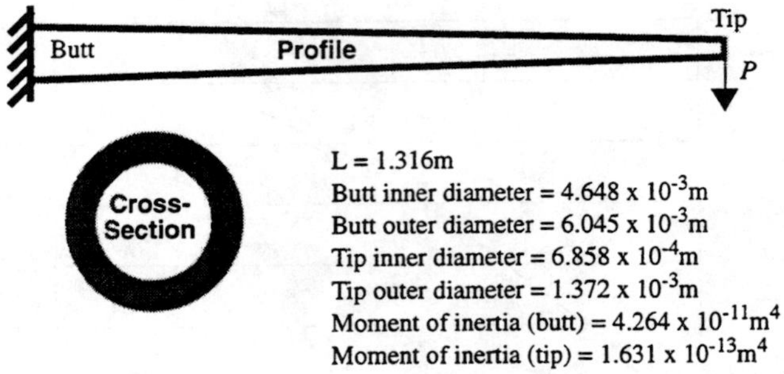

Figure 5: Physical properties of the tip section of a G. Loomis 9 foot, 5 weight GLX flyrod

For this example, the deflection path was determined by mounting the rod horizontally in a machining collett block. The block was clamped to a marking board that was covered with a 3.175 square mm grid. Weights were suspended from the end of the beam and the corresponding point of deflection was marked on the grid to determine the actual deflection path of the beam.

The predicted deflection path, obtained from the physical properties of the blank and Equation (10), is

$$y = 0.9864\sqrt{1-\frac{x^2}{1.732}} \tag{13}$$

The measured data and the path predicted by Equation (13) are plotted in Figure 6. The predicted data differs from the actual data by less than 2.05%, as seen in Table 4. The scattered nature of the error distribution is due to inaccuracies in the measured values. In general, the model does quite well and the results are within the accuracy suggested in its development. Note that despite radical differences between this example, the cases used to develop the model, and the previous example, each deflection path is represented accurately.

Table 3: Bernoulli-Euler Test Results

x (m)	0.988	0.974	0.956	0.913	0.867	0.822
y_{actual} (m)	0.142	0.208	0.268	0.372	0.454	0.517
y_{pred} (m)	0.143	0.209	0.271	0.376	0.459	0.524
error %	0.781	0.756	0.928	0.973	1.15	1.32

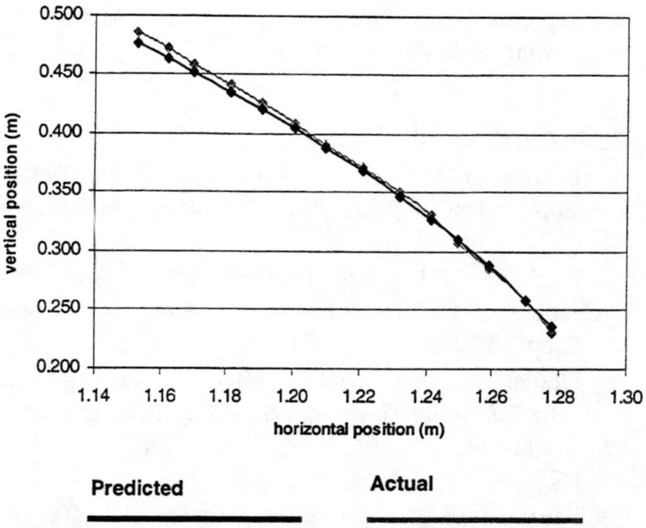

Figure 6: Predicted and Actual deflection path for Tapered Composite beam

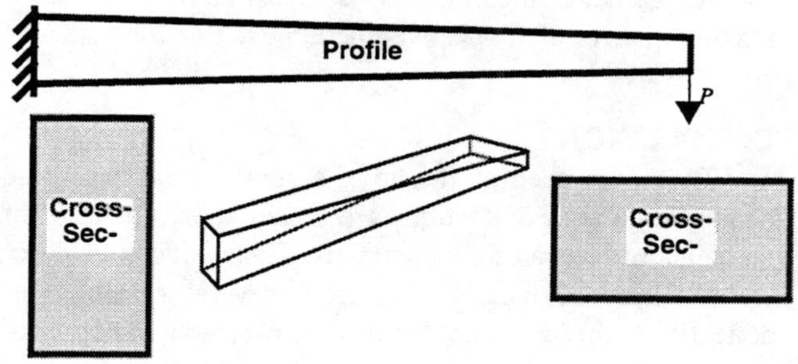

L = 1.00 m
Base width = 0.050 m
Base height = 0.100 m
Base moment of inertia = 4.167 x 10-6 m4
Tip width = 0.100 m
Tip height = 0.050 m

Figure 7: Physical properties and geometry of the morphed-rectangle beam

Morphed Rectangle

As a final example, consider the beam whose cross-section is a vertical rectangle on one end and a horizontal rectangle on the other (Figure 7), the beam tapering continuously between them. Although the shape of the cross-section is constantly changing and the geometry is quite different from the previous examples, the moment of inertia does continuously taper from the base to the tip. Since this is the case, the model should predict the deflection of this beam as accurately as it has the previous examples.

Given an I_{ratio} of 0.25 and a beam length of 1.0m, the equation predicting the deflection path is

$$y = 0.8768\sqrt{1-x^2} \tag{14}$$

Nonlinear finite element analysis was performed on a solid model of the beam. The base (the left end of the beam) was fixed in all degrees of freedom and a vertical deflection load was applied at the tip. The x locations for varying applied y deflections of the beam tip were recorded. These are summarized in Table 5, where they are compared with those predicted by Equation (14).

Table 4: Physical Validation Test Results

x (m)	1.28	1.27	1.26	1.25	1.24	1.23	1.22	1.21	1.20	1.19	1.18	1.17	1.16	1.15
y_{actual} (x10^{-1}m)	2.32	2.59	2.86	3.08	3.32	3.51	3.71	3.91	4.10	4.25	4.41	4.59	4.73	4.86
y_{pred} (x10^{-1}m)	2.36	2.59	2.88	3.10	3.28	3.47	3.69	3.89	4.05	4.21	4.35	4.52	4.63	4.77
error %	1.71	0	0.68	0.72	1.25	1.01	0.74	0.45	1.10	1.14	1.33	1.48	2.05	1.89

Table 5: Morphed Rectangle Test Results

x (m)	0.9857	0.9596	0.9190	0.8615	0.7835	0.6785	0.5376
y_{FEA}(m)	0.1500	0.2500	0.3500	0.4500	0.5500	0.6500	0.7500
y_{pred} (m)	0.1478	0.2466	0.3457	0.4451	0.5449	0.6441	0.7393
difference %	1.4440	1.3613	1.2296	1.0802	0.9363	0.9132	1.4295

As predicted, the model is also accurate for this case, with a maximum error of 1.44% over the large deflection path.

CONCLUSION

The path traced by the tip of a continuously tapered beam undergoing large deflection is a function of the ratio of the moments of inertia of the endpoints. This path can be described as an ellipse. Using the data from several carefully chosen deflections, the equation for this ellipse, as a function of the ratio of the moments of inertia, was found. Based upon these data and three other validation strategies, including comparison against the Bernoulli-Euler equation and physical experimentation, this model is believed accurate within 1.8% through 70 degrees of deflection.

In circumstances where the small deflection assumption is inappropriate, nonlinear finite-element analysis or an elliptic integral solution is typically required. This is particularly costly when the beam is tapered or the geometry is complicated. In such circumstances, the model developed here may be sufficiently accurate and is extremely easy to use.

ACKNOWLEDGEMENTS

The help of Brent Weight in gathering experimental data is gratefully acknowledged. The donation of materials from Orvis, Sage, Lamiglas, St. Croix, and G. Loomis fly rod manufactures is also acknowledged. This work is supported by the National Science Foundation (NSF) under an NSF CAREER Award under grant No. DMI-9624574.

REFERENCES

Baumeister, H.K. and Sebrosky, R.A., 1972, "Deflection and Slope in Tapered Beams," *Machine Design*, Vol. 44, pp. 65-67.

Bisshopp, K.E. and Drucker, D.C., 1945, "Large Deflections of Cantilever Beams," *Q. Applied Mathematics*, Vol. 3, pp. 272-275.

Flodin, J., 1957, "Deflection of Beams of Varying Moments of Inertia," *J. of American Society of Nautical Engineers*, Vol. 69, pp. 511-514.

Kemper, J.D., 1968, "Large Deflections of Tapered Cantilever Beams," *International Journal of Mechanical Science*, Vol. 10, pp. 469-478.

Ohnishi, H. and Matsuzaki, A., 1985, "The Large Deformation Analysis of a Graphite Fishing Rod," *Computers & Structures*, Vol. 21, No. 1/2, pp. 265-271.

Spolek, G.A. and Jefferies, S.R., 1982, "Analysis of Large Deflections of Fishing Rods," *Proceedings of the International Conference on Computational Methods and Experimental Measurements*, Washington D.C., pp. 637-648.

THE APPROXIMATE, HIGH FREQUENCY RESPONSE OF A PANEL WITH ATTACHED MASSES

Robert R. Reynolds
Department of Mechanical Engineering
University of Arkansas
Fayetteville, AR 72704
Email: rrr@engr.uark.edu

ABSTRACT

The high frequency response of a panel with attached masses is approximated using transient, asymptotic modal analysis (AMA). These relations are derived by applying asymptotic simplifications to classical solutions in both the time and frequency domains. The geometry consists of several masses that are attached to a panel to which impulsive excitations are applied. Predictions are made of the mean-squared, transverse displacement history and the localized responses near the added masses. These are compared to classical solutions and it is shown that for a few masses that are separated by more than a mean wavelength, the panel behavior can be estimated well by the AMA expressions. The approximate solutions are shown to require less than 20% of the computational time and memory of the classical solutions and are applicable to general forms of excitation.

The resulting capability can be used to estimate acoustic emissions from vibrating equipment and to model structure-borne sound and its effects on sensitive equipment.

INTRODUCTION

High frequency structural vibration is a source of workplace irritation and damage in many mobile electrical devices. While there are methods available to predict such vibrations when they occur in a steady-state form (Lyon, 1975; Dowell and Kubota, 1995; Kubota and Dowell, 1986; Kubota, 1990), there are few alternatives for transient excitation sources such as collisions or impacting components (Sepulveda and Thomas, 1996). Finite elements are extremely useful at low frequencies, but contain too much damping to be useful at high frequencies, say above 1000 Hertz (Hz).

There have been efforts to relax the ergodic assumptions of Statistical Energy Analysis (SEA) in order to provide some predictive capability for the transient regime (Lai and Soom, 1990; Dalton, 1992; Pinnington and Lednik, 1996). However, results vary depending upon the assumptions made and are usually order-of-magnitude estimates.

A reformulation of Asymptotic Modal Analysis (AMA) for transient vibration has been shown to be effective for simple elastic systems such as a panel (Reynolds and Dowell, 1998). In this paper, this method is extended to the more complex geometry of a panel with attached masses.

GOVERNING EQUATIONS

The classical equation of motion for a thin elastic panel may written in nondimensional form as

$$\frac{\partial^2 W}{\partial \tau^2} + C\frac{\partial W}{\partial \tau} + \left[\frac{1}{\gamma}\frac{\partial^4 W}{\partial \xi^4} + 2\frac{\partial^4 W}{\partial \xi^2 \partial \eta^2} + \gamma^2\frac{\partial^4 W}{\partial \eta^4}\right] = P(\xi,\eta,\tau) + \lambda_0\delta(\eta-\eta_0)\delta(\xi-\xi_0) + \lambda_1\delta(\eta-\eta_1)\delta(\xi-\xi_1) + \dots \quad (1)$$

where the nondimensionalized parameters are: the transverse displacement W=displacement/thickness; the coordinates, ξ=x/thickness and η=y/width (c.f. Fig. 1); time τ; the panel aspect ratio γ=length/width; the structural damping, C; the applied

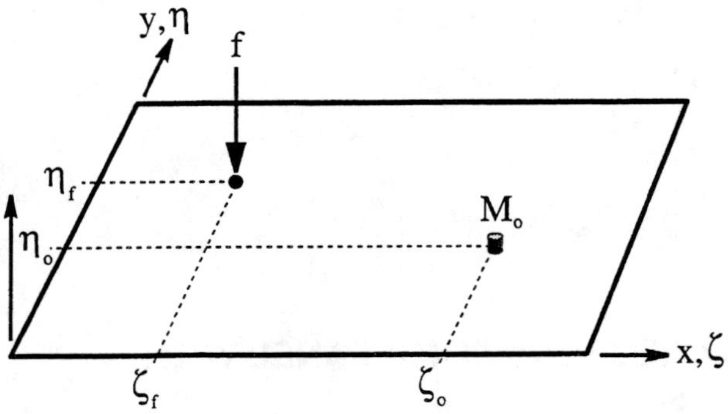

Figure 1. Geometry of a panel, load and a single, attached mass.

load P at (ξ_f, η_f); and λ_i are the loads applied by the added masses, M_i, located at (ξ_i, η_i).

A typical solution procedure is to apply Galerkin's method with a displacement of the form

$$W(\xi,\eta,\tau) = \sum_{n=1}^{\infty} q_n(\tau)\psi(\xi,\eta)\,. \tag{2}$$

This yields the single mode equation

$$\frac{\partial^2 q_n}{\partial \tau^2} + 2\zeta_n\omega_n\frac{\partial q_n}{\partial \tau} + \omega^2 q_n = Q_n \tag{3}$$

where ζ_n is the modal damping ratio, Q_n the generalized force, and ω_n is the modal natural frequency,

$$\omega_n = \left(\frac{\pi^2}{\gamma}\right)\left[n_x^2 + n_y^2\gamma^2\right]\,. \tag{4}$$

Here n_ξ and n_η are the modal indices in the coordinate directions.

Classical Solution of a Point Load

Now for the sake of specificity we restrict our attention to a localized load at the point (ξ_f, η_f), the applied pressure becomes $P(\xi,\eta,\tau) = f(\tau)\delta(\xi-\xi_f)\delta(\eta-\eta_f)$ (c.f. Fig. 1), and the generalized force can be written as,

$$\begin{aligned} Q_n &= \frac{\int_A (P + \lambda_0\delta_{\xi_0}\delta_{\eta_0} + \lambda_1\delta_{\xi_1}\delta_{\eta_1} + \ldots)\psi_n dA}{\int_A \psi_n^2 dA} &(5)\\ &= \frac{f\psi_{nf} + \lambda_0\psi_{n0} + \lambda_1\psi_{n1} + \ldots}{<\psi_n^2>}. &(6)\end{aligned}$$

This is used to determine the modal coefficients from

$$q_n = \int_0^\tau h_n(\tau-\tau')Q_n(\tau')d\tau'\,. \tag{7}$$

where h_n is the standard impulse response function,

$$h_n(\tau) = \frac{e^{-\zeta_n\omega_n\tau}\sin(\omega_n\sqrt{1-\zeta_n^2}\tau)}{\omega_n\sqrt{1-\zeta_n^2}}\,. \tag{8}$$

In this case, the modal coefficients are

$$\begin{aligned} q_n = \frac{1}{<\psi_n^2>}\{&h_n(\tau) * [f(\tau)\psi_{nf} + \lambda_0(\tau)\psi_{n0} + \\ &+ \lambda_1(\tau)\psi_{n1}] + \ldots\}\,. \end{aligned} \tag{9}$$

where $*$ indicates the convolution and the abbreviated notation $\psi(\xi_{ni}, \eta_{ni}) = \psi_{ni}$ is introduced for simplicity. The λ_i are found from the equations of motion for the masses, M_i,

$$\begin{aligned} \lambda_i &= -M_i\ddot{W}(\xi_i,\eta_i) &(10)\\ &= -M_i\sum \ddot{q}_n(\tau)\psi_{ni}. &(11)\end{aligned}$$

where the dots indicate derivatives with respect to time.

The $\ddot{q}_n(\tau)$ are determined from the eqn. (9) using Leibniz's Rule of differentiation,

$$\begin{aligned} \ddot{q}_n = \frac{1}{<\psi_n^2>}\{&\psi_{nf}[f(\tau) * \ddot{h}_n(\tau) + f(\tau)] \\ &+\psi_{n0}[\lambda_0(\tau) * \ddot{h}_n(\tau) + \lambda_0(\tau)] \\ &+ \psi_{n1}[\lambda_1(\tau) * \ddot{h}_n(\tau) + \lambda_1(\tau)] + \ldots\}\,. \end{aligned} \tag{12}$$

Substituting this into eqn. (11) results in a series of implicit, simultaneous equations for the λ_i in terms of the forcing, $f(\tau)$, and the modal impulse response functions, $h_n(\tau)$. To solve these equations, we employ the convolution theorem, $F(f*g) = \sqrt{2\pi}F(f)\cdot F(g)$ where F indicates the Fourier transform[1], and perform the algebra in the frequency domain where the relations will be *explicit*.

Now transform eqns. (11) and (12) and apply the convolution theorem and the transform relation $F[\ddot{h}_n(\tau)] =$

[1] The Fourier transform is used here rather than the Laplace for numerical simplicity. It is

$$G(\omega) = F[g(\tau)] = \frac{1}{\sqrt{2\pi}}\int_{-\infty}^{\infty} e^{-i\omega\tau}g(\tau)d\tau. \tag{13}$$

$-\omega^2\sqrt{2\pi}H_n(\omega)$. Using the notation $\Lambda_o(\omega) = F[\lambda_o(\tau)], H_n(\omega) = F[h_n(\tau)]$, and $F(\omega) = F[f(\tau)]$, this yields for the ith mass,

$$\Lambda_i(\omega) = -M_i \sum \frac{\psi_{ni}}{<\psi_n^2>} \times \left\{F(\omega)\psi_{nf}\left(1-\omega^2\sqrt{2\pi}H_n(\omega)\right)+ \Lambda_0(\omega)\psi_{n0}\left(1-\omega^2\sqrt{2\pi}H_n(\omega)\right)+ \Lambda_1(\omega)\psi_{n1}\left(1-\omega^2\sqrt{2\pi}H_n(\omega)\right)+\ldots\right\}. \quad (14)$$

For the entire system this is

$$\mathbf{A}\vec{\Lambda} = \vec{M} \quad (15)$$

where the elements of the symmetric $\mathbf{A}$ are

$$a_{ii} = 1+M_i\sum\frac{\psi_{mi}^2}{<\psi_m^2>}\left(1-\omega^2\sqrt{2\pi}H_m(\omega)\right) \quad (16)$$

$$a_{ij} = M_j\sum\frac{\psi_{mi}\psi_{mj}}{<\psi_m^2>}\left(1-\omega^2\sqrt{2\pi}H_m(\omega)\right) \quad (17)$$

and the elements of $\vec{M}$ are

$$m_i = -F(\omega)M_i\sum\frac{\psi_{mi}\psi_{mf}}{<\psi_m^2>}\left(1-\omega^2\sqrt{2\pi}H_m(\omega)\right). \quad (18)$$

Equation (15) is explicit for the Λ_i - the transformed Lagrange multipliers. We solve it and use these multipliers to determine the transformed modal coefficients (i.e. the transform of eqn. (9))

$$F(q_n) = \frac{\sqrt{2\pi}H_n(\omega)}{<\psi_n^2>}[F(\omega)\psi_{nf}+\Lambda_1(\omega)\psi_{n0}+ \Lambda_1(\omega)\psi_{n1}+\ldots]. \quad (19)$$

It is helpful to rewrite this as

$$F(q_n) = \frac{1}{<\psi_n^2>}[K_{nF}(\omega)\psi_{nf}+K_{n\Lambda_0}(\omega,\xi_0,\eta_0)\psi_{n0}+ K_{n\Lambda_1}(\omega,\xi_1,\eta_1)\psi_{n1}+\ldots] \quad (20)$$

where we have separated the effects of forcing and mass influences on the response by introducing the functions

$$\begin{aligned} K_{nF}(\omega) &= \sqrt{2\pi}H_n(\omega)F(\omega) \\ K_{n\Lambda_0}(\omega,\xi_0,\eta_0) &= \sqrt{2\pi}H_n(\omega)\Lambda_0(\omega) \\ K_{n\Lambda_1}(\omega,\xi_1,\eta_1) &= \sqrt{2\pi}H_n(\omega)\Lambda_1(\omega) \\ &\cdots \end{aligned} \quad (21)$$

Strictly speaking, the Λ_i are also functions of (ξ_i,η_i). Inverting the transforms (where $k_{n\Lambda_i}(\tau) = F^{-1}[K_{n\Lambda_i}(\omega)]$, etc.) and restricting our attention to a range of frequencies $\Delta\omega$ that contains ΔM modes we have the classical displacement solution

$$W(\xi,\eta,\tau) = \sum_{n=M}^{M+\Delta M-1}\frac{\psi_n(\xi,\eta)}{<\psi_n^2>}[k_{nF}(\tau)\psi_{nf}+ k_{n\Lambda_0}(\tau,\xi_0,\eta_0)\psi_{n0}+ k_{n\Lambda_1}(\tau,\xi_1,\eta_1)\psi_{n1}+\ldots]. \quad (22)$$

The classical, spatially averaged solution, $<W(\tau)>$, can be found by first squaring eqn. (22) (ignoring the cross terms as they are small) and spatially averaging this over the panel. This yields

$$<W^2(\tau)> = \sum^{\Delta M}\frac{1}{<\psi_n^2>}[k_{nF}(\tau)\psi_{nf}+k_{n\Lambda_0}(\tau)\psi_{n0}+ k_{n\Lambda_1}(\tau)\psi_{n1}+\ldots]^2 \quad (23)$$

and describes the temporal, panel response in an averaged sense.

Asymptotic Solution of a Point Load

If the frequency range $\Delta\omega$ contains enough modes with a sufficient modal density, then we may apply several simplifications to the classical solutions of eqns. (22) and (23). These simplifications eliminate the need to solve the eigenproblem and reduce the size of the summations. The result is an approximate solution that is simpler to evaluate and requires much less detailed information about the system.

We begin with the summation in the diagonal elements of $\mathbf{A}$ (c.f. eqn. (16)),

$$\sum_{m=1}^{M}\frac{\psi_{mi}^2}{<\psi_m^2>}\left(1-\omega^2H_m(\omega)\right). \quad (24)$$

Within our frequency range $\Delta\omega$, we note that $H_m(\omega)$ is *not* slowly varying over the range of the summation and cannot be removed from the summation. Thus, the asymptotic modal approximations used for the smooth panel (Dowell and Kubota, 1995; Reynolds and Dowell, 1998) cannot be applied here. However, two useful observations are:

- The scaling factor ψ_{mi}^2 ranges from zero to one throughout the frequency range, and
- The term $\left(1-\omega^2H_m(\omega)\right)$ is a frequency-dependent function in two senses. First, for any value of m, it is multiplied by and a continuous function of ω. Also, the m subscript indicates a dependence upon the natural frequency of the mth mode.

The first observation suggests that we may be able to eliminate ψ^2_{mi} by approximating its scaling effect on the H_m terms. If ΔM is large enough then we may only need to sum a few, say ΔM^*, representative terms of the complete summation. This will be true if H_m is a well behaved and smoothly varying function throughout the range of the summation.

This leads us to the second observation: the natural frequency content of H_m must be considered in any attempt to remove specific eigensolution references from the summation. It can be shown that H_m may be written

$$H(\omega) = \frac{1}{(\omega_n^2 - \omega^2) - j(2\zeta\omega_n\omega)} \tag{25}$$

where $j = \sqrt{-1}$ and ω_m is the natural frequency of the mth mode. Note that this is a smooth function of ω_m and ω. Thus we approximate the frequency dependence of the complete summation by a series of functions, $H_p(\omega)$, that are based on chosen frequencies, ω_p, distributed throughout $\Delta\omega$ rather than the actual natural frequencies themselves.

Hence we simplify our summation in three steps:

1. Extract $\psi^2_{mi}/<\psi^2>$ from the each term and scale the complete summation by its mean value, $\bar{\psi}^2_{mi}$.
2. Reduce the number of terms in the summation from ΔM to some fraction of it, $\Delta M^* = \alpha\Delta M$, where α is a chosen parameter (say, 0.1-0.2), and scale the sum accordingly.
3. Replace the natural frequency dependence of H_m with specified frequencies that span the frequency range $\Delta\omega$.

With these modifications our summation can be written as

$$\begin{aligned}\sum_{m=M}^{M+\Delta M} \frac{\psi^2_{mi}}{<\psi^2_m>}(1-\omega^2 H_m(\omega)) \\ \approx \frac{\bar{\psi}^2}{<\psi^2>}\left(\frac{\Delta M}{\Delta M^*}\right)\sum_{p=M}^{M+M^*}(1-\omega^2 H_p(\omega)) \\ \approx \frac{1}{\alpha}\sum_{p=M}^{M+M^*}(1-\omega^2 H_p(\omega))\end{aligned} \tag{26}$$

where $\Delta M^* = \alpha\Delta M$ chosen frequencies replace the natural frequencies in $H_p(\omega)$ and we observe that the spatial average of a mode squared is equal to the mean value of a summation of modes squared at any location (i.e. $\bar{\psi}^2(\xi_i,\eta_i) \approx <\psi^2>$). Relations for estimating ΔM are available in the literature for many continuous structures (Lyon, 1975).

The validity of the approximations in eqn. (26) may be seen by comparing both sides of the equation for a representative system. Figure 2 presents a comparison of the AMA summation on the right side of eqn. (26) and the CMA summation on the left side for a value of $\alpha = 0.1$ or 10% of the estimated number of modes in $\Delta\omega^2$. The transforms of the summations have been inverted so that the comparison can be made in the time domain where the results are most important. It should be noted that the AMA approximation required only 7 % of the computational time necessary for the CMA calculation.

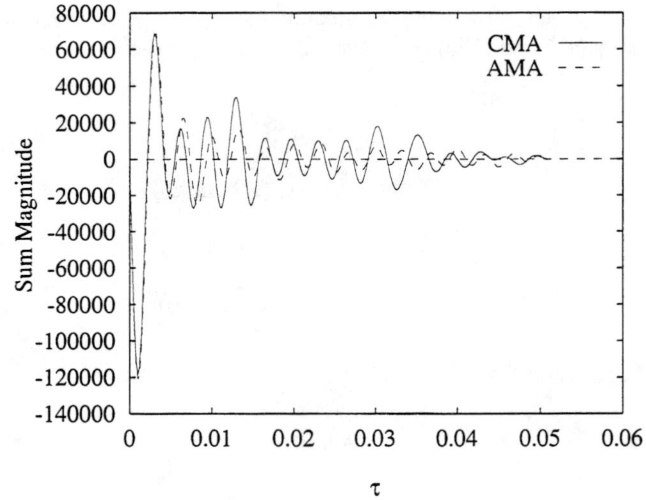

Figure 2. Comparison of the CMA summation on the left side of eqn. (26) to the AMA approximation on the right side. The AMA approximation only uses 10% of the estimated modes in the frequency range (i.e. $\alpha = 0.1$).

Now turn our attention to the off-diagonal terms of **A** in eqn. (17) and the elements of $\vec{M}$ in eqn. (18). The summations in these terms have the form

$$\sum_{m=1}^{M} \frac{\psi_{mi}\psi_{mj}}{<\psi^2_m>}(1-\omega^2 H_m(\omega))\,. \tag{27}$$

The simplifications applied to the diagonal terms are not applicable here since the eigenmode scaling is not a mode squared but rather a product of mode functions evaluated at, in general, different locations, $\psi_{mi}\psi_{mj}$. Clearly this product can be either positive or negative. Furthermore, the asymptotic limit of the product is zero

$$\lim_{\Delta M\to\infty} \psi_{mi}\psi_{mj} = 0\,. \tag{28}$$

Fortunately, our actual summation is over a finite number of modes and will be *nearly* zero. This is an important distinction. A zero value of this summation would imply that the response of

[2] Other parameters include $\nu = 0.359, \gamma = 1.21429, \zeta = 0.001, \omega_{min} = 769.64, \Delta\omega = 1172.93$.

the plate is independent of the added mass, M_i. As this cannot be true, a non-zero value of the summation in eqn. (27), however small, is crucial.

To approximate this summation we extract $\psi_{mi}\psi_{mj}$ from each term and multiply the sum by the mean value for the product over the frequency range, $\bar{\psi}_{ij}$. Again, we abbreviate the summation to $\Delta M^* = \alpha\Delta M$ terms that are based on chosen frequencies that span the frequency range $\Delta\omega$. Together these result in the relation

$$\sum_{m=M}^{M+\Delta M} \frac{\psi_{mi}\psi_{mj}}{<\psi_m^2>}\left(1-\omega^2 H_m(\omega)\right) \approx \frac{\bar{\psi}_{ij}}{\alpha<\psi^2>}\sum_{p=M}^{M+M^*}\left(1-\omega^2 H_p(\omega)\right) . \quad (29)$$

Here $\bar{\psi}_{ij} \not\approx <\psi^2>$, but rather must be estimated. For sufficiently large ΔM we make the continuous approximation

$$\bar{\psi}_{ij} = \frac{\sum_{i=M}^{M+\Delta M}\psi_{ni}\psi_{nj}}{\Delta M} \approx \frac{\int_{N_{\Delta\omega}}\psi_{ni}\psi_{nj}}{N_{\Delta\omega}} \quad (30)$$

where $N_{\Delta\omega}$ is the number of modes in the frequency range $\Delta\omega$. This is a lengthy but straightforward integration in mode-number space that yields a mean value for the cross product for the particular frequency range, $\Delta\omega$. This is discussed more thoroughly in the appendix.

Once the Λ_i are known, the $k_{i\Lambda_i}$ are determined and used with eqn. (22) for the panel displacement or eqn. (23) for $<W^2>$. Here we encounter further summations over the modes that can also be simplified. The k_{if} and $k_{i\Lambda_i}$ are in the time and may approximated using the asymptotic simplifications used for a smooth panel (Reynolds and Dowell, 1998). The summation becomes

$$W(\xi,\eta,\tau) \approx \frac{1}{\alpha<\psi^2>}\sum_{p=M}^{\Delta M^*}\psi_p(\xi,\eta)\left[k_{pF}(\tau)\psi_{pf} + k_{p\Lambda_0}^{AMA}(\tau,\xi_0,\eta_0)\psi_{p0} + k_{p\Lambda_1}^{AMA}(\tau,\xi_1,\eta_1)\psi_{p1} + \ldots\right] . \quad (31)$$

Although $k_{pF}(\tau)$ itself is not different in form than the analogous term in the classical solution of eqn. (22), the summation is abbreviated and the terms are evaluated at the fewer, chosen frequencies rather than the full range of natural frequencies.

Equation (31) contains a series of products of the modal functions at the forcing and mass locations, $\psi_{pf}\psi_{pi}$ which cannot be evaluated without the eigensolution. Hence we apply the same form of modal averaging used in eqn. (30) and focus our attention on the response of the panel near the locations of the masses or forcing. Other locations may be estimated using the average panel response discussed below. The panel asymptotic response near the excitation for NN added masses is then

$$W_{AMA}(\xi_f,\eta_f,\tau) = \frac{1}{\alpha}\sum_{p=M}^{M+\Delta M^*}\left[k_{pF}(\tau) + \sum_{i=1}^{NN} k_{p\Lambda_i}^{AMA}(\tau,\xi_i,\eta_i)\left(\frac{\bar{\psi}_{if}}{<\psi^2>}\right)\right] \quad (32)$$

and near the mass M_i will be

$$W_{AMA}(\xi_n,\eta_n,\tau) = \frac{1}{\alpha}\sum_{p=M}^{M+\Delta M^*}\left[k_{p\Lambda_n}^{AMA}(\tau,\xi_n,\eta_n) + k_{pF}(\tau)\left(\frac{\bar{\psi}_{nf}}{<\psi^2>}\right) + \sum_{i=1,\neq n}^{NN} k_{p\Lambda_i}^{AMA}(\tau,\xi_i,\eta_i)\left(\frac{\bar{\psi}_{in}}{<\psi^2>}\right)\right] . \quad (33)$$

The case of a coincident mass and force is easily handled by noting that $\bar{\psi}_{if} = <\psi^2>$.

Asymptotic, Spatially Averaged Solution

The asymptotic, spatially averaged solution can be found in a manner similar to that of eqn. (33). Squaring eqn. (31) and applying our asymptotic limits yields for NN added masses,

$$<W_{AMA}^2(\tau)> \approx \frac{1}{\alpha}\sum_{p=M}^{M+\Delta M}\left\{k_{nF}^2(\tau) + \sum_{i=1}^{NN}\left[k_{n\Lambda_i}^2(\tau,\xi_i,\eta_i) + 2k_{nF}(\tau)k_{n\Lambda_i}(\tau,\xi_i,\eta_i)\frac{\bar{\psi}_{fi}}{<\psi^2>}\right] + \sum_{i,j=1;i\neq j}^{NN} k_{n\Lambda_i}(\tau,\xi_i,\eta_i)k_{n\Lambda_j}(\tau,\xi_j,\eta_j)\frac{\bar{\psi}_{ij}}{<\psi^2>}\right\} . \quad (34)$$

These relations describe the panel response at the locations where a point load is applied or added masses are attached. They require none of the natural frequencies or modes of the panel and are valid for arbitrary forms of excitation. Furthermore, they require substantially less computational effort than the classical solutions of eqns. (22) and (23).

Care is required when selecting the frequency distribution for the asymptotic solution. Evenly spaced frequencies impose a prominent beat to the response and random frequency spacing is inappropriate for reasons not the least of which is repeatability. Experience indicates that a nominally even spacing with a superimposed perturbation (say, 20%-30%) is quite effective. Furthermore, the same frequencies should not be used for the diagonal

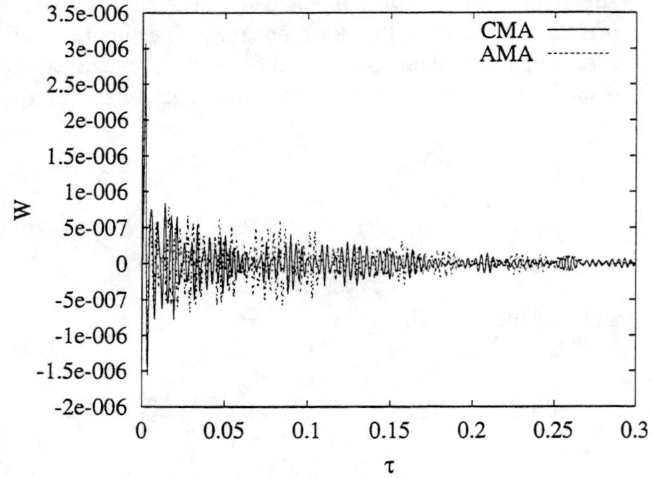

Figure 3. Comparison of the classical and asymptotic displacement predictions at the mass location (ξ_0, η_0). The AMA approximation only used 10% of the estimated modes in the frequency range (i.e. $\alpha = 0.1$).

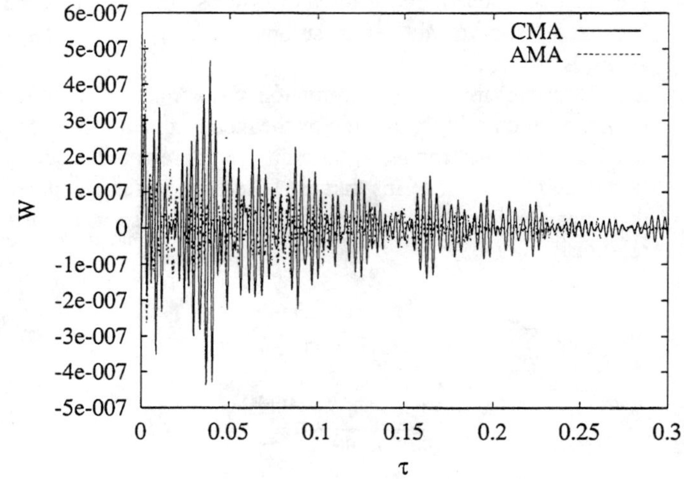

Figure 4. Comparison of the classical and asymptotic displacement predictions at the force location (ξ_f, η_f). The AMA approximation only used 10% of the estimated modes in the frequency range (i.e. $\alpha = 0.1$).

and off-diagonal elements of **A** as this can lead to a trivial solution.

Results

Let us consider the response of a relatively simple system as an example. We determine the transient response of an aluminum panel with the physical parameters, length = 0.864m, width = 0.711 m, ($\gamma = 1.214$), thickness = 0.00318 m, and modal damping $\zeta = 0.005$. The elastic modulus is 73 GPa, density is 2800 kg/m^3, and Poisson's ratio $\nu = 0.359$ (i.e. aluminum). A 100 gram mass ($M_0 = 0.00113$) is located at $(\xi_0, \eta_0) = (2/\pi, 2\pi)$ and the panel is excited by an abrupt strike from an impulse hammer at $(\xi_f, \eta_f) = (1/\pi, 1\pi)$. The response in the frequency range of 1000 – 1800 Hz. is shown in Figs. 3-5.

As Fig. 3 displays, the classical and asymptotic solutions near the mass location agree reasonably well. This is particularly true when one considers that the alternative methods of approximation (e.g. "transient" methods based on Statistical Energy Analysis) consider good agreement to be within an order of magnitude or two.

The comparison of displacement at the forcing locations in Fig. 4 is even closer and should be expected since the dominant excitation is due to the applied force (i.e. $k_{p\Lambda_i} \ll k_{pF}$). Clearly, this is where the maximum panel response will occur indicating that the AMA approximation is useful for maximum excursion estimates.

Figure 5 shows the spatially averaged responses. As these are useful for determining acoustic radiation, AMA appears to be a practical means of estimating sound pressure levels in a band-by-band form.

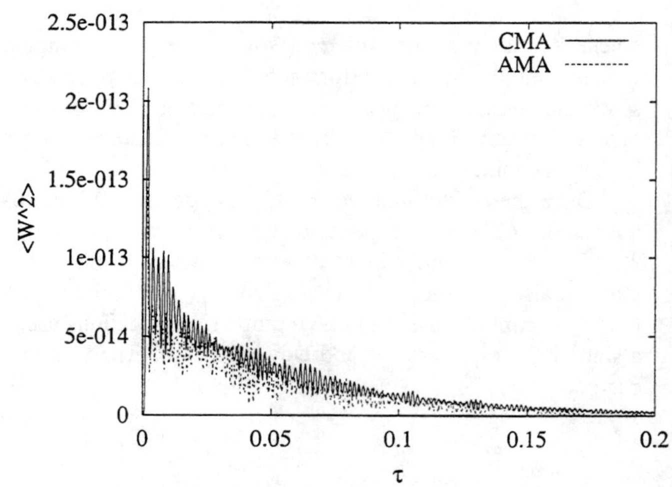

Figure 5. Comparison of the classical and asymptotic predictions of the spatially averaged, panel displacement, $< W^2 >$. The AMA approximation only used 10% of the estimated modes in the frequency range (i.e. $\alpha = 0.1$).

Analogous responses were calculated for multiple masses that showed similar agreement.

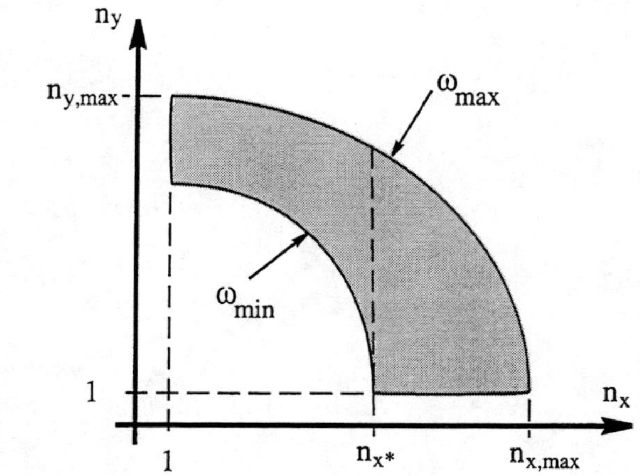

Figure 6. The frequency range $\Delta\omega$ as seen in mode number space, n_x-n_y.

Conclusions

Asymptotic simplifications were applied to the transient, classical solution of the transverse vibration of a panel with attached masses. The resulting transient, asymptotic modal analysis provides a reasonable approximation of the classical solutions in frequency ranges with significant modal content (i.e. high frequencies). Expressions were derived for the local panel response near the added masses and a point load as well as the spatially averaged panel response. These relations are much simpler to evaluate than their classical counterparts typically requiring 85-95% less computer memory and time.

Appendix: The Mean, Modal Cross Product: $\bar{\psi}_{ij}$

The mean value for the cross product of modes at different locations, $\bar{\psi}_{ij}$ was introduced in eqn. (30). We may estimate this using sinusoidal panel mode shapes by recognizing that away from the panel edges the boundary conditions are irrelevant and integrating over the mode numbers,

$$\bar{\psi}_{ij} \approx \frac{\int_A \sin(n_x\pi\xi_{ni})\sin(n_y\pi\eta_{ni})\sin(n_x\pi\xi_{nj})\sin(n_y\pi\eta_{nj})dA}{\int_A dA}. \tag{35}$$

We may recast the integrand into a more useful sum of Cosine functions and integrate on the region in mode-number space shown in Figure 6. We will perform the integral in n_y first since this can be done analytically. The limits of integration can be found using the frequency-mode number relation (4) and the appropriate values of ω. The domain has two portions: to the left and right of n_x^*. Thus our integration limits are

$$\left(\frac{\omega_{min}}{\gamma\pi^2} - \frac{n_x^2}{\gamma^2}\right)^{\frac{1}{2}} \leq n_y \leq \left(\frac{\omega_{max}}{\gamma\pi^2} - \frac{n_x^2}{\gamma^2}\right)^{\frac{1}{2}} \quad : n_x < n_x^* \tag{36}$$

$$1 \leq n_y \leq \left(\frac{\omega_{max}}{\gamma\pi^2} - \frac{n_x^2}{\gamma^2}\right)^{\frac{1}{2}} \quad : n_x > n_x^* \tag{37}$$

where n_x^* is determined from eqn. (4) with $n_y = 1$ and $\omega = \omega min$. The first integration in n_y yields

$$\bar{\psi}_{ij} \approx \frac{\int_{n_x}(\bar{\psi}_{ny,1} + \bar{\psi}_{ny,2})dn_x}{\int_{n_x}\int_{n_y} dn_y dn_x}. \tag{38}$$

where

$$(\bar{\psi}_{ij})_{ny,1} = \left\{\frac{\cos(n_x\pi\xi_i)\cos(n_x\pi\xi_j)}{\pi(\eta_j-\eta_i)(\eta_j+\eta_i)}\right\}[\Upsilon(\eta_j,\eta_i,\omega_{max}) - \Upsilon(\eta_i,\eta_j,\omega_{max}) - \Upsilon(\eta_j,\eta_i,\omega_{min}) + \Upsilon(\eta_i,\eta_j,\omega_{min})], \tag{39}$$

$$(\bar{\psi}_{ij})_{ny,2} = \left\{\frac{\cos(n_x\pi\xi_j)\cos(n_x\pi\xi_i)}{\pi(\eta_j-\eta_i)(\eta_j+\eta_i)}\right\} \times [\eta_i\cos(\pi\eta_j)\sin(\pi\eta_i) - \eta_j\cos(\pi\eta_i)\sin(\pi\eta_j) + \Upsilon(\eta_j,\eta_i,\omega_{min}) - \Upsilon(\eta_i,\eta_j,\omega_{max})] \tag{40}$$

where

$$\Upsilon(\eta_a,\eta_b,\omega_c) = \eta_a\sin\left(\eta_a\sqrt{\frac{\gamma\omega_c - (n_x\pi)^2}{\gamma^2}}\right) \times \cos\left(\eta_b\sqrt{\frac{\gamma\omega_c - (n_x\pi)^2}{\gamma^2}}\right). \tag{41}$$

Equation (38) must then be integrated over n_x and while it appears to be intractable analytically, the integrand is very well-behaved and can be done easily using a rudimentary numerical integration over the limits

$$\text{For } \bar{\psi}_{ny,1} : 1 < n_x < n_x^* \tag{42}$$

$$\text{For } \bar{\psi}_{ny,2} : n_x^* < n_x < \left[\frac{\gamma\omega_{max}}{\pi^2} - \gamma^2\right]^{\frac{1}{2}}. \tag{43}$$

The trivial case of coincident a point is simply $\bar{\psi}_{ii} = \bar{\psi}^2 \approx < \psi^2 >$.

ACKNOWLEDGMENT

The author appreciates the help of the Institute of Mechanics, Moscow State University, Moscow, Russia for acting as host during this work. Particular thanks to Prof. Sergei A. Shesterikov for his helpful comments and discussion. Support for this work was provided by NSF Division of International Programs under grant # 9803043, Susan Parris, Program Director.

REFERENCES

Dalton, E. C., 1992, "Ballistic shock response by an extension of statistical energy analysis," Proceedings, 63rd Shock and Vibration Symposium.

Dowell, E. H. and Kubota, Y., 1995, "Asymptotic modal analysis and statistical energy analysis of dynamical systems", ASME Journal of Applied Mechanics, 52, pp. 949–957.

Kubota, Y., and Dowell, E. H., 1986, "Experimental Investigation of asymptotic modal analysis for a rectangular plate," Journal of Sound and Vibration, 106(2), pp 203-216.

Kubota, Y., and Sekimoto, S., and Dowell, E. H., 1990, "The high-frequency response of a plate carrying a concentrated mass," Journal of Sound and Vibration, 138(2), pp 321-333.

Lai, M.L., and Soom, A., 1990, "Statistical energy analysis for the time-integrated transient response of vibrating systems," ASME Journal of Applied Mechanics, 112, pp. 206–213.

Lyon, R. H., *Statistical energy analysis of dynamical systems: theory and applications*, Cambridge, MA, The MIT Press, 1975.

Pinnington, R. J. and Lednik, D. "Transient statistical energy analysis of an impulsively excited, two oscillator system," Journal of Sound and Vibration, 189, pp. 249–264.

Reynolds, R. R. and Dowell, E. H., 1998, "The transient response of structures using asymptotic modal analysis", ASME Journal of Applied Mechanics, 65, no. 1, pp. 258–265.

Sepulveda, A.E., and Thomas, H.L., 1996, "Improved transient response approximation for general damped systems," AIAA Journal, 34(6), pp 1261–1269.

AUTHOR INDEX

PVP-Vol. 415
Recent Advances in Solids and Structures — 2000

Book Number: H01192